Lecture Notes
in Economics and
Mathematical Systems
Operations Research, Computer Science, Social Science

Edited by M. Beckmann, Providence, G. Goos, Karlsruhe, and
H. P. Künzi, Zürich

86

Symposium on the Theory of Scheduling and Its Applications

AF722659

WITHDRAWN

Edited by S. E. Elmaghraby

Springer-Verlag
Berlin · Heidelberg · New York 1973

Advisory Board

H. Albach · A. V. Balakrishnan · F. Ferschl · R. E. Kalman · W. Krelle · G. Seegmüller
N. Wirth

Professor Salah E. Elmaghraby
North Carolina State University
Grad. Program in Operations Research
Box 5511
Raleigh, NC 27607/USA

AMS Subject Classifications (1970): 60A30, 90B35

ISBN 3-540-06437-0 Springer-Verlag Berlin · Heidelberg · New York
ISBN 0-387-06437-0 Springer-Verlag New York · Heidelberg · Berlin

This work is subject to copyright. All rights are reserved, whether the whole or part of the material is concerned, specifically those of translation, reprinting, re-use of illustrations, broadcasting, reproduction by photocopying machine or similar means, and storage in data banks.

Under § 54 of the German Copyright Law where copies are made for other than private use, a fee is payable to the publisher, the amount of the fee to be determined by agreement with the publisher.

© by Springer-Verlag Berlin · Heidelberg 1973. Library of Congress Catalog Card Number 73-14483. Printed in Germany.

Offsetprinting and bookbinding: Julius Beltz, Hemsbach/Bergstr.

PREFACE

The theory of scheduling is receiving increased emphasis in research and practice for at least three good reasons. *First*, the management of large scale projects resolves itself, in the final analysis, into problems of scheduling interacting activities subject to limited resources. *Second*, a great deal of "fat" that used to exist in the past in production, distribution, and service systems is eliminated, thanks to tighter managerial controls in information systems, in financial management, in logistics, and in many other facets of industrial enterprises and military installations. Tighter scheduling methods are therefore called for. *Third*, the study of scheduling problems involves the study of combinatorial problems and optimization over discrete spaces which represent a radical, and interesting, departure from classical mathematics. This area of study has attracted a good number of distinguished researchers, engineers as well as mathematicians. There is a serious attempt to apply known number theory, and perhaps develop new theory, that would cope with the new problems. The computer enters the picture in novel and ingenious ways, which has not been possible before; etc.

To those working in the area, whether in theory or in practice, progress proceeds at an exhilarating pace, with new mathematical structures and computational approaches being continuously introduced to model and solve the problems in novel, and oftentimes ingenious ways. This requires rather frequent 'stock taking' of the progress achieved; which has been done at the rate of approximately once every five years over the past thirteen years. To this end a Symposium on The Theory of Scheduling and its Applications was held on the campus of North Carolina State University at Raleigh, between May 15-17, 1972. It was sponsored by the Office of Naval Research, and was attended by 46 people from several geographical regions of the United States and Canada, who represented a wide spectrum of background, occupation and contributions to the area of Scheduling.

The Symposium was organized in few general sessions and several discussion groups. The objective was to maximize interaction and cross-fertilization among participants.

Almost all papers presented at the Symposium had been previously distributed to the participants, and each paper presented in the general sessions was reviewed and discussed by two discussants. Their comments were conveyed in writing to the authors, who subsequently modified their paper in light of the criticisms presented to them. On the other hand, contributed papers in discussion groups received serious considerations from the audience, which was typically small in size and had opted to attend a particular group because of interest in the subject matter presented.

The upshot of all this is that while each paper bears the name of its author (or authors), it is, in fact, the distillation of several contributions since, in the majority of cases, the criticisms and contributions from discussants, or "from the floor", were incorporated in the revised versions of the author's manuscript. This "took care" of the discussants' remarks, and hence there was no justification to print them in the Proceedings. In a few instances (to be more specific, three cases), the discussant had a separate contribution to make; these are appended separately in the Proceedings following the paper. In any event, the discussants of a paper are identified, and to them go our thanks for the significant contributions they have made.

The Proceedings are organized in four sections according to the tenor of the contribution: Survey Papers, Applications, Theory, and Models of Processes. As the reader can imagine, it was sometimes difficult to draw the lines among these categories, especially between the last two. I take the blame for any error in designation, or apparent arbitrariness. The Index is organized by Section, and the papers in each section are arranged alphabetically. There is also a List of Participants, with the participant's contribution and the location of his paper (if any).

In an effort to bring out the volume without undue delay, authors were requested to submit their papers in typewritten form suitable for direct photographic reproduction. This procedure accounts for the variations in type style and arrangement.

Authors are, of course, solely responsible for the contents of their papers, as no editorial work was done, except regarding some minor format considerations.

The Symposium afforded the workers in this area a unique chance to get together, present their latest thoughts, exchange views, and in general assess the future of the area. The present thinking of the foremost contributors to the field are contained in these Proceedings. The serious reader will find not only the latest in "what's up" in scheduling, in both theory and practice, but also the trends of thought and applications. The collective lists of references in all the papers should be a valuable asset to any researcher.

Ideas and thoughts are significant by their presence as well as by their absence, because they elicit other ideas and thoughts. As the reader pores over these contributions, he should also reflect on the significance, if any, of the absence of papers applying the theory of queues; or the theory of dynamic programming, or heuristics, etc. On the other hand, there is an abundance of Models, but a relative scarcity of Applications (the ratio is almost two to one): Is this due to the dearth of applications, or to the scarcity of "applicators" who also write papers for presentation at Symposia? If the latter is closer to the truth, what can be done to fill that void?

A Symposium involving so many participants and requiring the expenditure of several thousand dollars cannot materialize without the significant contributions of a number of people. It isn't that they have "helped": they made it possible, and without their input the Symposium would not have been realized. They are, in a real sense, the co-organizers of the Symposium and the co-editors of these Proceedings. These are: Drs. G. E. Bennington and H. L. W. Nuttle, both of the Department of Industrial Engineering, N. C. State University, my secretary, Mrs. Lillian Hamilton, Mr. Eugene Starnes of the Division of Continuing Education at NCSU, and Mr. Richard Coppins, graduate student in the Operations Research Program at NCSU.

Most importantly, the profession owes a debt of gratitude to Dr. Thomas Varley, Director of the Operations Research Program, Office of Naval Research, for encouraging the idea of the Symposium and for the financial support that made it possible. I also owe a debt of thanks to Mr. Seymour M. Selig, Managing Editor, Office of Naval Research, for his continuous encouragement and support, especially relative to the various avenues of publication of the Proceedings.

Salah E. Elmaghraby
Raleigh, N. C.

CONTENTS

I. SURVEY PAPERS

II. APPLICATIONS

III. THEORY

IV. MODELS OF PROCESSES

LIST OF PARTICIPANTS

Participant	Discussant of Paper by	Also Author of Paper on Page No.
Ashour, Said	Fisher	206
Baker, Kenneth R.	Rau	201
Bennington, Gerald E.	-	1
Bradley, Gordon	-	224
Dessouky, Mohamed I.	Srinivasan & Thompson	-
Dudek, Richard	Bratley, Florian & Robillard	29
Elmaghraby, Salah E.	Giffler, and Srinivasan & Thompson	244
Emmons, Hamilton	Sidney	278
Fisher, Marshall	Elmaghraby & Mallik	294
Florian, Michael	Parker & Ashour	39,242
Fulcher, Doily	-	-
Ghare, P. M.	Bennington & McGinnis	-
Giffler, Bernard	Turksen & Shankar	44
Hodgson, Thom J.	Florian, Guerin & Bushell	56
Johnson, Gary	-	-
Kapur, Kailash C.	Rosenshine	324
Lawler, Eugene L.	Murty	164
Mallik, Arup	Parker & Ashour	244
Maxwell, William L.	Giffler, and Rau	-
Mitten, Loring G.	-	127
Murty, Katta	-	143
Nevins, Arthur	-	-
Nuttle, Henry L.W.	-	359
Panwalker, Shrikant	-	29
Parker, R. G.	Florian, Guerin & Bushell	206
Prabhakar, T.	Wiest	69
Ratliff, Donald	Bradley	-
Rau, John	Sidney	167
Robillard, Pierre	Bradley	242
Rosenshine, Matthew	Wiest	381
Rubin, David	Bratley, Florian & Robillard	-
Salvador, Michael	Rosenshine	83

Participant	Discussant of Paper by	Also Author of Paper on Page No.
Schrage, Linus	Mitten & Tsou	-
Schultz, George M.	-	-
Selig, Seymour M.	-	-
Shwimer, Joel	Emmons	391
Sidney, Jeffrey B.	Mitten & Tsou	393
Smith, Milton	Turksen & Shankar	29
Steinhoff, Harry W.	-	92
Swarc, W.	Elmaghraby & Mallik	-
Taha, Hamdy	Emmons, and Nuttle & Aly. . .	-
Thompson, Gerald L.	Kapur, and Giffler.	399
Turksen, Ismail B.	-	426
Varley, Thomas	-	-
Wiest, Jerome	Bennington & McGinnis	109
Zaloom, Victor	Fisher.	319,433

A CRITIQUE OF PROJECT PLANNING WITH CONSTRAINED RESOURCES

G. E. Bennington and L. F. McGinnis

North Carolina State University, Raleigh, North Carolina, USA

ABSTRACT

This paper surveys the current research in resource constrained project scheduling. Although CPM and PERT have gained wide acceptance and use, the problem of limiting resources used by each activity remains unsolved for practical-sized problems.

The past research follows three basic approaches. The problem may be formulated as an integer linear program and solved by standard integer programming techniques. A second approach is to directly employ some enumerative scheme for constructing an optimal schedule. Finally, the problem may be formulated in terms of minimaximal paths in a disjunctive graph and solved by network flow methods and implicit enumeration. The approaches will be compared and the essential difficulties of the several methods will be identified.

* This paper was prepared for presentation at the Symposium on Scheduling Theory and Its Applications on May 15-17, 1972 in Raleigh, N. C. This research was partially sponsored by ONR Contract N0014-70A0120002, ARO-D Contract DA-ARO-D-31-124-72-G106, and the National Science Foundation Traineeship Program.

1. Introduction

Project planning models based on network structure first appeared in 1958 under the acronyms PERT [30] and CPM [25]. These models have been further developed and refined, and have gained wide acceptance, primarily because of the relative ease with which they can be applied to large problems. PERT and CPM assume unlimited resources, which in many real problems is not a valid assumption. Many researchers in the past decade have addressed the problem of dealing with limited resources.

Project planning under resource constraints is one of a class of scheduling and sequencing problems which have been described as "combinatorial." Following Lawler [28], these are combinatorial problems because they deal with the optimal arrangement or ordering of activities. As the literature amply illustrates, these particular combinatorial problems have proved to be frustrating to deal with, and no general solution techniques of practical value have been developed.

There are several reasons why project planning under constrained resources has been such a frustrating problem:

(1) the problem is an extension of the PERT/CPM problem, which was readily solved through the theory of network flows,

(2) the problem is not difficult to visualize or state, and

(3) without exception, every solution procedure proposed has been computationally impractical, either because the model grows too large, or because the solution procedure is too lengthy, or both.

There are few versions of the constrained resources problem for which exact solution methods are known and almost all published results rely on some form of enumeration.

1.1 Problem Statement

The CPM problem can be represented on a graph G = (N,P) where N, the set of nodes, corresponds to the activities and P, the set of arcs, corresponds to the precedence relations. (N and P will also be used as the <u>number</u> of nodes and precedences, respectively. The intended meaning will be clear from the context.) An example of a project graph is given in figure 1.1. For any activity $i \varepsilon N$, let d_i be its duration, and t_i the starting time assigned to i by some schedule. Two schedules of particular interest are the CPM early start and late start schedules, i.e., those for which all activities start as early as possible and as late as possible respectively. The d_i and t_i for a CPM early start schedule are given in table 1.1 for the project of figure 1.1. The objective in the CPM problem is to minimize t_n, the starting time of the last or terminal activity, subject to the precedence constraints:

$$t_j - t_i \geq d_i \quad \text{if i precedes j, i.e. } (i,j) \varepsilon P$$

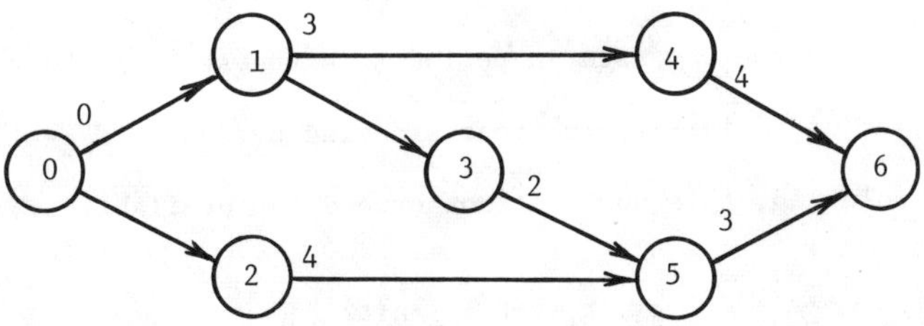

N = {0,1,2,3,4,5,6}
P = {(0,1),(0,2),(1,3),(1,4),(2,5),(3,5),(4,6),(5,6)}

Figure 1.1

Activity i	0	1	2	3	4	5	6
Duration d_i	0	3	4	2	4	3	0
Starting Time t_i	0	0	0	3	3	5	8

Table 1.1

This problem is easily solved since a project network is acyclic. The pairs $(i,j) \varepsilon P$ for which the constraint holds as an equality define the <u>critical path</u>, CP, i.e., delaying any activity on the critical path delays t_n.

When the resources required by a project are scarce, two problems can be defined. One is to minimize some function of resource consumption (e.g. peak consumption or the sum of the squared deviations from the mean consumption) subject to a fixed project duration and the other is to minimize project duration subject to fixed resource availabilities. It is the second problem that is treated in this paper.

In general, a project consumes several resources, which are available in varying amounts during the life of the project. The availability of resources will be written as $a(t) = (a_1(t), a_2(t), \ldots, a_m(t))$, where m is the number of resources and $a_i(t)$ is the amount of resource i available in period t.

Similarly, an activity may consume several resources, each at a varying rate, over its duration. The resource requirement for activity i will be denoted as $r_i(p) = (r_{i1}(p), \ldots, r_{im}(p))$ where $r_{ij}(p)$ is the amount of resource j required in the pth period of the activity's duration, i.e., period $t_i + p$.

The amounts of resources used in period t is the sum of the resource requirements for the set of activities in process during period t, $N_t = \{i \mid t_i \leq t \leq t_i + d_i\}$. In a feasible solution, this sum must not exceed the availability in any period, i.e.

$$\sum_{i \varepsilon N_t} r_i(t-t_i) \leq a(t)$$

Notice that the terms to be included in the above summation are dependent both on t <u>and</u> t_i.

A special case of the general project scheduling problem which has received considerable exposure in the literature is the machine sequencing problem. In this case, the precedence diagram has the special structure that all activities, other than the first and last (dummy) activities, have exactly one predecessor and one successor. (see figure 1.2) In addition, it is assumed that each activity requires only one unit of one type of resource (only one $r_{ij}(p)$ is positive and it is equal to one), and that all requirements and availabilities are constant.

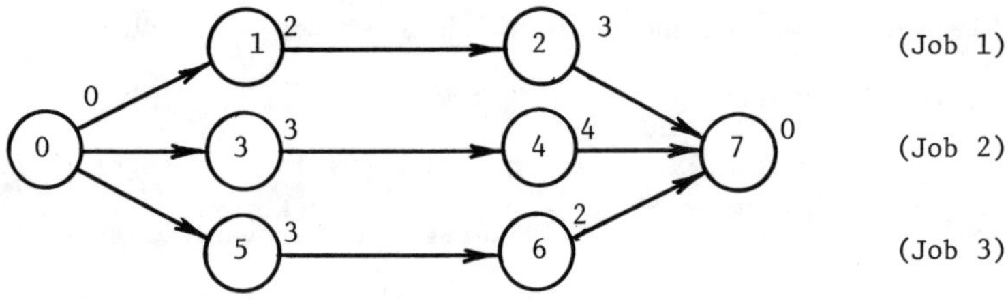

Nodes 0 and 7 are dummy tasks

Tasks 1, 3, 5 use machine A

Tasks 2, 4, 6 use machine B

Task durations are given adjacent to each node

Figure 1.2

The assembly line balancing problem is closely related to the project scheduling problem, since it can be represented on a similar network. It is not, however, just a special case, since the form of the resource constraints may be quite different.

These related problems are mentioned here because several of the papers surveyed are based on extending procedures originally proposed for them to the project scheduling problem.

1.3 Objectives and Organization

The objective of this paper is to critically examine, compare, and summarize the current approaches to project planning under constrained resources. Heuristic methods, while still the only viable solution techniques, will not be considered here, but are treated by Lambourn [27], Levy, Thompson, and Wiest [29], Mize [33], and Wiest[42, 43].

The research on resource constrained project scheduling falls under two broad classifications: approaches in which the problem is stated as a set of linear inequalities (possibly with integer restrictions), and approaches which deal with the combinatorial problem directly. Under the second classification, the approaches are further grouped as those which begin with a complete (possibly infeasible) schedule and modify it, and those which construct a schedule by means of a sequence of partial schedules.

A recent formulation by Fisher [16] presents an integer programming model which is solved by Lagrange multipliers. In the event of a "gap", a branch and bound algorithm is required. This paper appears elsewhere in these proceedings.

2. Integer Linear Programming Formulations

Efficient algorithms for finding the critical path in CPM problems can be derived by formulating the problem as a maximum cost flow problem and examining the dual linear program. Attempts to use a similar approach for the constrained resources problem have not been successful for several reasons.

In the CPM problem, integer data insures integer solutions because of unimodularity of the precedence constraint matrix. Since no formulation of the constrained resource problem has yet been proposed which maintains unimodularity, an integrality constraint must be specified. Also, the dual of the CPM problem has a structure which can be exploited to obtain an efficient one-pass algorithm. No straightforward formulation of the constrained resources problem has such a structure.

2.1 Models

Several integer linear programming formulations of the machine sequencing problem have been proposed [8, 31, 41], but were found impractical because a suitable code was not available. Recent developments in the theory of integer programming prompted Pritsker, Watters, and Wolfe [35] to develop a model for multi-project scheduling with resource constraints. The model allows great flexibility with respect to job-splitting, due dates, resource substitutability, and concurrency requirements, and is solved via a 0-1 integer programming code developed by Geoffrion. No attempt is made to identify and exploit any special structure.

Elmaghraby [15] and Bhadury [7] have developed an ILP model for the project scheduling problem, which is structured and has possibilities for solution by branch and bound. The basis for the model was developed by Elmaghraby for the problem of scheduling independent jobs through parallel processors, and is the result of applying decomposition to a larger, "almost" transportation model.

The model can be stated as:

$$\min \quad \sum_{k=1}^{T} k\, x_n \tag{2.1}$$

subject to

$$\sum_{k=1}^{T} x_{ik} = 1 \qquad i \in N \tag{2.2}$$

$$\sum_{i \in N} \sum_{k=1}^{T} x_{ik}\, v_{ik} \leq a_j\,(t) \qquad j=1,\ldots,m \tag{2.3}$$

$$-\sum_{k=1}^{T} kx_{ik} + \sum_{k=1}^{T} kx_{jk} \geq d_i \qquad (i,j) \in P \tag{2.4}$$

$$x_{ik} = 0,1$$

where:

$T =$ maximum project duration, and

$$x_{ik} = \begin{cases} 1 \text{ if activity } i \text{ starts in period } k \\ 0 \text{ otherwise.} \end{cases}$$

Constraints (2.2) require that each job must be done. Constraints (2.3) are the resource constraints (note that since $a_j(t)$ is a T period vector, these are actually mT constraints), where the vector v_{ik} represents the resource requirement by activity i if it starts in period k. The constraints (2.4) are the precedence constraints.

The model has P + mT + N constraints and NT variables. The number of variables can be greatly reduced by noting that x_{ik} is zero if k is less than the CPM early start time for i, or greater than the CPM late start time. It can be shown that this model is equivalent to the Pritsker, Watters, and Wolfe model, in the absence of job-splitting, due dates, resource substitutability or concurrency constraints. Both formulations require that T, the upper limit on project duration is specified *a priori*.

Elmaghraby gives a branch and bound algorithm for solving the machine sequencing problem which can be applied to the project scheduling model as well. The enumeration has at most N stages and at each stage, one of the x_{ik} is chosen to be non-zero. The resulting integer linear program is solved as a linear program to obtain a lower bound.

An extension of the approach used by Dantzig and Fulkerson [11] for allocating tankers to a fixed schedule is presented by Chapman [10] for the problem of allocating resource to a varying schedule. A computational procedure based on the transshipment algorithm is proposed, but few details of the model formulation and no computational results are given. The variables employed are

$$x_{hij} = 0, 1, 2... \quad \begin{matrix} h=1,...,m, \\ i=1,...,n+s+T \\ j=1,...,n+s+T \end{matrix}$$

the number of units of resource h assigned to the processing of activities i and j consecutively. Activities n+1,...,n+s are dummy sources, and n+s+1,...,n+s+T are dummy sinks, since some activities may not be processed consecutively with any other activity. A large number of the x_{hij} can be set to zero <u>a priori</u>, and the resulting tableau is well structured, although extremely large in comparison with the other models.

Greenberg [21] formulates the machine sequencing problem (see figure 1.2) in terms of a zero-one variable

$$x_{ij} = \begin{cases} 1, \text{ if i precedes j} \\ 0, \text{ otherwise} \end{cases}$$

for all $(i,j) \varepsilon P$ where i and j require the same machine. For each pair of tasks to be performed on a given machine, the constraints

$$t_i + d_i x_{ij} \leq t_j + M x_{ji} \tag{2}$$

$$t_j + d_j x_{ji} \leq t_i + M x_{ij} \tag{3}$$

$$x_{ij} + x_{ji} = 1 \tag{4}$$

are formed in addition to the precedence constraints for the tasks of a given job, where M is a suitably large constant. The algorithm proposed solves the resulting linear program without constraints (2), (3), or (4). A branch and bound procedure is employed, where branching consists of adding the constraints (2) and (4) for one branch and (3) and (4) for the other. The augmented problem is then solved as a linear program. Branching continues until all constraints, i.e. x_{ij}, have been examined. It should be obvious that the linear program solved at each node is nothing more than a CPM problem, which is better solved by other means. As will be seen in section 3, Greenberg's model is precisely that used in the disjunctive graph formulation.

3. Schedule Modifying Methods

Schedule modifying methods start with a schedule of the tasks that satisfies the precedence constraints. If none of the resource availabilities are exceeded or there is no schedule of shorter duration, the problem is solved. In general, at least one of the conditions will be violated. The method then modifies the existing schedule in an attempt to produce a feasible schedule, or one with a shorter duration. This modification procedure is then imbedded in an enumeration algorithm to find the optimal schedule.

The CPM problem is easily solved because the precedence relations have a natural interpretation as an acyclic network. The longest path in the network corresponds to the project duration and is easily computed via a simple recursive relationship. It is natural to attempt to exploit the similar network structure of the resource constrained scheduling problem. One such attempt [1, 38] results in the formulation of an equivalent problem of finding a minimaximal path in a disjunctive graph.

As a means of describing this method, consider the simple machine sequencing problem associated with the graph G = (N,P) in figure 3.1. N^j is the set of tasks requiring machine type j, e.g. $N^1 = \{1,4,5\}$. If there were no limit on the number of each machine type being used at one time, this could be solved as a CPM problem.

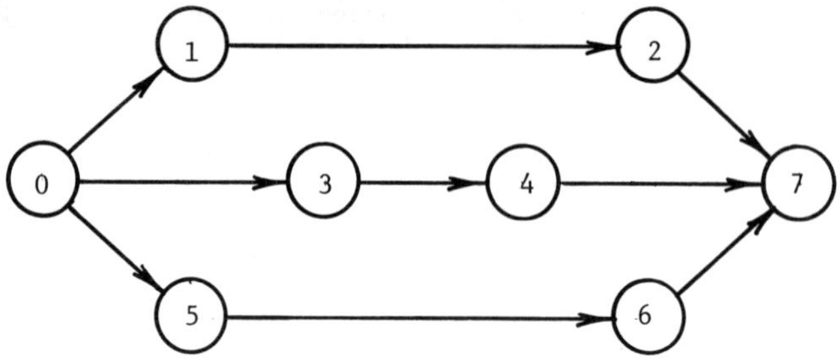

$N^1 = \{1,4,5\}$ $N^2 = \{2,3,6\}$

Figure 3.1

Alternatively, if there were some way to specify an ordering for the tasks requiring each machine type, the CPM algorithm could still be used. For example, the precedence relations in figure 3.2 order the set of tasks, N^1, requiring machine type 1, so that only one may be processed at a time regardless of the number of machines of type 1 available.

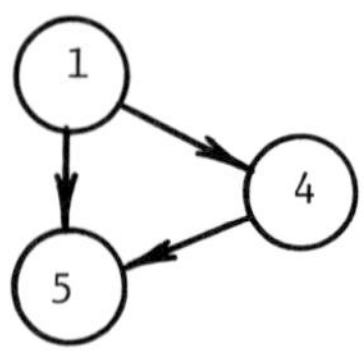

Figure 3.2

The difficulty is, of course, that although it is not unusual to have several precedences prescribed for tasks requiring a given machine type, (due to changeover, etc.) it would be most unusual to have a complete ordering. To avoid this difficulty, a "psuedo-precedence" could be specified for each pair of tasks in N^j for each machine type j. Since for a given pair of tasks, say 1,5 in figure 3.1 it is not known _a priori_ whether in an optimal schedule, 1 will precede 5 or vice versa, the psuedo-precedence can be represented as the pair, $\{(1,5), (5,1)\}$ in figure 3.3, with the understanding that at most one of the arcs will be included in any solution. Since the pair serves to separate tasks 1 and 5, it is referred to as a disjunctive pair, or d.p.

Figure 3.3

There is a d.p. for every pair of tasks requiring a given machine type, provided there is no path already in G connecting them. The set of all disjunctive arcs is denoted by W. If the original precedences, P, are referred to as conjunctive arcs, then combining the conjunctive and disjunctive arcs yields a disjunctive graph, D = (N,P∪W). Figure 3.4 illustrates the disjunctive graph corresponding to the machine sequencing problem of figure 3.1. Obviously D cannot be solved as a CPM problem, since each d.p. forms a cycle, and the longest path is unbounded.

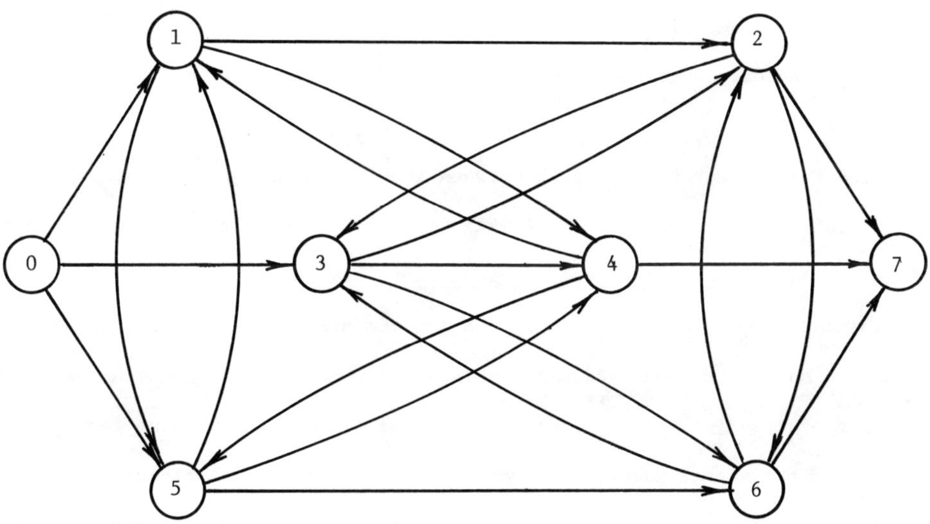

Figure 3.4

The d.p. between nodes in N^j represent conflicts between tasks competing for machine type j, which are resolved by choosing one of the arcs from each d.p.

Making this choice for each d.p. associated with each machine type results in a set of disjunctive arcs called a resolution. If some, but not all of the d.p. are "resolved", a partial resolution is obtained. Sequences of partial resolutions are used in several solution methods. For each complete or partial resolution, R, a conjunctive graph $D_R = (N, P \cup R)$ can be defined. If R is a complete resolution and D_r is acyclic, the CPM solution for the conjunctive graph will be feasible for the machine sequencing problem.

The machine sequencing problem has now been transformed into an equivalent problem of specifying a complete resolution R^o, for which the resulting critical path in $D_{R^o} = (N, P \cup R^o)$ is the minimum for all resolutions. Since each d.p. in a resolution may be in either of two states or orientations, depending on which arc is chosen, the equivalent problem is referred to as finding a two-state minimaximal path in a disjunctive graph.

The project scheduling problem is not as easily formulated in terms of a disjunctive graph, since there may be more than one unit of each resource available and several activities requiring the same resource may overlap in a feasible schedule. To generalize the method to project scheduling, some preliminary definitions are needed. Note that the assumption of constant availability is required, i.e., $a_j(t) = a_j$.

Two tasks which have no path between them in the conjunctive graph are unrelated, and form a stable set. The maximum number of unrelated tasks in any set N^j is called the coefficient of internal stability for j, or CS_j. Since two unrelated tasks may overlap, CS_j is the maximum number of tasks requiring machine type j which may be processed simultaneously, hence is a bound on the number of machines of type j that are required. For the case of the machine sequencing problem, with several machines of each type, a partial resolution is feasible if $CS_j \leq a_j$ for j = 1,...,m. For example, in figure 3.5 a partial resolution of the disjunctive graph of figure 3.4 has $CS_1 = 2$, $CS_2 = 1$, and $\{1,5\} \subseteq N^1$ is the only stable set, i.e., the only pair of activities which may overlap.

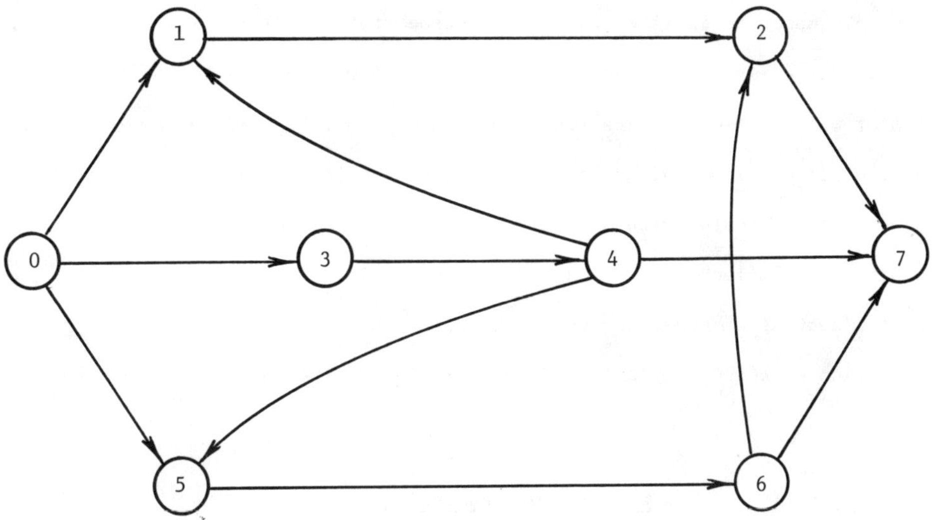

Figure 3.5

The notion of coefficient of internal stability is generalized to project scheduling by considering all internally stable subsets, B, of jobs in N^j, and defining the <u>generalized coefficient of internal stability</u>, GCS_j, as

$$GCS_j = \max_{B \subseteq N^j} \sum_{i \in B} r_{ij}.$$

A partial resolution is feasible if $GCS_j \leq a_j$ for $j = 1,\ldots,m$. The total requirement for resource j may be less than GCS_j, depending on the schedule, but cannot exceed it.

For two activities, i and k, in an internally stable set, neither of the arcs in the d.p. between i and k has been chosen. This means that, for the project scheduling model, a d.p. may be in one of three states, either one of the arcs is chosen, or neither arc is chosen. Thus the project scheduling problem can be transformed into a problem of finding a <u>three state minimaximal</u> path in a disjunctive graph.

Let $|N^j| = k_j$. There are at most $d = \sum_{j=1}^{m} \binom{k_j}{2}$ d.p. in D; hence there are at most 2^d resolutions to the two state problem (not all are feasible) and at most 3^d resolutions to the three state problem. It is easy to show that an upper bound on the number of feasible resolutions to the two state problem is $\prod_{j=1}^{m} (k_j!)$.

No simple upper bound is known for the three state problem.

In the example of figure 3.4, there are 64 resolutions to the two state problem of which at most 36 are feasible. There are 729 possible resolutions to the three state problem. Although the increase from 64 to 729 may not appear critical, for a similar problem with 3 machine types, and 5 tasks per machine, the number of possible resolutions is 2^{30} and 3^{30}, respectively. There are only $(2/3)^{30}$, or about 1/2060 as many possible solutions to the two state problem, which gives an indication of added complexity of the combinatorial aspect of the three state problem.

3.1 Disjunctive Graphs and Machine Sequencing

From the comments in the previous section, it appears that the three state problem should be more difficult to solve than the two state problem, since it has many more possible resolutions. Indeed, the majority of the published research has been on the two state or machine sequencing problem. These results will be discussed as a preface for describing the project scheduling problem.

Several of the algorithms [17, 18, 21, 40] are quite similar to the permutation method proposed by Schrage [29] (see Section 4) in that the partitioning proceeds by specifying partial resolutions, which correspond to Schrage's partial permutations.

As was noted in section 2.1, Greenberg's [21] mixed integer linear programming model is actually a statement of the disjunctive graph model. It turns out that the linear program that he proposes to solve at each node of the branch and bound is nothing more than a CPM problem, and the bounds could be obtained as the longest path for the corresponding partial resolution.

Florian and Trepant [17] describe the branching scheme in terms of a cut in the disjunctive graph. The bounds are obtained by solving a sequence of subproblems, one for each machine that performs some terminal task (the last task for some job).

Sussman's [40] branching scheme resembles that of Florian and Trepant, in that the partial resolutions are constructed in order of "earliest conflicts". The bounds are the critical path length for the graph with the unresolved d.p. deleted.

Balas [1,2,3,4,5] and Raimond [36,37] give algorithms based on implicit enumeration of feasible solutions. Some resolution is specified (perhaps arbitrarily) as a starting point for the enumeration. A candidate set of disjunctive arcs can be identified, which are the only ones that need be considered when searching for resolutions which yield a shorter critical path. For a given resolution, subsequent resolutions are generated by reversing the disjunctive arcs in the candidate set, one arc at a time. In order to eliminate redundancy in the enumeration, a state vector can be associated with each resolution, indicating which arcs have already been reversed and in what order, so that backtracking can be used.

The proposed algorithms differ principally on two points. First, the algorithms in [1,2] and [36,37] consider resolutions that generate cycles in D_R, which cannot be feasible, whereas in [3,4,5] it is assumed that the starting resolution is feasible, and only feasible resolutions are examined. Second, the bounding procedures employed differ, but since few computational results are given, the several bounds cannot be compared. Balas [6] essentially interprets [5] in terms of degree constrained subgraphs, which are a representation of the enumeration.

3.2 Schedule Modification and Disjunctive Graphs

Although Balas [3,4] defined the disjunctive graph model for project scheduling in 1969, only one algorithm, due to Gorenstein [20], has been proposed for its solution. The most apparent obstacle to developing an algorithm is in devising a method for determining the generalized coefficient of stability, GCS_j. Essentially, the maximal internally stable subsets of N^j (i.e., those having the largest number of unrelated elements) must be identified for each resource j, in order to find the GCS_j.

Gorenstein determines the GCS_j by a maximum flow algorithm based on the Konig-Egervery theorem and the Dilworth Chain Decomposition theorem applied to a bipartite graph defined on $N^j \times N^j$. The enumeration is a three state version of

[5] and [37], using bounds based on the critical path length, and the GCS_j to determine feasibility.

The GCS_j are not used as bounds, rather they only indicate if a given resolution is resource feasible. Also, although $GCS_j \leq a_j$ implies resource feasibility, the converse is not necessarily true. A simple example serves to demonstrate that using GCS_j as a criterion may result in <u>not</u> recognizing an optimal solution when it first occurs.

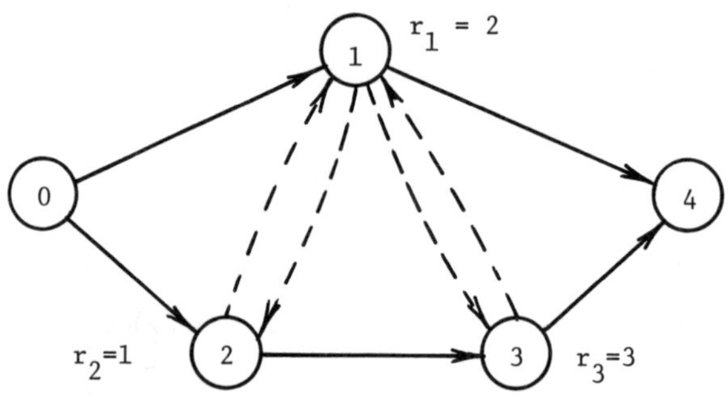

Figure 3.6

If all activities have a duration of 1 period, and $a_1 = 3$, the CPM schedule is resource feasible and optimal. However, $GCS_1 = 5$ for the resolution with all of the d.p. deleted, so the null resolution, although optimal, would not be recognized. Of course, for this example, the optimal solution is found after one branch, but clearly a serious problem exists in using the GCS_j criterion.

The problem may be resolved by comparing the skyline (see Section 5) to the resource limit to determine feasibility. Since the skyline can be determined as the schedule is generated, it requires very little additional computation, in contrast with the GCS_j calculations. If the GCS_j criterion is satisfied, then by definition, the skyline criterion will be satisfied. However, as illustrated above, the skyline criterion may detect a feasible solution even though the GCS_j criterion does not; thus it <u>always</u> dominates the GCS_j criterion.

4. Schedule Construction Methods

The general approach in the algorithms of this section is to "build up" to a feasible solution. For this reason, they will be referred to as schedule construction methods. As with many of the other methods, the schedule construction methods were originally proposed for the machine sequencing problem. A few of the results have been generalized to the project scheduling problem.

One way to obtain a feasible schedule to the machine sequencing problem is to use an ordering or permutation of the jobs to determine the starting times. The permutation is selected so that if $(i,j)\varepsilon P$ is a precedence between tasks i and j then i precedes j in the ordering. Given this ordering, the tasks are scheduled as soon as the appropriate resources are available. Notice that although task i may be considered before task j, $t_j < t_i$ is possible because of resource availabilities. This procedure is the basis for almost all of the heuristic scheduling rules.

Schedule construction using permutations was first used by Giffler and Thompson [19] in enumerating all feasible active schedules (a schedule is active if all jobs start as early as possible for the given permutation). For example, for the problem of figure 1.1, the permutation (1,2,4,3,5,6) generates the CPM early-start schedule, as give in table 1.1. Brooks and White [9] used this enumeration in a branch and bound algorithm, but made no attempt to show that the procedure generated schedules non-redundantly.

W. Karush [24] proved that a similar method could be used to solve the constrained resource project scheduling problem, provided that the resource requirements, $r_i(t)$, are non-increasing. He also gives an example that shows how the method fails if some $r_i(t)$ is an increasing function of time. (i.e., the optimal solution need not be active in the permutation sense).

Several theorems for constructing permutations for a more general case of the machine sequencing problem where additional precedences between tasks of different jobs may be specified are given by Schrage [39]. A feasible permutation is defined as one for which, if $(i,j)\varepsilon P$, then j does not precede i in the permutation. Schrage shows that a feasible active schedule and the permutation obtained by a lexicographic ordering of the vectors $[t_i,i]$ are equivalent (t_i is the start time from the feasible active schedule). From this result, a redundant permutation is one for which i precedes j in the permutation, but $t_j<t_i$ in the corresponding feasible active schedule. It follows that redundant permutations need not be considered.

An activity, i, is <u>dominant</u> in position K of the partial permutation J(1), J(2),...,J(K-1) if it can be started in position K and finished before any other activity could start if it were in position K. If $\sigma_j(K)$ is the (feasible) start time for j if it is in position K, this result is stated as

$$\sigma_i(k) + d_i \leq \sigma_j(K) \quad \text{for all j not in the partial permutation.}$$

Similarly, i is <u>dominated</u> in position K if there is some activity, j, which is dominant in position K. The immediate results are:

i) A complete permutation constructed from J(1),...,J(K-1) for which i is dominant in position K is redundant if J(K)≠i, and

ii) A complete permutation constructed from J(1),...,J(K-1) for which i is dominated in position K is redundant if J(K)=i.

Schrage describes an implicit enumeration scheme for generating the feasible permutations, and two weak bounds for directing the search for the optimal permutation. The results may be applied to the general project scheduling problem only in the case of non-increasing resource requirements. It should be noted that in general, it may be very difficult to determine $\sigma_j(K)$, which diminished the usefulness of the dominance results.

Another approach to schedule construction is to bypass the artifice of job permutations, and work directly with schedules. An algorithm by Mason and Moodie [32] utilizes a direct approach in an enumeration scheme for minimizing cost of resource consumption in the project scheduling problem. This model minimizes the costs associated with increasing project duration, increasing resources utilization, and decreasing resource utilization. Although it is possible to fix the resource availability, ignore the leveling costs, and concentrate on minimizing the project duration, this would destroy the cost bounds used in the branch and bound algorithm.

Johnson [23] develops a branch and bound algorithm based on a modification of a heuristic given by Moder and Phillips [34]. The heuristic constructs a feasible schedule as follows:

(1) Given a partial schedule S_k at time t_k, consider the set of all unscheduled activities, whose immediate predecessors have been completed, $C = \{i \varepsilon N,\ i \not\varepsilon S;\ t_j + d_j \leq t_k,\ (j,i) \varepsilon P\}$.

(2) Order the set C by decreasing maximum remaining path length $(t_n - t_i - d_i)$.

(3) Schedule as many activities from C to start at t_k as there are resources available, in the order give by (2).

Let $S_{k+1} = S_k + \{\text{activities scheduled in step (3)}\}$.

(4) Determine t_{k+1} as $\min\limits_{i \varepsilon S_{k+1}} \{t_i + d_i \mid t_i + d_i > t_k\}$. Go to step (1).

Steps (1) - (4) are repeated until all activities have been scheduled.

Johnson's modification is that, instead of considering single activities, sets of activities are considered, i.e.

(2´) Form all subsets of C for which the total requirement by the subset does not exceed the remaining availability. Suppose there are n_k subsets.

(3´) Form n_k new partial schedules by appending each of the subsets from (2) to S_k.

Now each of the new partial schedules, S^i_{k+1}, $i = 1, \ldots, n_k$, must be considered in turn. A rule is given for determining the order in which to consider the new subsets, which is intended to lead quickly to a good solution.

Two bounds are calculated for each partial schedule. The critical path length yields one bound, and the other is obtained by considering the resource requirements. For a given partial schedule, S_k, any resources that are idle in periods prior to t_k will be idle in any complete schedule constructed from S_k. Thus a bound on duration can be obtained by adding the amount of these idle resources to the total requirement by the project, and dividing the sum by the constant availability.

An algorithm based on assembly line balancing models has been presented by

Davis [12,13] and Davis and Heidorn [14]. The model requires that each activity, i, be decomposed into d_i unit duration activities, as illustrated in figure 4.1.

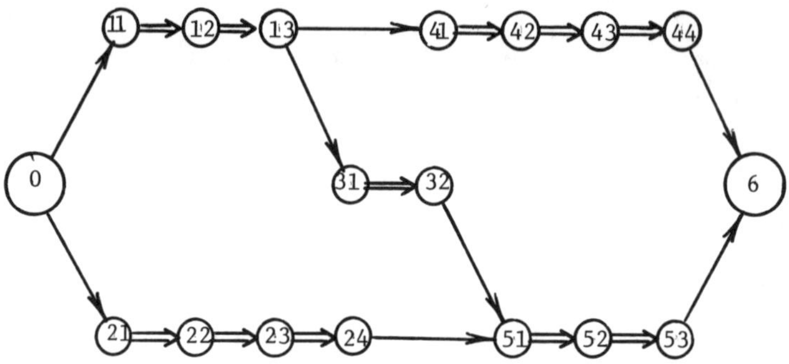

Figure 4.1

Since, in the original problem, no activity can be interrupted once it has been scheduled, in the decomposed problem a "must immediately precede" relation is defined and represented by double arcs in figure 4.1. Davis employs an algorithm by Gutjahr and Nemhauser [22] to generate feasible schedules for the decomposed problem. The result of the algorithm is a sequence $\{S_1, S_2, \ldots, S_T\}$, where S_t is the set of activities to be done in period t. A feasible solution is one for which the sequence does not violate either the precedence or resource constraints, i.e.:

(1) if $(i,j)\varepsilon P$ then $i\varepsilon S_k$ and $j\varepsilon S_p$ with $k<p$;

(2) if $(i,j)\varepsilon P$ and i must immediately precede j then $i\varepsilon S_p$ and $j\varepsilon S_{p+1}$ for some p;

(3) the resources used in each period $\sum_{j\varepsilon S_t} r_j(t) \leq a$ for $t=1,\ldots,T$.

An optimal solution minimizes the project duration T. Note that a constant resource availability is required for this model.

The sequence $\{S_t\}$ is not constructed directly, but is determined from an equivalent shortest path problem in the following manner. A <u>feasible subset</u> of N, $F_m \subseteq N$, is a set of activities such that, if $i\varepsilon F_m$ then all predecessors of i are in F_m. If all feasible subsets of N are constructed, then the project scheduling problem can be transformed into an equivalent shortest path problem in a special network.

The network is constructed from the set of feasible subsets as follows:

1) let each feasible subset be represented as a node in the network
2) construct an arc from node F_k to node F_m if the following conditions are satisfied:

   i) $F_k \subseteq F_m$
   ii) $j \varepsilon F_m$, then all predecessors of j are in F_k
   iii) $i \varepsilon F_k$, i immediately precedes j, then $j \varepsilon F_m$
   iv) the resource requirement by activities in $F_m - F_k$ does not exceed the availability, a.

The network is called the A network and a straightforward procedure for its construction is given in [12] with rules for eliminating some feasible subsets. Some of the elimination rules depend on selecting a "target" duration, and their effectiveness naturally depends on the accuracy of the estimated duration. Davis also gives a dynamic programming algorithm for finding the shortest path in the A network, which is applied simultaneously with the subset generation procedure, so that only one arc into each node (feasible subset) need be saved.

Davis describes a method for determining how many feasible subsets will be generated for any particular problem, and shows that this number is directly related to the number of tasks and inversely related to the number of arcs. From this, the easier problems would be those having many precedence constraints and a large total slack.

Note that the feasible subsets defined by Davis are completely analagous to the partial schedules generated by Johnson's algorithm. The algorithms differ greatly in the manner in which they use the information about feasible subsets or feasible partial schedules in searching for an optimal schedule.

5. Bounds

Although several algorithms apparently have successfully utilized dominance rules for reducing the necessary enumeration, the bounds used have been weak ones.

There are essentially only two types of bounds: those which ignore resource constraints, and those which ignore precedence constraints.

In the schedule modifying algorithm, only bounds of the second type are used. The bound is determined from a CPM duration. In the schedule construction algorithms either type bound may be used. The first type is computed by finding the CPM start times for all unscheduled activities, given the start times already assigned to activities in the partial schedule. The second type is computed by totaling the requirement by unscheduled activities, and comparing this to the duration necessary, at the given availability, to achieve this total.

Zaloom [44,45] has investigated the possibility of deriving a stronger bound by simultaneously considering both resource and precedence constraints. The basis of this stronger bound is the project skyline or resource requirement profile for a given schedule. The cumulative requirement profile is compared to the cumulative availability to derive a lower bound on the amount of project delay. Since this bound requires little more computation than the CPM bound, it is clearly an improvement.

6. Summary

Only five algorithms give any computational results. In three cases, the results are of such limited scope as to make comparison on a relative basis meaningless. Computing times for the algorithms will not be given, because the authors feel that the size of problem that can be solved, rather than time to solution, is the critical measure.

The model of Pritsker, Watters, and Wolfe [35] is limited in usefulness by the state of the art in integer programming. The solvability of an integer programming problem decreases fairly rapidly as the number of constraints increases. Since the number of constraints in this model varies as mT, even if the optimal duration is known *a priori* (e.g. from a hueristic solution), the number of resource types that could be considered will be small. Only one example, having eight activities, is cited.

The approach by Elmaghraby [15] and Bhadury [7] is, in a sense, subject to the same limitations, since the solvability of a linear program depends on the number of constraints. The number of nodes that would have to be examined in the enumeration proposed is very much an open question. The method was tested for the problem of scheduling a network of activities with due dates. The penalty costs were incorporated in the objective function and were used to obtain a cost-related bound. Fourteen problems were solved, ranging in size from 5 to 20 activities. Only six problems required branching (i.e. in 8 problems the initial LP solution was integer) and no problem required more than two branches.

The disjunctive graph algorithm [20] has been applied to variations of 7 machine sequencing problems where more than one machine of each type is available. The largest problem had no more than 40 tasks; the exact number cannot be determined. Problem size in terms of the number of d.p. ranges from 61 to 9. In two cases an optimal solution was not obtained, in one variation of the 61 d.p. problem, and in a 24 d.p. problem.

By far the most comprehensive computational experience is reported by Johnson [23]. Table 6.1 shows the relevant results, where failure to complete a problem was due to excessive running time.

Number of Activities	20	30	40	50	60	70	100
Number of Problems	50	79	18	12	6	4	1
Per Cent Completed	100	93	72	92	67	50	0

Table 6.1

The networks for each problem size were randomly generated, and Johnson indicates that they do, in fact, resemble real project networks. Johnson further asserts that problems of 50 or more activities can be considered "realistic." Although only one resource was considered, the extension to multi-resource problems is discussed.

Davis [12, 13, 14] also gives extensive computational results, which are summarized in Table 6.2. Failure to complete a problem was caused by exceeding

Number of Activities	22	27
Number of Problems	33	32
Per Cent Completed	85	62.5

Table 6.2

internal storage capacity when generating the feasible subsets. The problems all had networks similar to that in Figure 6.1, where there are several ranks of nodes, and each node has arcs only to nodes in adjacent ranks.

The results indicate that the procedures have not been used successfully for any problems of a realistic size. Further, the different algorithms each have characteristics that will inherently limit their extension to problems of a significantly larger size, say 100 activities.

The models of Pritsker, Watters, and Wolfe [35]; and Elmaghraby and Bhadury [15, 7] grow in size so rapidly that the resultant integer programming problems would quickly become unmanageable. If Gorenstein's [20] algorithm is used with the GCS feasibility criterion, the CPM bounds will be so weak as to be useless. The number of feasible subsets generated by each partial schedule is critical in Johnson's algorithm [23]. As the number of activities increases, the set C will become larger and the number of feasible subsets will become unmanageable. Finally, Davis' [12] algorithm will be unable to generate all of the feasible subsets for a much larger problem.

It appears that little progress has been made on the resource constrained project scheduling problem since Johnson's work in 1967. Thus, the problem remains unsolved for practical problem sizes. This may be caused by the difficulty of including <u>both</u> the precedence and resource constraints in a mathematical formulation that maintains sufficient structure to aid in the solution.

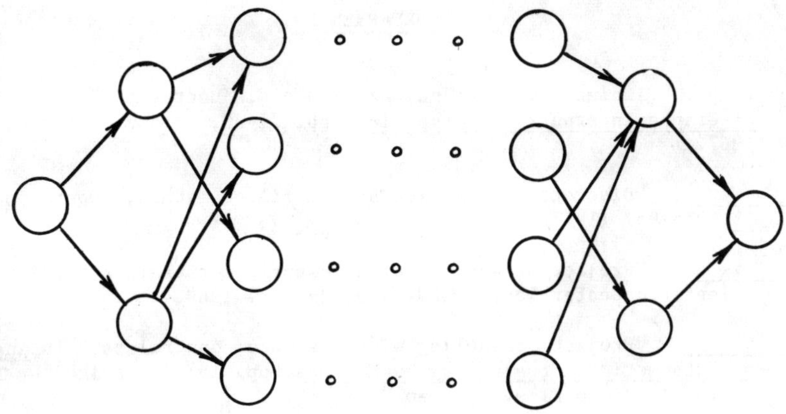

Figure 6.1

REFERENCES

1. Egon Balas, "Finding a minimaximal path in a disjunctive PERT network," Theory of Graphs International Symposium, pp. 21-30, Gordon & Breach, New York, 1967.

2. __________, "Discrete programming by the filter method," Operations Research, 15, 915-957 (1967).

3. __________, "Project scheduling with resource constraints," IBM New York Scientific Center Report 320-2960, January 1969.

4. __________, "Project scheduling with resource constraints," In Applications of Mathematical Programming Techniques, pp. 187-200 edited by E. M. L. Beale, American Elsevier, New York, 1970.

5. __________, "Machine sequencing via disjunctive graphs: an implicit enumeration algorithm," Operations Research, 17, 941-957, (1969).

6. __________, "Machine sequencing: disjunctive graphs and degree constrained subgraphs," NRLQ, 17, 1-10 (1970).

7. Bikash Bhadury, "An experimental investigation of a model for sequencing N jobs on M machines in parallel," unpublished thesis, North Carolina State University (1970).

8. E. H. Bowman, "The schedule-sequencing problem," Opns. Res., 7, 621-624 (1959).

9. G. H. Brooks and C. R. White, "An algorithm for finding optimal or near-optimal solutions to the production-scheduling problem," J. Ind. Eng., 16, 34-40 (1965).

10. C. B. Chapman, "Allocation of resources to variable timetables," Opl. Res. Q., 21, 81-90, (1970).

11. G. B. Dantzig, and D. R. Fulkerson, "Minimizing the number of tankers to meet a fixed schedule," NRLQ, 1, 217-222 (1954).

12. E. W. Davis, "An exact algorithm for the multiple constrained-resource project scheduling problem," Ph.D. dissertation, Department of Administrative Sciences, Yale University, 1968.

13. __________, "Computational experience with a new multiresource algorithm, in Project Planning by Network Analysis, pp. 256-260, edited by H.J.M. Lombaers, North Holland Publishing Co., Amsterdam - London, 1969.

14. __________, and G. E. Heidorn, "An algorithm for optimal project scheduling under multiple resource constraints," Mgt. Sci., 17, B803-816 (1970).

15. S. E. Elmaghraby, "The sequencing of n jobs on n parallel processors with extensions to the scarce resources problem of activity networks," Proceedings of the Inaugural Conference of the Scientific Computation Center and the Institute of Statistical Studies and Research, pp. 230-255 Cairo, 1969.

16. M. Fisher, "Optimal Solution of scheduling problems using LaGrange multipliers: Part II," Proceedings of Symposium on Theory of Scheduling and Its Applications, Raleigh, N. C., May 15-17, 1972.

17. M. Florian and P. Trepant, "An implicit enumeration algorithm for the machine sequencing problem," Department d'informatique, Universite de Montreal (1969).

18. _____________, and G. McMahon, "An implicit enumeration algorithm for the machine sequencing problem," Mgt. Sci. 17, B782-792 (1971).

19. B. Giffler and G. L. Thompson, "Algorithms for solving production scheduling problems," Opns. Res., 8, 487-503 (1960).

20. S. Gorenstein, "An algorithm for project (job) scheduling with resource constraints," Opns. Res., 20, 835-850 (1972).

21. H. H. Greenberg, "A branch and bound solution to the general scheduling problem," Opns, Res., 16, 353-361 (1968).

22. Allen L. Gutjahr and G. L. Nemhauser, "An algorithm for the assembly line balancing problem," Mgt. Sci. 11, 308-315 (1964).

23. T. J. R. Johnson, "An algorithm for the resource constrained project scheduling problem," Ph.D. thesis, Massachusetts Institute of Technology, 1967.

24. W. Karush, "On scheduling a network of activities under resource constraints over time," SP-2654, System Development Corporation (1966).

25. J. E. Kelley, Jr., "Critical-path planning and scheduling: mathematical basis," Opns. Res., 9, 296-320 (1961).

26. ____________, "The critical path method: resource planning and scheduling," in Industrial Scheduling, pp. 347-365 edited by J. F. Muth and G. L. Thompson, Prentice-Hall, Englewood Cliffs, N. J. (1963).

27. S. Lambourn, "Resource allocation and multi-project scheduling (RAMPS), a new tool in planning and control," Computer Journal, 5, 699-719 (1966).

28. E. L. Lawler, Notes on Combinatorial Optimization, manuscript, 1969.

29. F. K. Levy, G. L. Thompson, and J. D. Wiest, "Multi-ship, multi shop workload-smoothing program," NRLQ, 8, 37-44 (1962).

30. D. G. Malcomb, J. H. Roseboom, C. E. Clark, and W. Fazar, "Applications of a technique for research and development program evaluation," Opns. Res., 7, 646-669 (1959).

31. A. S. Manne, "On the job-shop scheduling problem," Opns. Res., 7, 219-223 (1959).

32. A. Mason and C. L. Moodie, "A branch and bound algorithm for minimizing cost in project scheduling,: Mgt. Sci. 18, B158-173 (1971).

33. J. H. Mize, "A hueristic scheduling model for multi-project organizations," unpublished Ph.D. thesis, Purdue University, August, 1964.

34. J. J. Moder and C. R. Phillips, Project Management with CPM and PERT, Reinhold Corporation, New York, New York, 1964.

35. A. A. B. Pritsker, L. J. Watters, and P. M. Wolfe, "Multiproject scheduling with limited resources: a zero-one programming approach," Mgt. Sci. 16, 93-108 (1969).

36. J. F. Raimond, "An algorithm for the exact solution of the machine scheduling problem," I.B.M. New York Scientific Center 320-2930 (1968).

37. ___________, "Minimaximal paths in disjunctive graphs by direct search," I.B.M. J. of Res. and Dev., 13, 391-399 (1969).

38. B. Roy and B. Sussman, "Les problems d'ordonnancement avec constraintes disjonctives," SEMA, Note D.S., No. 9 bis decembre 1964.

39. Linus Schrage, "Solving resource constrained network problems by implicit enumeration-nonpreemptive case," Opns. Res., 18, 263-278 (1970).

40. B. Sussman, "Scheduling problems with interval disjunctions," Tech. Rep. No. 70-5; Operations Research House, Stanford University (1970).

41. H. Wagner, "An integer programming model for machine scheduling," NRLQ, 6, 131-139 (1959).

42. J. D. Wiest, "Some properties of schedules for large projects with limited resources, Opns. Res., 12, 359-418, (1964).

43. __________, "A hueristic method for scheduling large projects with limited resources," Mgt. Sci., 13, 359-377, (1967).

44. Victor Zaloom, "An examination of the expected minimum time to complete a resource constrained project," Ph.D. thesis, Department of Industrial Engineering, University of Houston (1970).

45. __________, "On the resource constrained project scheduling problem," AIIE trans., 3, No. 4 (1971).

SEQUENCING RESEARCH AND THE INDUSTRIAL SCHEDULING PROBLEM*

S. S. Panwalkar, R. A. Dudek, and M. L. Smith

Texas Tech University, Lubbock, Texas, USA

Types of industrial scheduling problems were investigated by personal visits to plants and by questionnaires mailed to scheduling departments. Information on problem sizes, job flow, optimization criteria and job similarity was obtained. Results indicate that most of the present procedures in theoretical research cannot handle average industrial problems. Also most commonly used objective criteria differ from industrial goals. There is a definite need for better communication between sequencing researchers and scheduling practioners.

Sequencing research from 1954 to the present time has been directed toward solution of many different problems related to the order of processing jobs on machines. Variations in problem definitions arise from criteria to be minimized or maximized and restrictions on precedence, technological requirements both within jobs and among jobs.

Criteria that have been considered include minimization of total elapsed time or makespan, minimizing average time in the shop, minimizing job lateness or lateness penalties, minimizing machine set-up costs, minimizing inprocess inventory costs, and others. Concerning precedence restrictions, the common problem

*This research was sponsored by the National Science Foundation Grants GK 1156 and GK 2869.

definitions are flow shop with no passing of jobs, job shops, and resource constrained networks. A special case is a shop with only one machine.

Research in job sequencing at Texas Tech University has concerned several of the above variations in problem definition. Although several algorithms have been developed and publicized, the lack of little known practical application of the research results has led to a reexamination of the scheduling problems found in industry and the sequencing research conducted at Texas Tech. One possible reason for a lack of application of sequencing research is that sequencing procedures solve problems that are rarely found in industrial scheduling environments. To gather information on the scheduling problems of industry two surveys were conducted by mail, and several plant visits were conducted.

Each survey was designed to collect data on a specific aspect of scheduling in addition to general information. The first survey was conducted in 1969 and concerned the similarity or patterns of processing times.(2) This survey was sent to 250 companies throughout the United States; 75 replies were received.

The second survey was designed to collect data on the importance of set up times in schedule determination.(1) This survey was conducted in 1971 and was sent to 108 companies; 40 replies were received. Since the majority of companies in the 1971 survey were the same as those surveyed in 1969, answers to the general questions were combined.

Results from General Questions

1. When a number of jobs are being scheduled for several operations in a specific shop, how many jobs and machines are generally involved at a given time?

Jobs	Machines	% of Total Responses
1-10	1-10	19
11-20	1-10	3
	21-30	5
	31-40	3
21-40	1-10	3
	11-20	5
	31-40	3
41-80	31-40	5
	more than 40	8
more than 80	1-10	14
	21-30	3
	31-40	3
	more than 40	27

2. Are most of the machines of a different type?

Answer	% of Total Responses
Yes	60
No	40

3. What type of job flow is in the shop?

Answer	% of Total Responses
Flow shop	20
Job shop	57
Combination or other	23

4. Rank in importance (1 - highest/4 - lowest) the following criteria for a scheduling decision.

A. Minimizing total processing time
B. Meeting due dates or minimizing penalty costs
C. Minimizing total set-up time or cost
D. Minimizing in-process inventory costs

Criterion	Rankings	1	2	3	4
		No. of Responses for Each Ranking			
A. Minimizing total processing time		6	14	9	11
B. Meeting due dates or minimizing penalty costs		33	2	1	4
C. Minimizing total set up time or cost		4	12	15	9
C. Minimizing inprocess inventory costs		2	8	11	19

A single measure of importance of the criteria was obtained by multiplying the highest ranking (1) by 4, the second ranking (2) by 3, etc., these subtotals were added for each criterion resulting in the following:

Criterion	Total
1. Meeting due dates	144
2. Minimizing total processing time	95
3. Minimizing total set-up time or costs	91
4. Minimizing inprocess inventory costs	73

Results from the Questions on Scheduling for Set Ups

5. In scheduling jobs, is it important to minimize set up times (costs) in order to achieve a lower total cost?

Answer	% of Total Response
A. Yes, most of the time	55
B. Yes, sometimes	30
C. Rarely	15

6. What percentage of the situations are set up costs mainly due to:

A. Time and labor
B. Material
C. Other

	Combinations of Set Up Cost Sources					
A. Time	100%	75-95%	50%	25%	--	--
B. Material		5-25%	50%	25%	20%	--
C. Other		5-10%	--	50%	80%	100%
% of Total Responses	45	42	3	3	3	5

7. Sequence dependence of set up times occurs with what percentage of jobs?

% of Operations With Sequence Dependence	% of Total Responses
100	13
75	15
50	20
25	23
0	30

8. What percentage of jobs require similar set ups?

Percentage of Jobs Requiring Similar Set Ups	% of Total Responses
100	13
75	23
50	18
25	38
0	10

9. Is the schedule arranged to allow jobs using similar set ups to be processed together?

Answer	% of Total Responses
Yes	78
No	22

The questions on ordered processing times concerned relationships of processing times among jobs and machines. An ordering of processing times according to jobs could be due to relative sizes or complexity of the jobs. Similarly, an ordering of processing times according to machines can result from the different natures of the operations performed on the machines. If one operation is complex for all jobs and another is simple for all jobs, the former machine normally will have larger processing times than will corresponding jobs on the latter machine. A plot of elemental values from a processing time matrix for a three job, three machine ordered problem is shown in Figure 1.

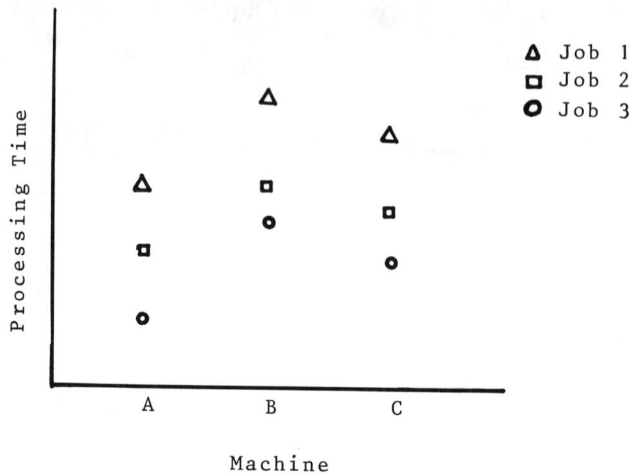

Figure 1

One of the two surveys was designed to investigate the prevalence of ordered matrix problems in industry. Questions and results follow:

10. Are all of the products in the shop similar in terms of manufacturing operations involved?

Answer	% of Total Responses
Yes	40
No	60

11. Is there an approximate similarity in trend for process times for similar products?

Answer	% of Total Responses
Yes	77
No	23

12. What is the predominate trend in processing times for similar products?

Answer	% of Total Responses
Progressively increasing	14
Progressively decreasing	23
Similarly fluctuating	63

Discussion of Survey Findings

The survey findings provided several suggested reasons for the lack of application of sequencing research to the industrial scheduling problem. These results indicated there was little in common between the problem approached by many sequencing algorithms and the problem that actually exists.

The number of jobs scheduled at any time usually exceeds the computational capability of most sequencing algorithms. Only 19% of the responses indicated problem sizes with no more than 10 jobs and 10 machines. Therefore, many sequencing procedures, while perhaps approaching the correct industrial problem, are incapable of handling the prevalent problem size. The responses suggest the use of heuristic or approximating procedures rather than optimizing due to problem size.

Considerations that were not fully investigated were availability of computers with adequate capability for sequencing to the scheduling organizations. While some companies did have large computers, many did not.

Most companies considered the meeting of due dates to be the most important criterion. This same finding was reported during the ten plant visits without exception. Minimization of total processing time and minimization of set up time were given approximately equal importance after meeting due dates, and minimizing inprocess inventory was considered least important. From the plant visits, it was found that almost every company had primary and secondary criteria. Of first importance was the meeting of due dates, and when that was accomplished, the most common secondary objective was to minimize set up times. It appeared that inprocess inventory levels generally were not considered in routine scheduling decisions; rather inprocess inventory levels became important every year or two when higher management ordered a reduction in the level. Thus, it appears that inprocess inventory levels are important but are not very important to those individuals making routing scheduling decisions.

The criteria most commonly used in sequencing research were not extensively used in industrial scheduling. Although minimization of total time was considered important in scheduling, caution should be used in equating this criterion with the make-span of sequencing research. One reason for this is that most industrial problems are dynamic in that incoming jobs are incorporated in the schedule before the previously scheduled group of jobs is completed; thus the make-span criterion is inapplicable to the dynamic situation. There is a possibility that the minimization of total processing time criterion was interpreted as meaning either minimization of time in the shop for each job or minimization of time required in processing only; the latter objective would be achieved primarily by proper selection of tooling, material, and machine speeds and feeds, etc., rather than by job scheduling.

From the results on criteria, it appears that sequencing algorithms provide little assistance to industrial schedules.

There are very few algorithms that can handle either due dates or set up times, and there are even fewer algorithms that can handle both due dates and set up times or any other combination of criteria.

Responses to the questions related to problems with ordered processing times indicated the ordered problem does exist in industry. It is quite probable that other types of specially structured problems are common in industrial scheduling. One approach to sequencing research might be the development of solution procedures for these special types of problems. In some cases characteristics of the problem might be utilized so that efficient solution techniques can be developed.

Conclusion

Significant advances in theoretical developments and solution techniques have resulted from the efforts expended in sequencing research. It does appear that most of the research has been basic in nature. Application of this research and development of procedures for common industrial scheduling problems has lagged the theoretical research.

It may be desirable for many researchers to link up with the industries to better understand their problems and thus to gain insight into solution procedures needed in industrial environments. Some of the problems that one might face in this approach are the difficulty in collecting accurate data, making proper assumptions and comparing solution techniques, to name a few.

While both basic and applied research are necessary, there should be a balance between the two. In the theoretical research it may be desirable to investigate problems with composite criterion rather than a single one such as make-span, machine idle time, etc. Such work can eventually lead to more practical applications.

References

1. Prunzel, J.F., "Setup Times and Setup Costs in Sequencing Problems," Unpublished Master's Thesis, Texas Tech University, 1972.

2. Smith, M.L., "A Critical Analysis of Flow Shop Sequencing," Unpublished Doctoral Dissertation, Texas Tech University, 1968.

THE ENGINE SCHEDULING PROBLEM IN A RAILWAY NETWORK
PART I

M. Florian, G. Guérin, and G. Bushell

Université de Montréal, Montréal, Québec, Canada
Université de Québec, Québec, Canada
Canadian National Railways

Summary

In this paper we present the mathematical version of a scheduling problem that is faced by any railway company that employs several engine types to provide power for its trains. Usually, a railway employs engines of several types that differ in their tractive effort capability and horsepower rating. There are two distinct but related aims in reducing the engine costs to the railway. One is to select the mix of engine types that gives the lowest capital investment and operating cost for the trains operated by the railway. The other is to provide a scheduling method that assigns available engines to trains on a short time horizon (e.g. a week ahead). We describe the practical context of the former problem below.

A railway runs many of its trains to a predetermined schedule that is publicly available to buyers of their service. The convenient graphical representation of the scheduled trains is the widely used <u>time space</u> diagram. This diagram is a network that is plotted to scale using two coordinates : the horizontal coordinate is time and the vertical coordinate is distance. Thus, a train is specified on the abscissa by its departure time and arrival time and on the ordinate, by the originating and terminating cities. Train arcs are diagonal arcs in the network, when this convention is used. Horizontal arcs are added to the network to connect the terminal node of a diagonal arc to initial nodes of other diagonal (train) arcs that originate at the same city (station).

The flows associated with each diagonal arc are the engines that are assigned to the corresponding train, and with horizontal arcs engines that are idle and waiting to be reassigned to another train that originates at the same city. Empty or "dummy" trains may also be indicated on the time space diagram. Figure 1 gives an example of a time space diagram for 3 cities and 4 trains. The "smaller" arcs such as (2,3) and (4,5) represent the movement of engines within the railway terminal, starting at the time they arrive in a yard and ending at the time they are ready to be reassigned. Similarly, arc (11,12) represents the flow of engines that stays with the continuation of train 3 from city B to city A.

Each train (x,y) has a motive power requirement T(x,y) that is determined by the weight and length of the train. Every engine type i, for train (x,y) has a tractive effort rating $s_i(x,y)$ that is related to engine design and steepness of grades over which the train must be pulled. In addition, every engine type i, for train (x,y) has a horsepower rating $w_i(x,y)$ that is related to engine design and speed that the train must maintain in order to meet its scheduled arrival time. The operating constraint is to provide a train with sufficient engines in order to meet its motive power requirement T(x,y) with sufficient tractive effort and horsepower. Strictly speaking, $s_i(x,y)$ and $w_i(x,y)$ are not constants but vary with the mix of engine types that are assigned to a particular train. But, because they do not vary greatly when the majority of engine types are mixed together, they are considered to be constant in this formulation.

In addition, the engines must circulate in the railway network such as to prevent imbalances that may cause short supply at certain cities. The situation arises typically in the case where more tonnage is moving in one direction than in another (e.g. from east to west in the geographical sense).

We give a mathematical formulation of the problem of selecting the mix of engine types that gives the lowest capital investment and operating cost for the trains operated by the railway and outline a method of solution based on the decomposition method of Benders.

We have a small amount of computational experience with the method utilizing the weekly train schedule for the Atlantic Region of the Canadian National Railways. The time space diagram for this schedule involves 718 nodes and 986 arcs, with 216 of these arcs representing train movements.

The application of the algorithm with 2 engine types required 10 Benders type iterations, that were done in 155 seconds of CDC 6600 central processor. The same problem with 5 engine types was attempted. We obtained a solution where the difference between the lower and upper bounds provided by the Benders method was of the order of 1% after 6 iterations and 300 seconds.

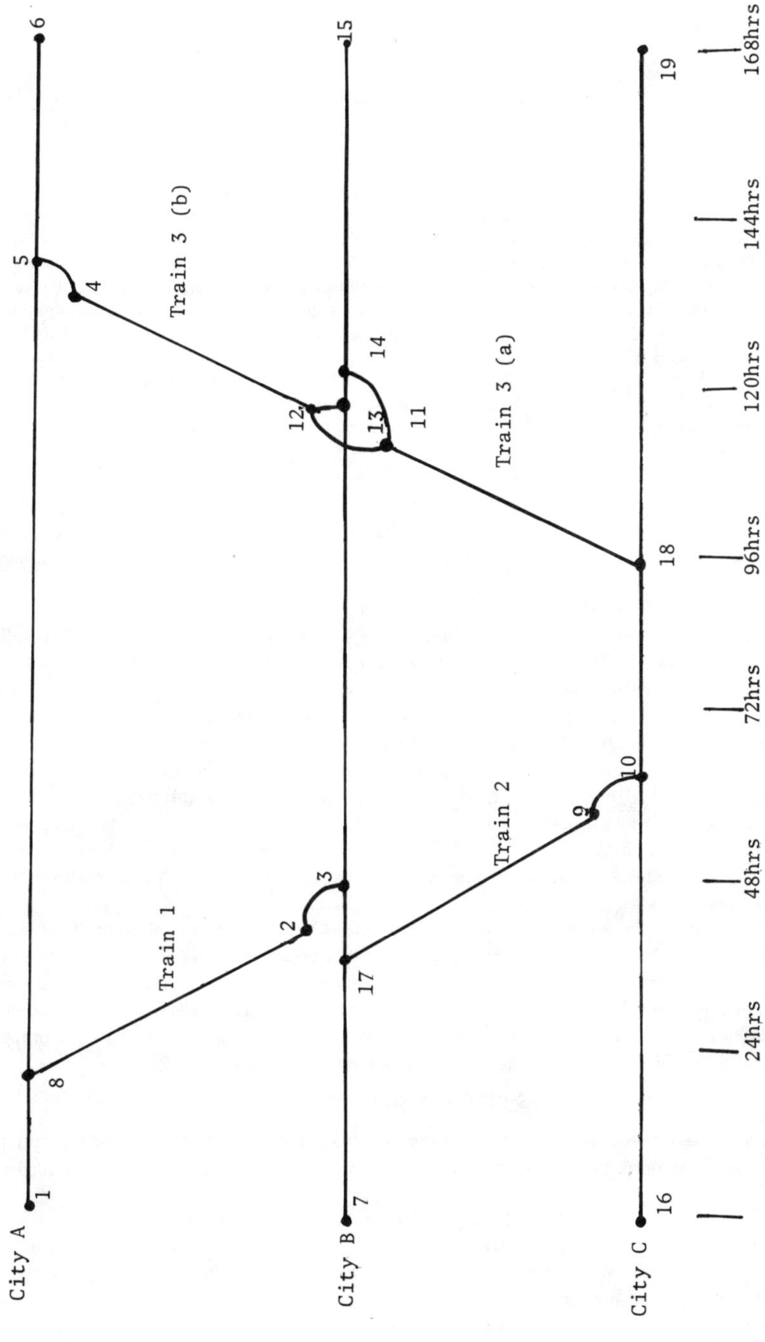
City A
City B
City C
Train 1
Train 2
Train 3 (a)
Train 3 (b)
1
2
3
4
5
6
7
8
9
10
11
12
13
14
15
16
17
18
19
24hrs
48hrs
72hrs
96hrs
120hrs
144hrs
168hrs

FIGURE 1

DETAIL SCHEDULING MODELS AND SYSTEMS

B. Giffler

Industrial Consultant
North Caldwell, New Jersey, USA

This paper is a summary account of Chapters 20 and 21 (pp. 432-514) of the book PRODUCTION MANAGEMENT MODELS AND SYSTEMS (copyright Nov. 1972, The Management Seminars Institute, Inc., Box 414, Caldwell, N.J.). The chapters describe detail scheduling models and systems the author has developed to schedule and control resource-limited, on-going production systems: e.g. factories, data processing centers, etc. The emphasis of the work is on the design of man-machine systems in which a computer will generate detail schedules that the production system's human controllers would then review and, if necessary, modify – by repeating the scheduling run with additional, special instructions. The models have an algebraic formulation which, the author claims, has helped simplify and speed the numerical calculations to generate and evaluate detail schedules, and which has also helped him recognize directions in which these models could be generalized without having to abandon their relatively simple mathematical structure.

THE DETAIL SCHEDULING PROBLEM

To detail schedule a production system is to determine when the operations of a given set of jobs will (or can or should) be performed, and the specific facilities that will perform them. The detail scheduling problem (that is considered in this paper) is to determine a particular detail schedule, namely the one that will satisfy the requirements of the problem.

Requirements of detail scheduling problems are of two kinds. One is called the *technological requirements* and refers to the constraints that must be satisfied by a *feasible schedule,* that is, a schedule that will produce the products of the jobs in a technically correct manner. The requirements of the *other kind* could be called the *esthetic requirements.* These requirements are used to single out particular feasible schedules that we would be inclined to describe as acceptable, good or optimal schedules – depending on how we would feel about the quality of the schedules we will have singled out.

An esthetic requirement can, in short, be whatever we want it to be. For example, we could require that the optimal schedule (that is, the one that will be singled out) be the feasible schedule that minimizes the time-span to complete all jobs in the problem's definition. This happens to be a quite difficult requirement to implement, except in special cases: e.g. there are not too many operations that we could not enumerate *all* the feasible schedules, or there are just two or three facilities and each job must take the facilities in the same order.

It is not necessary, however, that the esthetic requirement be difficult to implement. That is, there is nothing to prevent our defining an acceptable, or even an optimal schedule as the schedule that will be produced when the scheduling problem's indeterminancies are resolved by a specified loading procedure. This way of defining an optimal schedule is very attractive from an engineering point of view because very good schedules can be produced in this manner and with not too much difficulty. This is, in fact, the way we will be defining and solving detail scheduling problems in this paper.

EXAMPLES OF LOADING PROCEDURES

Loading procedures have two parts. One will determine the next facility for which a loading decision will made. The other part (it is often called the *loading rule* or *discipline*) will determine what the loading decision will be.

Loading procedures in our models will select as the next facility at which a loading decision will be made, the one that could start a next operation before any other facility – where the point in time at which a facility could start a next operation is taken to be the maximum of the earliest point in time at which the facility will be able to start a next operation and the earliest point in time at which a next ope- will be ready to start at that facility. We call this point in time the facility's NET (an abbreviation for "next event time"). Thus it follows that the facility at which the next loading decision will be made, will be the one that has the smallest NET. We call this the "facility of the min NET".

While it will always be possible to start an operation at min NET, it does not follow that the model, we will be using, will always do that. Note that the facility of the min NET is not described as the one at which the next operation will start – but only as the one at which the next loading decision will be made.

The point is that there are cases in which we will want to use loading rules that can make decisions *not* to start the next operation at the facility of the min NET — or, at least, not to start it at the current value of the min NET. To understand how and why we might want to make decisions not to start operations, it will be necessary that we understand how our models make loading decisions.

To begin, a loading rule will have two parts. One will define the conditions that will *qualify* an operation to be started next. The second part will determine which (if any) of the qualified operations will be selected to start next. If operations are qualified if (and only if) their jobs will be on-queue at min NET, we can be sure an operation will be started at min NET. It is, however, possible to qualify operations so long as their given next-preceding operations have started in which case the loading rule could select an operation to start that would not be ready to start until *after* min NET.

The loading procedure will not accept a decision to start an operation after min NET but will, instead, advance the facility's NET to the point in time at which the selected operation will be ready. This, of course, may change the value of the min NET and trigger one or more loading decisions before the min NET will return to the original facility. And when it does return, the loading rule may (in the interim) have qualified additional operations at the original facility of the min NET and may have made such other changes in the production system that the operation that was the original selection to start may no longer be the selection.

The loading rules in our models are implemented by a common logic. It is to select as the operation that is to start next, the qualifying operation that has the greatest (or smallest) priority number — depending on whether we prefer to use priority numbers that vary directly (or inversely) with the operations' actual priorities. Thus the differences in the actions of the loading rules will depend on the rationales that are used to determine the operations' priority numbers.

The rationales that are used to determine the operations' priorities can be almost anything we want them to be, but with one firm exception. It is that they must not use information that could not be obtained without knowing or presuming the outcome of loading decisions that had not yet been made. They can, thus, be numbers that were assigned to the operations when their jobs were released, or the values of characteristics of the production system that were determined by the loading decisions that had already been made.

One of the simplest loading rules we can use is called the FIFO (or *first in, first on*) rule. This rule will assign qualified operations priority numbers that are times their jobs joined the queue at the facility of the min NET and will select as the next operation to start, the one that has the smallest priority number. Another simple, but usually more efficient loading rule is called the FOFO (or *first off, first on*) rule. This rule will assign each qualified operation a priority number that is the nominal time at which the operation would (if selected) allow its facility to start a next operation and will select as the operation to start next, the one that (again) has the smallest priority number.

Another simple loading rule is called the SST (or *set-back start time*) rule. To use this rule we must first set-back schedule the operations, that is, determine the latest possible point in time at which each operation must start if their jobs' due-out times are not to be slipped. The loading rule will treat the operations' set-back start times as their priority numbers and will select as the operation to start next, the one that has the smallest priority number, that is, the one whose latest possible start time did (or will) occur first.

It is not necessary that we use rules with single rationales as in the above examples, or that we use the same rule for all facilities, operations and conditions of the production system. Thus we could, for example, use the FOFO rule at facilities at which the operations' processing times tend to vary to a considerable extent, and the FIFO or SST rules at the other facilities.

We could also use loading rules that will make an exception for very urgent operations, for example, operations whose on-time completion is extremely urgent — or even vital. It can be accomplished by identifying these operations' jobs in the scheduling input so that the scheduling program system could then assign their operations a priority class number of 1; and all other operations, a priority class number number of 2. A loading rule could then be defined that would give each operation whose priority class number is 1, a priority number that would be the number 1 followed by the operation's SST-number; and that would give all other operations a priority number that would be the number 2 (their priority class number) followed by their FOFO-numbers. If we now select operations to start that have the smallest priority numbers, we will automatically select an urgent operation if an urgent operation qualifies, and will select the most urgent of the urgent operations if two or more should qualify. If no urgent operation qualifies, we will select the the operation that would be the first to make way for the next urgent operation that will qualify.

We prefer to use simple loading rules like FOFO, FIFO, SST and their combinations. The reason is that detail scheduling problems' objectives and constraints are often vague or incomplete–and are also subject to unanticipated changes. We feel therefore that the systems we design to solve these kinds of problems should be man-machine systems, that is, systems that human controllers can instruct and over-ride when necessary. Obviously, if we use complicated loading rules, the controllers may not be sure how the scheduling system will react to their instructions. They may also be uncertain as to what the system could be trusted to do on its own.

It would be different, of course, if we were to design a system that did not need human controllers. But that would be like designing an automobile that did not need a driver. Such a vehicle could conceivably perform a few maneuvers on a highway. But who would want to ride in it?

DYNAMIC vs. STATIC SCHEDULING PROBLEMS

A detail scheduling problem can be static or dynamic. If it is a static problem, it would be implied that the problem's definition will not change before the problem's solution could be executed. We will therefore have no reason to regret or want to change that solution, that is, assuming we do not make mechanical errors in solving problems or become dissatisfied with a solution's theoretic basis.

If the detail scheduling problem is dynamic, it would be implied that the problem's definition will change before its solution could be executed. In this case we may have to be continually updating the problem and its solution. This is an important consideration when we design scheduling systems – because, obviously, the cost to run the system will depend on the frequency with which we will have to be updating last solutions.

It should be noted that dynamic scheduling problems must be solved as though they were static problems. The reason is that we will not know how the problems will change. Obviously if we did, we could take the changes into the problems' definitions – in which case they would no longer be dynamic. The fact, however, that the dynamic problem must be solved as a static problem does not mean we should ignore the problem's dynamic aspects. That is, we should bear in mind that these problems' definitions will change, and we should want to develop at least a "feel" for the nature of the changes that can occur – so we can design systems that will take these things into account. We should not want, for example, to sink increasing amounts of time and money into solutions of detail scheduling problems unless we are confident these expenditures will pay their way.

THE DETAIL SCHEDULING MODEL

We discuss in this section the main features and assumptions of a half-dozen scheduling models and systems we have developed for scientific and commercial application. This discussion will also bring in the kinds of the technological constraints that can be accomodated, the nature of the problems' indeterminancies and the manner of their resolution.

To begin, we assume the jobs will be known in advance, and that each will have a given due-out date or time. It is, further, assumed that the jobs' methodizations will also be known, but that these methodizations will not be so complete and detailed that there could be only one feasible schedule. For example, we will normally find that the methodizations will not prescribe the specific units of facility-types that are to perform each operation, or the specific sequence in which each unit is to perform the operations it will be its lot to perform. We will also frequently use models in which operations may be interrupted, but in which the particular operations that will be interrupted will not prescribed or the points at which operations will be interrupted. We will also frequently use models in which jobs can have alternate methodizations, and would not expect to find in these cases that the methodizations that will ultimately be adopted had, in fact, been prescribed.

One important feature of the scheduling models has already been described. This is the loading procedure, that is, the logical process that is used to determine which facility will be loaded next and the specific operation it will perform. In the remainder of this section we will be considering other features of the models, and also extensions and variations in the loading procedures that are needed to support them.

The Operations

We assume that each operation of each job will have a given or determinable *release time* (that is, an earliest possible point in time at which that operation would be ready to start), that each will be addressed to a specific facility-type and will have to be performed by a single facility (or unit) of that facility-type, and that each will have a given, nominal *duration* which, in our models, is taken to be the interval of time it will keep its facility from starting a next operation. We also assume that operations may be additionally addressed to a specific unit of their facility-types, or to a specific subset of its units.

The preceding characterization of an operation is that of a "single, continuous slug of work" that will be performed by a single facility. This does not, however, prevent our defining "slugs of work" that would require the joint services of two or more facilities. But to do that, it will be necessary that we create an operation for the work that would be done by each facility and that we state that these operations are to be performed *in concert,* that is, that the start times of these operations are *fixed* with respect to the start times of the other operations of their concert. [How concerts are specified and executed is described in a following section.]

Every operation, in our models, will not necessarily have to be performed. The reason is that jobs can have alternate methodizations in which there will be subsets of operations of which only one operation will be performed in each subset, or (more generally) in which there will be sets of subsets of operations of which only one subset will have to be performed in each set. The choice of the alternate methodization that will be adopted, is made by the loading procedure. Briefly, each alternate operation would be treated as any other operation – but only until another operation is started that would preclude its performance. The adopted methodizations will, in short, be the ones that could get started first.

The Facilities

Every facility must be a member of exactly one facility-type. This is a set of facilities that can perform any operation that would be addressed only to the facility-type, and that would perform them at approximately the same rate. The facilities of a facility-type must, in short, be interchangable with respect to the operations that would be addressed only to the facility-type. This interchangeability, however, does not refer solely to what a facility could physically be expected to do, but rather to *that plus how the facility will itself be used by the production system.* For example, two otherwise identical machines would probably not be treated as units of a same facility-type if they were (say) located in different buildings.

It is important to note that units of a same facility-type do not have to be *generally* interchangeable. That is, it is possible that individual units or even subsets of the units of facility-type could perform operations (or perform them at a rate) that the other units of its type could not. This situation can be tolerated because operations can be *additionally* addressed to specific units or subsets of the units of their facility-types. The preceding subsets are referred to as *variants* of their facility-types. The identical machines (in the preceding paragraph) could, for example, be treated as units of a same facility-type, by treating the units in each building as separate variants.

Operations may be directly or indirectly addressed to the specific units or variants of their facility-types. They would, for example, be directly addressed if the actual codes of the units or variants were specified in the scheduling problems' definitions. They could, however, also be indirectly addressed–by specifying that they are to be performed by the same unit or variant that performed some other, given operation, or by a unit or variant that has the same or a corresponding code. The other operations should, of course, be operations that would be started before the indirectly addressed ones.

The indirect addressing of the scheduling models has many uses. One is to subdivide very long operations into a linear sequence of suboperations that will keep returning to the same unit that will perform the first suboperation. Subdividing long operations in this way makes it possible for more urgent operations to "play through" the long operations at prescribed break-points, and with no danger that the suboperations would resume on different units – which, it is assumed we do not want.

Another interesting use of the indirect addressing is to channel all operations of particular jobs to a particular assembly-line or building. For example, we could have two buildings that have similar machines so that we could assign particular jobs to either building. It could be done by making one facility-type of all units that would be formed into one facility-type if they had been in the same building, and then making separate variants of the units in each building. We could, further, assign all variants of all facility-types that are in a same building, a common code that could be (say) *Building A;* and all variants in the other building, a common code that could be *Building B.*

We could now assign jobs to one building or the other by directly addressing their operations to the variants that have the code of the building to which we want their jobs to go. We could also let the jobs find their own building. This could be done by addressing the first operation of each job just to its facility-type and then *indirectly* addressing all other operations of the job to whichever variant will perform the first operation.

The terms *facility* and *facility-type* refer to the functions facilities perform, and not specifically to the machines or other substances that are performing the functions. This distinction can be very important when we have machines that can perform quite dissimilar functions at different times – and sometimes even at the same time. For example, a traffic light at an intersection could be two facility-types: one that allows east-west traffic to use the intersection part of the time, and another that allows north-south traffic to use it at other times. Another case in which the distinction between a facility and its substance can be important occurs with machines that have multiple spindles, or their equivalent. In these cases we can often vary the functions of individual spindles so that at some times all spindles may be functioning as identical units of a facility-type, and at other times as units of different variants of a facility-type.

The use of machines that can perform different and/or multiple functions is simulated by creating a facility for each function and then scheduling the facilities of the functions that are to be inoperative at a given time to be idle at those times. How facilities are scheduled to be idle is the subject of the next subsection.

The Operating Schedule

An operating schedule is a statement of the times and/or logical conditions that will determine when the facilities of a production system will be operating, that is, available to perform operations. These actions are implemented by scheduling the facilities to perform pre-defined *down operations* when they are *not* scheduled to be operating. The action of a down operation will depend on its *type,* and on the status of the production system when it must be executed.

Each down operation is a one-operation job, and is similar in many respects to ordinary operations that may happen to be the only operation of their jobs. For example, each type of down operation will be addressed to a specific facility-type and have a nominal duration. It can also be addressed to a specific unit or variant of its type. One important difference, however, is that each down operation *must* have an assigned *earliest permitted start time* which is referred to as the down operation's *nominal start time.*

We have used (or experimented) with eight different types of down operations to date. Four are of *forcing type* which means they must start at their nominal start times – even if they have to use force. These four forcing downs are of two types. Two of the four downs, in particular, are *interrupting downs* which means they will interrupt an in-process operation if that is the only way they can start at their assigned start times. The other two forcing downs are *inhibiting downs* which means they will prevent the starting of operations that would otherwise have to be interrupted. One each of interrupting and inhibiting downs is *unrestricted* which means it could be performed by any unit of its facility-type or variant. The other interrupting and inhibiting downs are *restricted* which means they must be performed by the specific unit of their facility-type or variant to which they will have been assigned in the scheduling problems' definitions.

The four other down operations are *non-forcing* which means they will wait for an in-process operation to be completed. Two of the non-forcing downs are of *fixed duration* which means they will hold their facilities for the full amount of their durations – whether they start at their nominal start times or not. The two other non-forcing downs are of *variable duration.* These downs' durations will be automatically reduced by the length of time they will have had to wait for an in-process operation to be completed. One each of the fixed and variable duration downs is unrestricted; one each is restricted.

Fixed duration downs are loaded the same as ordinary operations, except that these downs will have an automatic super-priority. Variable duration downs will be loaded similarly, except that in these cases the loading procedure will sometimes have to reduce their durations. Forcing downs are also loaded as ordinary operations that would have a super-priority, except when they will have to use force. How they are loaded on those occasions is described in a following subsection.

THE NEXT-PRECEDE STATEMENTS

Each operation of each job will next-precede one or more other operations. If the operations are not terminal ones, they will next-precede operations in their same jobs. If they are terminal, they will next-precede a common dummy operation.

The pair of a next-preceding operation and the operation it next-precedes defines a *next-precede relation*. The next-preceding operation of the pair is called the *referent* of the relation. The other operation is called the *next-successor* or the *next-following operation.*

The typical next-precede relation is often referred to as the *(i, j) th* next-precede relation where i and j are the indices of operations that are the respective referents and next-successors of next-precede relations. If there is an *(i,j) th* next-precede relation, it would be implied that *i must start at the same time or before j* because it is a technological constraint that these operations be performed in one-two squence or because it was a decision of the loading procedure that they be performed in one-two sequence by a same facility.

The fact that a given operation starts at the same time or before another operation does not imply that the given operation *next-precedes* the other operation. It could be, for example, that the given operation *simply precedes* the other operation, that is, that the operations must be performed in the indicated order, but not necessarily in one-two sequence. It could also be that the relative start times of the operations are an accident of the numbers of the scheduling problem, specifically of the given numeric values of the minimal time-intervals that must separate the starts of referents and their next-successors.

The facts of each next-precede relation are documented in a *next-precede statement.* This statement will identify the referent and next-successor of the relation. It will, however, also name the type of the facility that must perform the referent, the unit or variant of the facility-type (if they are given), and will state how the referent's and the next-successor's performances are to be time-phased. One of the referent's next-precede statements will also indicate how it is to be time-phased to the operation its facility will perform next. One of the referent's next-precede statements may also contain information the loading procedure will use to determine the operation's priority.

THE TIME-PHASING REQUIREMENTS

Each next-precede statement in which there is a real next-successor, will have two time-phasing statistics. One is called the statement's TIJ and is the minimal interval of time that must elapse after the referent starts that the next-successor can start. The other time-phasing statistic is called the statement's TIL and is the minimal length of time the referent must be actively processed (that is, without an interruption) in order that the next-successor can start the indicated TIJ time-units after the referent. The values of TIJs and TILs can be any non-negative numbers including zero. However if it *is* zero, we would assume it is a *numeric zero* – that is, that it is not the *logical zero* that would deny the time-phasing statistic's existence.

TILs are used to determine if an operation will have to be forced and, if it must be interrupted, which of the operation's TIJs will then have to be adjusted. Specifically, an operation will be forced if it will not let its facility start a forcing down at its assigned start time without force – or, in other words, if the forcing down would have to start before the operation had been processed as long as its maximal TIL. If the forcing down if of inhibiting type, the forced operation will be prevented from starting until the forcing down has started. If the forcing down is of interrupting type, each TIJ whose corresponding TIL was breeched by the forcing down, will be automatically increased by the duration of the forcing down.

A referent's TIL can also be its TIJ with respect to the operation its facility will perform next. That is, it can be the minimal length of time it will keep its facility from starting that operation. The value of this TIL is referred to as the operation's *duration.* In many applications this will be the operation's maximal TIL – and, when that is the case, we would be inclined to make it an assumption (or ground-rule) of the scheduling model.

Often in applications it will be reasonable to set each of an operation's TILs equal to its duration, and to set the TIJ in each of the operation's next-precede statements (that is, the statements in which the operation is the referent) equal to this duration plus a time-allowance to move the "work" to the facility that will perform the next-successor of the statement. It is not necessary, however, that we use the TIJs and TILs in this rigid manner. That is, we can use TIJs that are greater *or less* than their corresponding TILs, or that are even numeric zero. We can also set TILs equal to numeric zero, which would be the same as to say that the interruption of the referents will not affect the starting of their next-successors.

If Operations Must Be Forced

If an operation must be forced, it will be the loading procedure that will do the forcing. And when the loading procedure does force an operation, it will know that it will have to force that operation before it has actually been started. Thus, if the forcing operation is an inhibiting down, it will be able to prevent the starting of the operation; that is, it will not have to "unstart" it. Similarly if the forcing operation is an interrupting down, it will be able to adjust the forced operation's TIJs before the affected next-successors will have started.

The loading procedure will automatically re-start each interrupted operation when the interrupting down's duration has elapsed. It will not, however, check if the resumed operation would have to be interrupted a second time because that can be complicated – and because that capability is not normally required. It is, however, easy to determine (after the fact) that an operation would have had to have been interrupted twice (to satisfy the operating schedule) and to report this as an "error condition".

Two procedures are used to determine if an operation must be forced. One is called the *naive logic.* This logic assumes that each impending *unrestricted* forcing down is addressed to the facility of the current min NET – which means that the operation that would be selected to start next, will be forced if any of its TILs will keep it in-process beyond the point in time at which the next-impending forcing down must start. The naive logic is the proper logic to use if all forcing downs are *restricted* or if it is the practice to use these downs to bring down all units of a facility-type simultaneously – or, in other words, if the forcing actions could not possibly be avoided short of re-scheduling operations that had already started.

Forcing actions can often be avoided when forcing downs are unrestricted and it is not the practice to bring all units of a facility-type down at the same time. The loading process that can do this is called the *sophisticated logic.* It is called into play only when an operation would be forced by the naive logic and the forcing down is unrestricted.

The sophisticated logic starts by determining if another unit of the facility-type will be available to start the next-impending forcing down without force, that is, if it is not currently scheduled to be performing an operation when the forcing down must start. If the forcing down cannot be passed off to another unit, it would mean that the forcing action cannot be avoided. However, the fact the forcing down could be passed off would not mean a forcing action can be avoided. *That* could only be established if it could be shown that the operation will not now be in conflict with the *new* next-impending forcing down. We will, however, have recreated the circumstances of the original problem (that is, to determine if a selected operation will clear a next-impending forcing down) and could deal with it by repeating the previous procedure – except that, if the new next-impending forcing down must be passed off, we must remember not to pass it off to the unit to which the previous forcing down was passed. The sophisticated logic will continue in this vein until it will establish that the operation can avoid a forcing action – or it will have run out of units to which it could pass a forcing down.

It should be noted that the sophisticated logic does not actually load the forcing downs to the facilities to which they were passed. That is, all it wants to do is to establish if the forcing downs *could be* performed at their assigned start times. Also, though it will only pass a forcing down to a facility that could currently start it without force, it will not require that they be actually started without force. That decision would be made when the min NET moves to those facilities.

Another point to note about the sophisticated logic is that, when it must interrupt an operation, it will always interrupt it with the last impending forcing down it could not avoid. It will, thus, tend to interrupt the operations (it must interrupt) either late or early in their performance – rather than at any point with equal probability. This tendency can frequently be turned to our advantage.

SPECIFYING THE PERFORMANCE OF CONCERTS

A concert is a set of two or more operations whose relative start times are given in advance and must be realized exactly. An example would be a processing step in the manufacture of a product that had to be performed by a machine and a machine-operator and in which it was necessary to treat the machine and its operator as separate facilities.

The work of the machine and the work of its operator could be scheduled as a concert by making a separate operation of the work that would be performed by each facility, and then specifying which operation is to start first and *exactly* how long after that the second operation must start. For example, if we want the two operations to start together, we could specify that either operation must start first and that the other one must start exactly zero time-units later. We will not normally have to worry about when the operations will end as each one can be assigned an independent duration. Thus, if we would want the two operations to end at the same time, we would simply give them the same durations.

It would be quite easy to specify a concert if we could create next-precede statements that would have TIJs that would have to be realized exactly – rather than as lower bounds. For example, in the previous paragraph we could make a concert of the separate operations that would be performed by the machine and the operator by creating a next-precede statement in which the machine's and the operator's operations would be the respective referent and next-successor, and then setting the TIJ of the statement equal to numeric zero – provided, of course, our models could realize that TIJ exactly.

Unfortunately our models do not have TIJs that must be realized exactly, and so we must find some other way to achieve that effect. The other way, we have devised, utilizes two features of the models that have not yet been discussed. One is called the *hold feature* and makes it possible to assign operations artrarily large durations so they could (conceivably) hold their facilities forever. This feature can be used to simulate systems that have facilities that can only be used once. However, we are not interested (at the moment) in that use of the feature, but rather in its use in conjunction with another feature of the models which is called the *release feature*.

The release feature is used to release a holding operation's hold on its facility a specified interval after a designated operation (called the *releasing operation*) starts. Any operation can be used to release a held facility. The only requirement is that we must be sure the releasing operation will not start before the holding operation.

The releasing action is implemented by creating a *quasi next-precede statement* in which the releasing and the holding operations are the respective referent and next-successor. The fact that the relation is a quasi relation would be indicated by the presence of an appropriate code. The TIJ of this quasi statement is the interval of time after the releasing operation starts, that the holding operation's hold will be lifted – that is, the point in time at which the held facility will be available to start a next operation. The TIL of the quasi statement would be used to indicate if the releasing operation will have to be forced and, if it must be interrupted, if the interruption will affect the time time at which the held facility will be released. The TIL, in short, will perform the same functions TILs perform in ordinary next-precede statements. It can, in fact, even double as the releasing operation's duration.

To make a numerical example, let us say we want to schedule the two operations, i and j, as a concert that we want i to start exactly 5 time-units before j, and that we want i and j to have durations of 10 and 20 time-units respectively. This concert could be specified with just two next-precede statements. One would be an ordinary next-precede statement in which i and j would be the respective referent and next-successor, in which the TIJ would be 5 time-units, and the TIL would be an arbitrarily large number.

The other next-precede statement would be a quasi statement in which j and i would be the respective referent and next-successor. The TIJ of the quasi statement would be 5 time-units, that is, the difference between the values of the duration of i and its TIJ. The TIL of the quasi statement would be the ordinary duration of j, that is, 20 time-units.

Note how the concert works and, in particular, that we do not force j to start exactly 5 time-units after i, but rather force i to capture and hold its facility until j can capture its facility – but, at the same time, making sure that j will not be able to capture a facility until at least 5 time-units have elapsed since the start of i. The principle, we are illustrating, can be used to specify the performance of concerts that have any number of operations.

The main weakness of our models' way of specifying and scheduling concerts is that concerts can sometimes be stalled for long periods and ther is no way, when that happens, to "lend out" a holding fa-facility–even if the facility could be returned without delaying the concert. The system's controllers, however, could spot these possibilities in print-outs of schedules and their evaluations, and could implement them by repeating the scheduling exercise with appropriate special instructions.

THE SCHEDULING ALGORITHM

The problem to determine the schedule that will be generated by a loading procedure is formalized and solved in terms of a mathematical system, called *star algebra.* This system is postulationally equivalent to the conventional matrix algebra we would get if it was required that all entries of matrices be non-negative. That is, star algebra will obey the postulates of conventional matrix algebra—except the one that says that all matrices have an additive inverse.

The objects of star algebra are matrices of numbers that include the positive and negative integers, plus zero and an additional number called *iota.* Iota is *numeric zero* (or "zero magnitude"). Its symbol is the greek letter ι. The ordinary (or round) zero is the symbol for *logical zero.* Why we need separate symbols for the two kinds of zero will soon be apparent.

The rule for adding star algebraic matrices is the same as in conventional matrix algebra, except for the way we will add the numbers that are the entries of the matrices. Specifically, the sum of two numbers is simply the greater number. Logical zero, in this connection, is taken to be less than other number. Iota, on the other hand, is exceeded only by the positive numbers. The symbol for star algebraic addition is the asterisk or star—which is why the system is called *star algebra.* The additive identity in star algebra can be any matrix that has logical zeros everywhere.

The rule for multiplying star algebraic matrices is also the same as in conventional matrix algebra, except for the rule that will determine the product of two numbers. The rule is defined, below, for any two numbers a and b. The symbol for star algebraic multiplication is the symbol for a musical sharp. The conventional addition, that is indicated in the definition, is not a formal operation of the algebra; it is simply a way to explain how the magnitude of certain star algebraic products is to be determined.

$$
\begin{aligned}
a \mathbin{\#} b &= 0 && \text{if } a \text{ or } b \text{ is } 0\\
&= a + b && \text{if } a \text{ and } b \text{ are not } 0\\
&= \iota && \text{if } a \text{ equals } -b\\
&= b && \text{if } a \text{ is } \iota
\end{aligned}
$$

The identity matrix in multiplication can be any square matrix that has iota everywhere on its main diagonal, and logical zero in all other positions.

If the problem to generate a detail schedule is *determinate,* we could express the essential *givens* of the problem with just two star algebraic matrices. One would be the square *schedule matrix S* whose $(i, j)\,th$ entry is 0 if the $i\,th$ operation does not next-precede the $j\,th$ operation, and is the minimal interval of time that must elapse after i starts that j can start if i *does* next-precede j. The other matrix is the column vector T_0 whose $i\,th$ entry is the given earliest permitted start time of the $i\,th$ operation if the operation has a given earliest permitted start time, and is logical zero if it does not.

The matrix we want to determine is the column vector T whose $i\,th$ entry is the set-forward start time of the $i\,th$ operation (or, equivalently, the greatest of the lower bounds on the start time of that operation) and that is determined by the following star algebraic equation

$$T = T_0 * T_1 * T_2 * \cdot\cdot$$

where

$$T_k = S' T_{k-1} \qquad k > 0$$

Solving the above equation will not identify the specific facility that will perform each operation. However, if we know the first operation each facility performs and the facilities to which the other operations are addressed, we could easily trace out the specific facility that will perform each operation.

If the scheduling problem is *indeterminate,* it will be assumed that its indeterminancies will be confined to the matrix S. For example, if the indeterminancies are with respect to the operations each facility will perform and the order in which they will perform them, they could be expressed by defining an S matrix that will document each possibility so that we could subsequently save the possibilities that were exercised, and discard the others.. To be specific, if operations a, b, and c could be performed by a same facility and in *any* order, we could show this in S by giving the *(a,b) th, (b,a) th, (a,c) th, (c,a) th, (b,c) th,* and *(c,b) th* entries the values $t_{ab}x_{ab}$, $t_{ba}x_{ba}$, $t_{ac}x_{ac}$, $t_{ca}x_{ca}$, $t_{bc}x_{bc}$, and $t_{cb}x_{cb}$ where t_{ij} is the duration of the *i th* operation when it is next-followed by the *j th*, and x_{ij} is a variable that will be set equal to iota if j is selected to next-follow i and that will be set equal to logical zero in all other cases.

The values of the x_{ij} will be determined by the loading procedure. For example, if it selects a to start first, it will have determined that $x_{ba} = x_{ca} = 0$. If it then selects b to next-follow a, it will have determined that $x_{ac} = x_{cb} = 0$, and that $x_{ab} = \iota$. When it selects c to next-follow b, it will determine that $x_{bc} = \iota$. The loading procedure can be likened, in short, to a mathematical function in that it will determine the values of dependent variables (the x_{ij}) from the given or determined values of other variables.

The scheduling algorithm determines the entries of the vector T by updating the matrix S following each loading decision, and then using the updated S to determine and accumulate the new vector $T_k = S'T_{k-1}$. These matrices must be updated after each loading decision because we will need the information in these matrices to determine the facility and time of the new min NET. We will also need this information to determine the operations' priorities, that is, when the priority is a function of the current status of the production system.

It is important to note the "ready availability" of the above information in the current S and sum of the T_k. It should be noted, for example, that the fact that an operation started can be spotted from the fact that its column in the current S will be null if we delete (or pay no attention to) the rows of operations we will *know* have already started, that the time the operation started will be the value of its entry in the current sum of the T_k, that a *last-started* operation of a facility can be spotted from the fact that its row in the current S has an entry that has a *coefficient of* x and that these entries will occur in the columns of still-unstarted operations the facility might yet perform, that the time a facility will be able to start an operation it might yet perform will be the star algebraic product of its last-started operation's start time and the coefficient of x in that operation's row in the current S that is also in the column of the next-following operation, that a currently unpreceded operation can be spotted from the fact that its column in the current S will be null (assuming we skip the rows of started operations) except for one or more entries that have coefficients of x, that the time at which a currently unpreceded operation will be ready to start will be the value of its entry in the current sum of the T_k, etc. We do not want to give the impression that the preceding information is developed *exactly* as described. Actually, there are numerous ways to speed and compact the numerical analysis of the matrices, and which we try to use whenever possible. We will not normally, for example, perfrom empty calculations in which we would be multiplying by logical zero, or perform (or repeat) calculations that we would know in advance could not produce a new greater lower bound on an operation' start time.

Most of the methods we have developed to compact and speed the calculation of detail schedules were stimulated by the mathematical formulation of the problem. This formulation, for example, helped us recognize (and appreciate) the logical equivalence of these problems to other numerical problems (e.g. problems to generate material and resource requirements) that have conventional matrix algebraic formulations and for which efficient numerical methods had already been developed. This break-through (if we may call it that) has helped us develop computer-program systems that will generate and evaluate detail schedules at a very rapid rate. This is important because it reduces the cost of the computations, and because it makes it possible to design man-machine systems in which the production controllers can actively participate in the scheduling process.

THE SCHEDULING SYSTEM

Detail scheduling algorithms are implemented by computer-program systems that will normally be imbedded in a *total production planning, scheduling and control system.* This total system will usually consist of three inter-related subsystems. One, the planning system, will perform general production planning functions and will, in addition, produce the next-precede statements that will make up the bulk of the technological constraints the resultant schedules must satisfy.

Another of the three subsystems is called the "monitor"—because it is used to compare the actual and scheduled performances of operations. It will, however, also unload next-precede statements in the current input file as operations are completed, and will statistically forecast and control the operations' performance-times.

The third subsystem is called the *scheduling subsystem.* This is the system that will generate the detail schedules and that will evaluate them *as schedules.* That is, it will determine the scheduled percent utilization of the facilities, scheduled average delays of jobs, etc.

The scheduling subsystem takes three kinds of input. One consists of the next-precede statements that define the products' methodizations, and that will ordinarily be generated mechanically. Another kind of input will define the numbers and types of the facilities in the production system, and will give the times and/or logical conditions that will determine when these facilities will be operating, that is, available to perform operations.

The third kind of input is referred to as the *special instructions*, and is used by the production system's controllers to direct and over-ride the scheduling system and algorithm. The instructions have three main uses. One is to synchronize a generated schedule with the conditions that did (or will) prevail in the production system at the point in time from which the schedule will be projected. Another use is to temporarily correct the current input file: e.g. to incorporate last-minute changes.

The third use of the special instructions is to introduce changes that will improve a last-generated schedule. These changes would be introduced by the production system's controllers after they have had a chance to examine a last-generated schedule and to analyze its evaluative statistics. The controllers can use the special instructions to change the assignments of operations to facilities and/or the order in which the facilities will perform them . . to change the alternate methodizations that were adopted . . to postpone the processing of non-critical jobs if they are getting in the way of the more critical ones, or to advance their processing if it is more important to soak up excess capacity . . to force facilities to wait for particular, urgent jobs . . to lend out facilities of temporarily stalled concerts, or to even disband concerts that are not going well . . and so on.

MECHANICAL GENERATION OF NEXT-PRECEDE STATEMENTS

The next-precede statements that define the jobs' methodizations, will normally have to be generated mechanically – because they will usually be very numerous, and because there will usually be a considerable time-pressure to get these statements into the current input. The information needed to mechanize the generation of this input will often be found in summary statements of the products' methodizations in the company's engineering data base. These summary statements will generally identify the operations that must be performed to make each product and the order in which they must (or will usually) be performed. They will also identify the type of the facility that must perform each operation (they may also cite specific units or variants when necessary), and the time one should allow to set-up the facility and to process each unit of the product.

Other essential input will come trom the job orders. This will include the quantity of the product that must be made and the time at which the order must be completed. They may also include the jobs' due-in times and priority-class designations.

The problem to mechanically generate the next-precede statements may be complicated by the common failure of engineering data bases to describe the products' *real* methodizations, that is, the methodizations that will (or could) be used by the production controllers after the products' jobs have been placed in process. The engineering methodizations, for example, will rarely specify products' alternate methodizations, that jobs to make particular products may have to be subdivided for certain operations and/or combined for others, or even that a particular operation could be performed by more than one facility-type. Fortunately, we have found that the deficiencies of the engineering methodizations can be mechanically corby and within the scheduling subsystem, and without having to change the engineering data base itself – other than to incorporate the coded clues the program system will need to know how to deal with particular jobs and operations. This aspect of the scheduling system's design is discussed in Chapter 21 (pp. 475 to 500) of the cited reference.

THE SCHEDULING OUTPUTS

Outputs of scheduling systems are of three types. One is the schedule, itself. Another consists of the evaluative statistics that will be used to determine if the schedule is satisfactory and if (and how) it may be improved. The third category of the output consists of the *transmittal cards* that are the medium of the data-gathering system that will collect and channel the observations needed to monitor and update the scheduling system.

The bulk of the output will consist of the schedule and the evaluative statistics that will characterize each operation's scheduled performance. This will include the identification of the operation and the facility that is to perform it, the time at which the operation will start, if and how long the operation will be interrupted, if an how long the operation will have to wait for its facility, if an how long the facility will have been waiting for a next operation, and if and how long the operation's scheduled start could be postponed without forcing another operation's scheduled start to be postponed.

The time an operation has to wait for its facility is called (in the scheduling outputs) the operation's *delay*. The time the facility had been waiting for a next operation is called the operation's *minus delay*. It can also be called the facility's *idleness* – provided we understand it is only the idleness since the last point in time at which the facility was able to start a next operation. The time an operation's start can be postponed without forcing another operation's start to be postponed is called the operation's *slack*. This is not, however, the slack statistic encountered in PERT/CPM scheduling. It is, if anything, more related to the *free float* statistic in PERT/CPM. The analogy, however, should be advanced with caution because *slack* in detail scheduling is a dependent variable whose values will be determined by the loading decisions, while in PERT/CPM it will be fixed and implicit in the problem's givens.

The above output can be brought out in any one or more of the following sequences:

a) in numeric sequence of the operations' identifications within facility-types;

b) in the order of the operations' start times within facilities and/or facility-types;

c) in the order of the operations' start times within jobs.

The output can be produced in any or all of the above sequences without having to formally sort the output. This is an important consideration because sorting can be time-consuming and could, thus, diminish the scheduling system's effectiveness as a man-machine system.

The output will also include evaluative statistics that are not specific to the performance of individual operations. They are called *summary statistics* and include: totals and average levels of operations' delays by facility-types, jobs, and time-periods . . idleness of facilities by facility-types and time-periods . . status of queues (e.g. how many jobs are waiting? what is the average delay? etc.) by facility-types and time-periods . . and so on.

CONCLUDING REMARKS

We have developed a half-dozen detail scheduling systems, to date. The first three were developed at IBM; the fourth was developed at a former division of ITT; the fifth was developed at C-E-I-R for the Wolverine Tube Division of Calumet & Hecla. The most recent system was developed for the Travelers Insurance Co. in contract with BERNARD GIFFLER ASSOCIATES. The systems were programmed for the IBM 704, 7094, and 360/65 and for the Honeywell H-200.

The first three of the above systems were used to look for flaws in the systems' design and logic, and to evaluate their potential usefulness in a variety of applications which included a steel mill and a hospital facility. The fourth system was used to investigate the scheduling implications of the design and operating procedures of production systems which included a missile range, an air transport fleet, and the command center of a command/control system.

The last two of the above systems were specifically designed to detail schedule and control existing, on-going production systems. One is being used by the Wolverine Tube Division to detail schedule and control its factories. The other (it is not yet installed) will detail schedule and control the Data Processing Center of the Travelers Insurance Co.

An idea of the systems' capacity and run-times can be gained from our experience with the third of the above systems that was developed at IBM, and which was programmed for the IBM 7094. This is a computer that has an internal memory of 32,768 36-bit addressable words and a cycle time of 2.18 micriseconds. Most of its instructions (e.g. add, load, store, transfer, etc.) are executed in two cycles. This system can schedule approximately 1000 operations per minute and includes the time to write the print images of the output magnetic tape for off-line printing. The main emphasis of this system's design, however, was not on speed, but rather to be able to solve very large problems in one piece. For example, W. Gutschow (then at IBM) wrote a version of this same program that was two to three times faster – by giving up a little of its capacity.

A NAVAL AIR REWORK FACILITY SCHEDULING PROBLEM

Christopher Haas and Thom J. Hodgson

University of Florida, Gainesville, Florida, USA

In the following paper we describe an independent job shop production sub-system within the Jacksonville Naval Air Rework Facility, Jacksonville, Florida. The various job flows within the shop are identified. The mechanisms controlling job flows are quantified. The <u>controllable</u> costs of operating the shop are identified and the scope of control open to management is discussed. Finding a method of control for the system is left as a challenge for the reader.

I. INTRODUCTION

The Naval Air Rework Facility (NARF) at Jacksonville Naval Air Station is an industrial plant with some 3,000 employees. Top management officials are Naval officers and the remainder are civilians. The civilian work force includes engineers, planners, administrators, and mechanics representing some 35 different highly skilled trades.

The NARF mission includes maintenance engineering and heavy (depot) maintenance of military aeronautical items ranging from complete aircraft to spare components. By policy, NARF maintenance on all items is a selective process which involves diagnosis of item condition, updating with latest required changes, and limited maintenance rework to recondition the item for another period of service.

The NARF is organized by support services (Engineering, Quality Assurance, Planning, etc.) and the Production Department. The Production Department, with some 2/3 of the 3,000 civilian workers, is made up of over 100 different shops. These shops, the hardware producing elements of the plant, are organized: some by product (radio shop), some by process (cleaning shop), and some by function (assembly). The shops form an interrelated network through which products flow as the maintenance process proceeds.

The NARF product workload consists of about 50% aircraft. The aircraft are are examined (diagnosed), and disassembled as required to permit shop component processing. As a matter of production policy, the processed components are used to reassemble the aircraft, which is then flight tested. About 15% of the NARF workload consists of aircraft enginines (which enter the plant as such) and are overhauled and returned to the Navy supply system for issue and reuse. Another 15% of the workload is made up of miscellaneous spare aeronautical components (the F/J program) which have been in use and are sent to NARF for required maintenance. After being reconditioned by NARF, these are returned to the Navy Supply System for reissue and reuse. The remaining 20% of NARF workload is of a miscellaneous nature including such unplanned items as aircraft repair (repair of damaged aircraft), customer service (on demand), and field modification.

II. THE HYDRAULIC SHOP

This paper deals with a specific NARF shop (the Hydraulic Shop). This shop is typical of many other shops in NARF. Specifically, the shop under study belongs to a class of shops each of which has the following common characteristics:

1) The bulk of processing on jobs entering each such shop is confined to that shop. This means that a shop functions as a repair station rather than a disassembly, routing, or reassembly shop.
2) A shop of this class is manned by a substantial number of mechanics whose interchangeable skills make it possible to work on a variety of jobs in backlog.
3) Jobs in these shops have individual work content which is small relative to the total load in the shop. Since work content is small it is also generally true that a single worker at a time processes each job.
4) Shops in this class are not highly dependent on equipment capacity as limiters.

From a production control viewpoint, the shop is composed of three basic elements: the workable backlog area, the nonworkable backlog area, and the processing area (see Figure 1). Jobs arrive at the shop and normally are entered into the workable backlog, to await processing by an available worker. If, for some reason, a job cannot be processed "as is," it is entered into the nonworkable backlog. Usually jobs will be placed in the nonworkable backlog because of the nonavailability of a component part required for processing or technical data required for

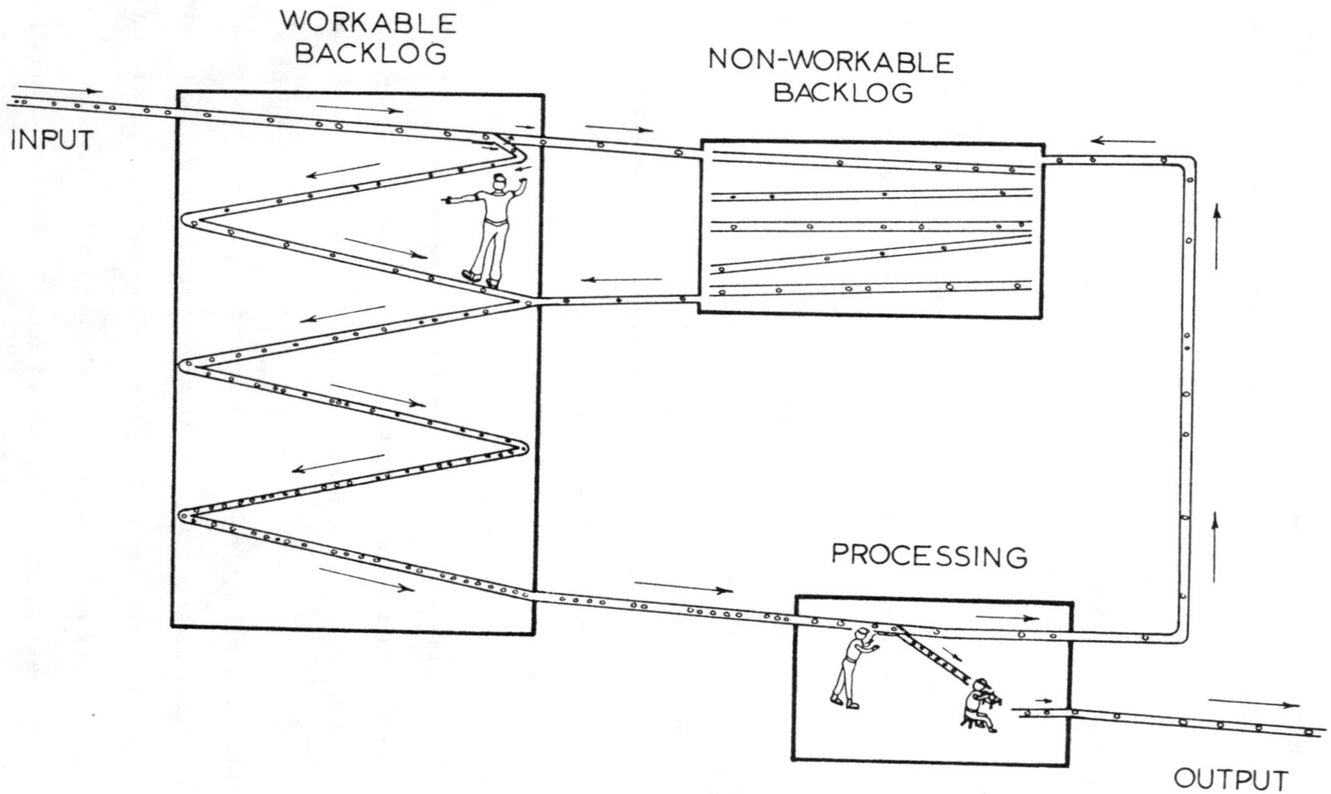

SCHEMATIC OF THE SHOP

FIGURE 1

processing. Once the nonavailability is satisfied, the job returns to the workable backlog.

Jobs move from the workable backlog to the processing area as needed. Sometimes, during the initial disassembly of a job, it is determined that an additional component part(s) will be necessary to complete the job. If the component part(s) is not available, the job is entered into the nonworkable backlog (see Figure 1). When processing is completed on a job, it leaves the shop.

Management has at least two methods by which they can exercise control over the shop on a week to week basis. First, they can increase the shop capacity by moving qualified workers into the shop from other shops in the NARF, or they can decrease the shop capacity in the same manner. Second, they can increase shop capacity by working the existing shop manpower overtime. Management control of the shop is exercised with the objective of performing the mission of the shop, and incurring the least possible cost. Clearly, these two objectives are somewhat in conflict. Consequently, any rational control policy is formed by trading off mission performance with cost of operation.

The use of overtime increases the operating cost of the system. In addition to an increased labor rate, overtime work usually is performed at a lower efficiency rate than that of straight time work. It would seem initially that increasing the shop capacity by bringing in more workers would be preferable to overtime, but that also results temporarily in increased costs. First, there is an administrative cost for moving workers. Second, and more important, workers moved into a shop take time to learn their jobs and, consequently, work at a reduced rate for a period immediately after assignment to the shop (assuming that the worker is qualified, this period of lowered efficiency may last one to three weeks).

The amount of work in the system is of concern to management. On one hand, if the workable backlog is depleted, the workers in the shop will be idle. Another possibility is that if the workable backlog nears depletion, the workers' efficiency will decrease, thereby resulting in an application of Parkinson's Law. In any event the effect on the system is the same. The efficiency of the system is reduced.

On the other hand, if backlogs are too large, management also has reason for concern. Large backlogs result in increased flow time for jobs in the system. If a job has come from an aircraft being reworked at the NARF, an excessive delay of the job in the hydraulic shop will cause the processing of an aircraft to be delayed. This will result in a real cost for NARF. If a job is for the Navy supply system, an excessive delay of the job will cause a depletion of the Navy inventory which may result in the grounding of a fleet aircraft. In either case, the result is a reduction the number of operational aircraft available to the Navy.

From conversations with NARF personnel, it would seem that the function depicted in Figure 2 is a reasonable estimate of the form of the deferral cost for jobs. There is essentially no cost incurred for delay up to a certain point. But once that point is reached, a cost is incurred (approximately linear) which continues until the job is processed through the hydraulic shop.

Consequently, one would expect that an optimal control policy for the hydraulic shop would be one that:

1. maintained the backlogs at levels high enough to avoid depletion but not so high as to cause any significant delay (deferral) costs, and
2. accomplished the above with an absolute minimum of excess labor costs.

It is not clear, however, what control policy will, in fact, do this in the most expeditious fashion.

The Application of Control

The use of overtime and the movement of manpower within the NARF are not methods of control which are applied in a continuous fashion. Men cannot be transfered back and forth on a daily basis between shops. If a man (or men) is transfered into a shop, he (or they) would expect to remain in that shop for a period of time. It has been our experience that the application of overtime in an industrial system also is not applied on a daily basis.

The Input Process Mechanism

From Figure 1, it is seen that a job enters the shop and immediate determination is made as to whether the job is workable or nonworkable. Given that "m"

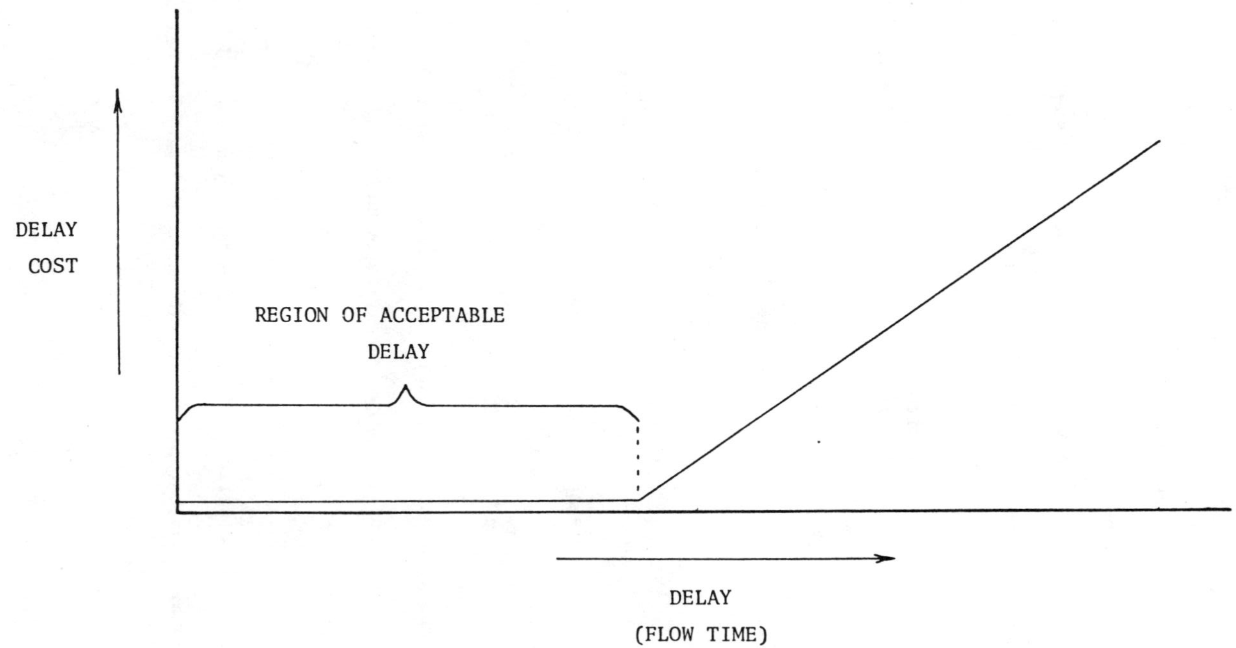

DELAY COST .VS. FLOW TIME

FIGURE 2

hours of work enter the shop it is necessary to know the probability that "w" hours go into workable backlog (this is equivalent to "m-w" hours going into workable backlog). The following definitions are needed.

$p_1(j|m)$ = The probability that a group of jobs having "m" standard man hours of work content has exactly "j" jobs.

$p_2(k|j,m)$ = The probability that "k" jobs are nonworkable, given a group of "j" jobs with a total work content of "m" manhours.

$f_1(w|k,j,m)dw$ = The probability that "w to w + dw" man hours go to nonworkable backlog, given "k" nonworkable jobs are selected from a group of "j" jobs with a total work content of "m" manhours.

$f_2(w|m)dw$ = The probability that "w to w + dw" manhours go to nonworkable backlog, given "m" hours of work enter the shop.

It can easily be seen that

$$f_2(w|m) = \sum_{j=1}^{\infty} \sum_{k=0}^{j} f_1(w|k,j,m)\ p_2(k|j,m)\ p_1(j|m). \qquad (1)$$

In order to calculate $f_2(w|m)$, it is necessary to know the form of its component distributions.

$\underline{p_1(j|m)}$

A sample of one thousand jobs processed in the hydraulic shop showed that the distribution of standard hour content is approximately exponential [3] (see Figure 3). The number of jobs j in a group having a total standard hour content of m can be thought of as a realization of a process where jobs are selected from an infinite supply of jobs with exponentially distribution standard manhour content. Jobs are selected until the total manhour content of the jobs is greater than or equal to m. The distribution of the number of jobs selected under this process is easily seen to be Poisson (offset by one).

$$P_1(j|m) = \frac{m^{j-1}e^{-m}}{(j-1)!} \qquad (2)$$

$\underline{p_2(k|j,m)}$

The workable-nonworkable selection process can be assumed to be a stationary binary process. Consequently the distribution of the number of jobs (k) going non-

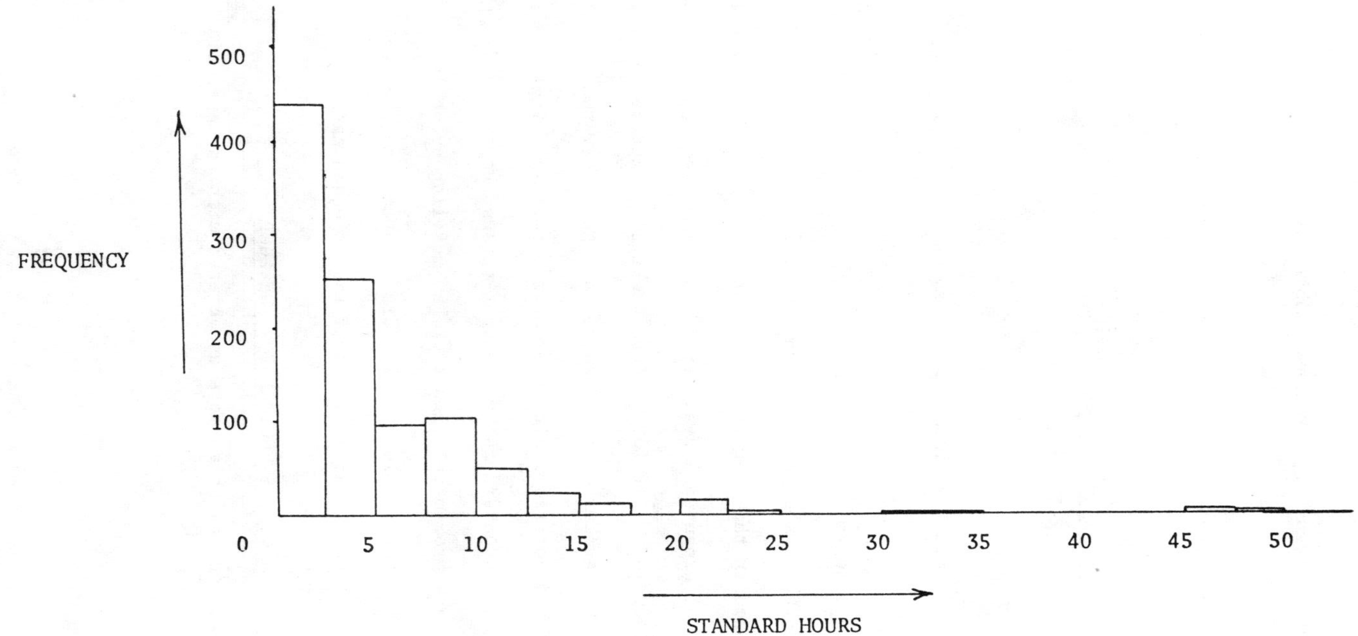

FREQUENCY HISTOGRAM FOR STANDARD HOURS/JOB

FIGURE 3

workable from a group of j jobs is binomially distributed.

$$P_2(k|j,m) = \binom{j}{k} p^k(1-p)^{j-k}, \quad \text{where} \tag{3}$$

p is the probability that a given job goes nonworkable.

$f_1(w|k,j,m)$

The conditional distribution of manhours entering nonworkable has been shown to be [3],

$$f_1(w|k,j,m) = \frac{(j-1)!w^{k-1}(m-w)^{j-k-1}}{(k-1)!(j-k-1)!m^{j-1}} \tag{4}$$

Nonworkable Delay Mechanism

The distribution of the delay in nonworkable can be assumed to be approximately Erlang 2 with a mean delay of ~8 days [3] (see Figure 4).

The Process (or Output) Mechanism

Shop data shows that the number of standard manhours of work output for a given number of manhours expended in the shop is approximately normally distributed with mean equal to the number of expended hours. Figure 5 summarizes the results of chi-squared tests for goodness of fit for various numbers of expended hours. As one might expect, the larger the number of expended hours, the better the normal density fits (another example of the central limit theorem in practice). Figure 6 is a graph of the variance of the standard manhour output verses the expended manhours. The function is seen to be approximately linear. For modeling purposes, it can be assumed that the output distribution is normally distributed, and that the variance of the output distribution is a linear function of the expended manhours.

III. A COST STRUCTURE FOR THE SHOP

There are three basic controllable costs associated with the operation of the shop:

1. Labor costs
2. Personnel transfer costs
3. Delay costs

Labor costs consist of direct labors, both straight-time and overtime. The actual rate is a direct function of the direct labor hourly rate, the number of men

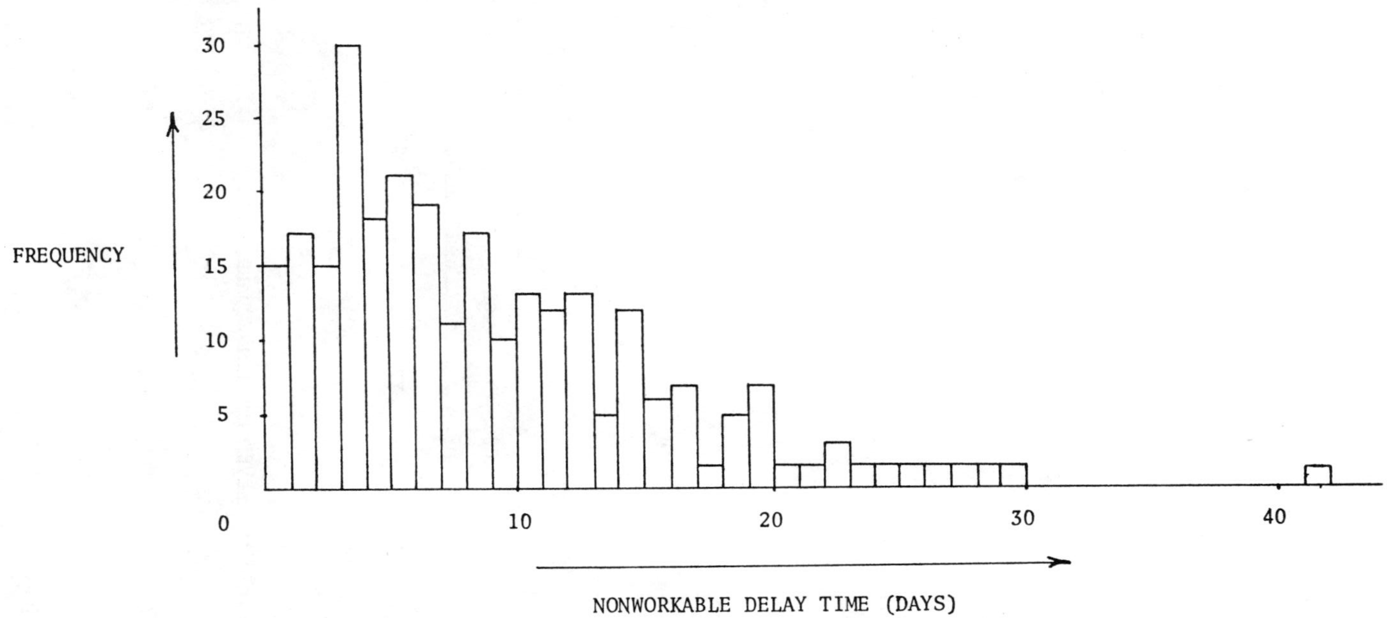

FREQUENCY HISTOGRAM FOR NONWORKABLE DELAY TIME

FIGURE 4

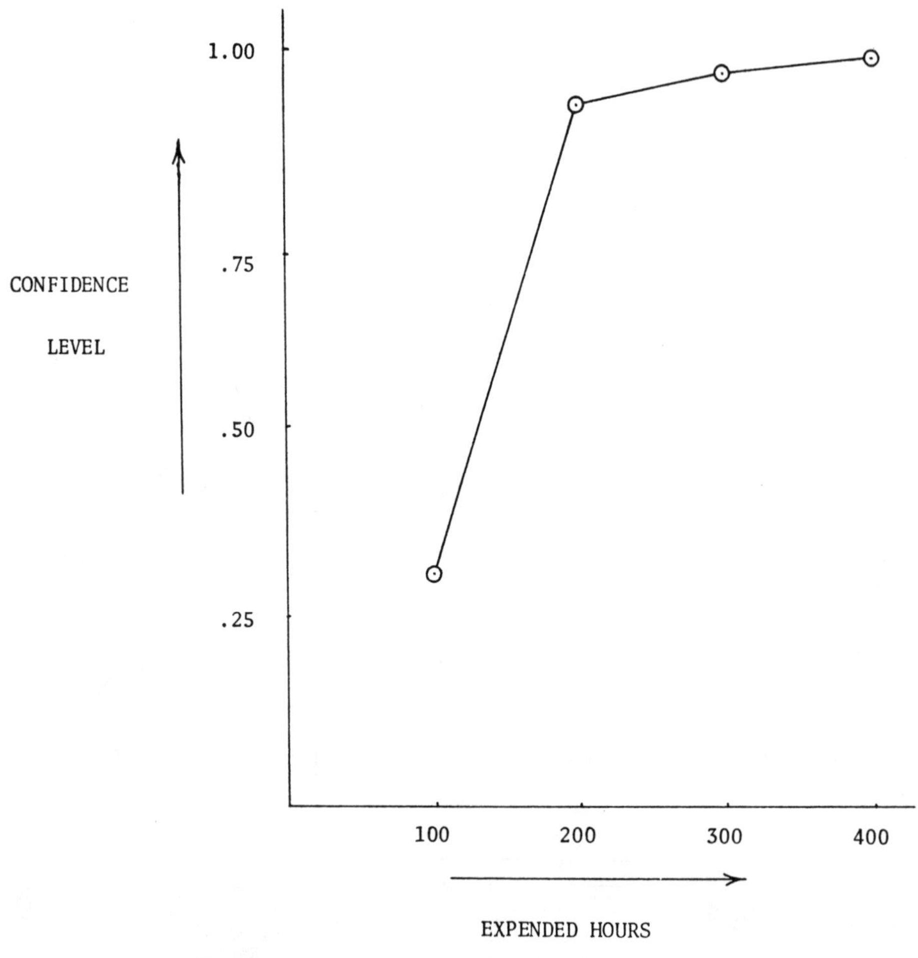

CONFIDENCE LEVEL FOR CHI-SQUARED TEST

.VS.

EXPENDED HOURS

FIGURE 5

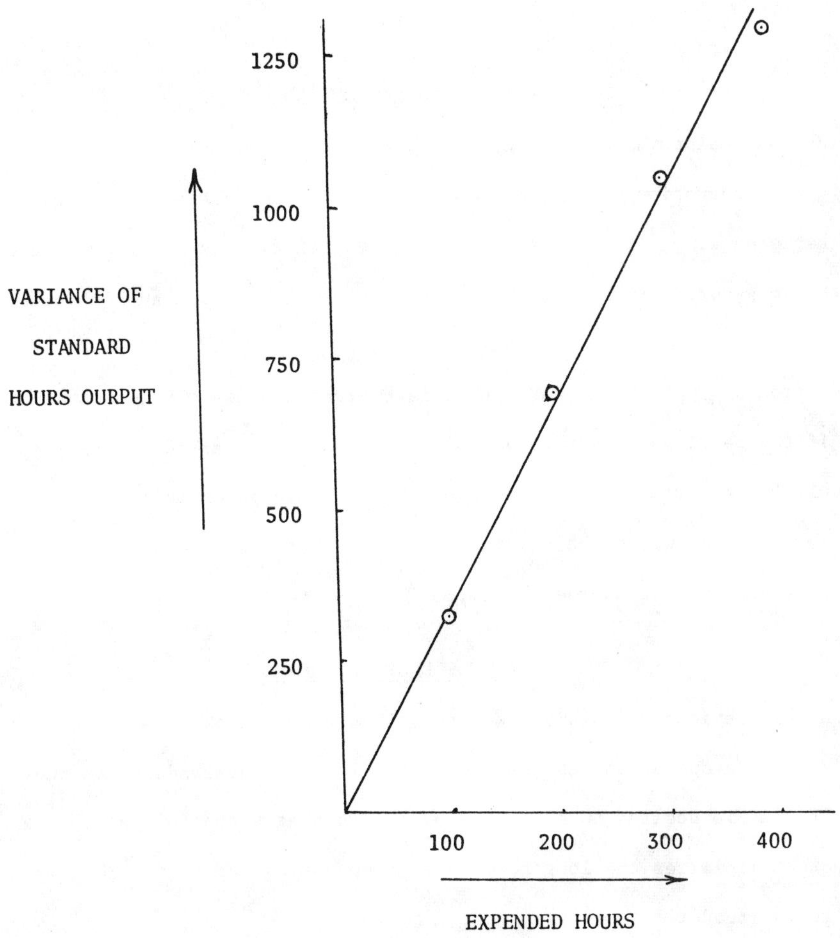

VARIANCE OF STANDARD HOURS OUTPUT

.VS.

EXPENDED HOURS

FIGURE 6

in the system, and the amount of overtime worked. Personnel transfer costs consists of the administrative costs associated with transfering A from one department to another.

IV. THE EXISTING INFORMATION AND CONTROL SYSTEM

The NARF maintains and operates a "transactor" system for the control of the shops. Each operation on a given job (i.e., entry to workable backlog, entry to nonworkable backlog, etc.) is entered in a central file as it is completed. Status reports are generated which summarize the status of workable and nonworkable backlog.

Control of the various NARF shops is exercised presently in a heuristic fashion. If the backlogs appear to be too large, then the various control options available to NARF management (described earlier) can be exercised. There is, however, no method by which the various controls options can be evaluated in terms of their long term effect on the system.

V. A CHALLENGE TO THE READER

In the previous discussion, we have given a description of a shop within the NARF. The shop operation was described as virtually independent of other shops in the NARF. It is desired to model the system so that a control policy can be obtained that minimizes the long term cost of operating the shop. That is the challenge to the reader.

Present work on a model to do this centers around a Markov decision model of the shop. That research will be reported in detail at a later date.

References

[1] Feller, William, An Introduction to Probability Theory and its Applications, Wiley, New York, 1968.

[2] Howard, R. A., Dynamic Programming and Markov Processes, MIT Press and Wiley, New York, 1960.

[3] Haas, Christopher, and Hodgson, Thom J., "Preliminary Scheduling Models for Control of the NARF Shops," Report submitted to ONR under contract N00014-68-A-0173(NR-047-101), January, 1972.

SOME SCHEDULING APPLICATIONS IN CHEMICAL INDUSTRY

T. Prabhakar

UNION CARBIDE CORPORATION
Kanawha Valley Regional Computer Center
South Charleston, West Virginia, USA

Some scheduling applications in Union Carbide Corporation are discussed. These include production and distribution scheduling, inventory scheduling, vehicle scheduling, and project scheduling. The solution techniques used include linear programming, mixed integer programming, dynamic programming, simulation, network analysis, and heuristic approaches.

1. INTRODUCTION

This paper briefly discusses some scheduling applications in Union Carbide Corporation. The problems dealt with in these applications are typical of some of the scheduling problems encountered in the chemical industry. Unlike machine scheduling problems, these need not always be associated with sequencing. There are, however, instances where scheduling problems with sequencing aspects are encountered; for example, sequencing and scheduling of products on reactors with setup costs, the setup costs being a function of the reactor characteristics and the product sequence. The scheduling models which serve as an overall planning tool ignore the sequencing aspects which are normally solved locally, say, at the plant level by the plant personnel.

The primary purpose here is to show some of the types of scheduling problems encountered in the chemical industry and the techniques used in solving them. While discussing a scheduling application, therefore, the emphasis is on a clear definition of the problem with a brief discussion of the solution approach.

The scheduling models discussed here include a production and distribution scheduling model (Section 2), a transportation scheduling model (Section 3), a product sequencing and scheduling model (Section 4), an inventory scheduling model (Section 5), a vehicle scheduling model (Section 6), and two project scheduling systems (Section 7).

2. A PRODUCTION AND DISTRIBUTION SCHEDULING MODEL

Union Carbide has several facilities for the manufacture and distribution of a major line of products whose annual sale is over a billion pounds. Based on the sales demands at the various marketing areas, the production and distribution schedules are set up so as to either minimize the total production and distribution costs or maximize the profit derived from earnings from sales less the production and distribution costs. A high level of coordination and effort go into the planning of these schedules. The planning tool is a large and complex linear programming (LP) system which has a very highly detailed representation of the production process and the distribution network. The model is run once a month for monthly (tactical) planning and once or twice a year for annual (strategic) planning. The LP model is solved by the IBM's Mathematical Programming System (MPSX). The auxiliary programs required for LP model generation and report writing are written in PL/I and ASI-ST*. The model is run on an IBM 370/165 computer system.

The production process primarily involves two stages: reaction and processing. The product coming out of a reactor can be a finished

---

*ASI-ST is a proprietary software developed by Applications Software, Inc., Torrance, California.

product which may need only packaging, or an intermediate product which, on mixing with some additives and further processing, is converted to a finished product. An intermediate product can be transported to some other location to be used as a feed to a processor. The product coming out of a processor is usually a finished product. There are usually multiple reactors and processors at a given production facility. The reactors are grouped into various types depending upon the engineering design and the type of catalyst handled by them. The processors are also grouped into various types depending upon the type of function performed. A product can usually be made/processed on more than one reactor/processor. There are, of course, certain technological considerations restricting product-reactor and product-processor combinations.

A reactor/processor may handle more than one product during a given scheduling period. Whenever there is a product change-over on a reactor or processor, there may be a certain amount of unwanted product produced before the reaction or processing is stabilized at the new operating conditions. This material, instead of being a single product, is usually a mixture of two or more products. As much of these materials as possible is converted to utility products and disposed of. Since the utility products are less attractive in the market, one of the major objectives of the production strategy is to minimize their formation during the formation of the high-valued products.

The distribution network is comprised of plants, warehouses, and demand centers or marketing areas. A warehouse is an intermediate distribution place which receives a product from one location, stores it, repackages it (if necessary), and ships it to another location. A marketing area is a final distribution point from where the product is shipped to individual customers. There are basically two classes of transportation: bulk and packages. The bulk transportation is further classified depending upon the vehicle used, such as railway cars, trucks, and tankers. The transportation by package is also further classified depending upon the type of packaging, such as bags and drums.

The independent variables in the LP model are comprised of the production and distribution activities. The restrictions are comprised of the production capacities, raw material availability, inventory limitations, and sales demands.

The output from the LP solution is translated into a set of management reports describing the production and distribution activities and financial analyses. These reports are analyzed by the Operations Team comprised of Production, Distribution, Marketing, and Operations Managers. If there are any discrepancies in the results, the model is run again. Finally, the information on the production schedules is transmitted to the Department Heads located at various production facilities.

The production and distribution LP model for the high-valued products has a very large distribution network consisting of the producing centers (sources), warehouses and terminals (transshipment points), and market areas (sinks). A transshipment point can be a sink (with inventory accumulation) or a source (with inventory depletion). The routes (arcs) connecting the various locations (nodes) are directed and uncapacitated. There is a transportation cost associated with each one. Since there can be several routes from a source to a sink, the LP solution will naturally pick up the least cost route. This is what is currently being done. However, instead of unnecessarily enlarging the LP with the various combinations of routes, the shortest route in the sense of least cost can first be established between a source and a sink for each product through a shortest path algorithm. Incorporation of such an algorithm is planned for some time in the future. The LP can then use a simplified distribution network which consists of only the least cost routes.

There is a second LP model for optimizing the production and distribution schedules for the utility products. This has all the essential characteristics of the first LP model and, in addition, it has a certain priority order on the demands for products. For example, a certain market area demand should first be satisfied completely, if possible, before satisfying the demand at some other market area. This requirement is met by assigning a binary (0,1) decision variable (x_i) to each demand and setting up predecessor relationships between these variables such as $x_i \geqslant x_j \geqslant x_k$. The normal LP problem is thus converted to a mixed (zero-one) integer programming (MIP) problem. The MIP problem is solved using the IBM's MPSX/MIP system.

3. A TRANSPORTATION SCHEDULING MODEL

A certain major chemical product and its required raw materials are manufactured at several production facilities of Union Carbide. The raw material is shipped, if necessary, from its origin to the place of production of the finished product. The supply of this raw material is fixed. The finished product is shipped to customers for sales or to warehouses for inventory or to other Union Carbide production facilities for making further derivatives. There are high and low limits imposed on the production and inventory of the finished product. The scheduled arrival and departure times of the ships are known beforehand. Given the raw material availability, product demands, production capacities and inventory limitations, the problem is to minimize the net transportation and inventory costs and to specify, if possible, the bulk movements by barges in integers.

The above problem is solved through a mixed integer programming (MIP) model covering eight time periods. The LP matrix generation and report writing are done through programs written in PL/I. The MIP model is solved using the IBM's Mathematical Programming System Extended (MPSX) with its mixed integer programming facilities. This model is run once a month to establish the production-distribution schedule and to provide a barge and ship schedule guide.

4. A PRODUCT SEQUENCING AND SCHEDULING MODEL

This is a deterministic simulation model written in GPSS (General Purpose Simulation System) and run on an IBM 370/165 computer system. The model is run once a month to establish a daily production schedule at a plant for a period covering three months.

The production facility represented in the model is comprised of about seven reactors and three processors. About thirty intermediate products are made on the reactors which are then mixed with some additives and processed on processors to yield about sixty finished products. A product can be made (processed) on one or more reactors (processors). The technological considerations restrict the products to certain types of reactors and processors. Whenever there is a product change-over on a reactor or processor, there is always a certain amount of substandard material produced. The amount of the substandard product made, however, depends upon how widely the two products differed in the properties, such as specific gravity and melting point. A primary objective in scheduling is to minimize the total substandard produced which is accomplished by minimizing the total turnaround time due to product change-over.

Given a total number of hours available for reaction or processing during a given time period, the reactor and processor rates, product turnaround time, product demands, and product sequence on each reactor and processor, the model determines the feasibility of a given schedule. Different product schedules originating from different product sequences are simulated and the most satisfactory schedule, determined by the least total change-over time, is selected.

5. AN INVENTORY SCHEDULING MODEL

Hydroelectric power is generated at a cetain place through a set of reservoirs and power stations. There are uncertainties regarding the inflow to these reservoirs, future power requirements, and future buying and selling prices of power. Under these circumstances, the problem is to decide for the current time period whether to buy or sell power or do neither and if buy or sell, how much to buy or sell.

The problem is solved through a dynamic programming approach. The planning period is first divided into, say, n time periods. The storage capacity of the reservoir system, the generating capacity of the power stations, the penalty for not meeting the power requirements, and the "recovery value" of the power stored in the form of dammed water in the reservoir at the end of the planning period are known. Also, for each time period, forecasts of selling prices, purchase costs, limitations on sales and purchases, probability distribution of power requirements, and probability distribution of inflow are known.

The stored power is the state variable in the dynamic programming model. The decision variables are buy and sell decisions. The return function is the expected total variable costs which include purchase costs, credit for sales, and penalty costs (economic measure of not meeting requirements). The basic recursive function used for optimal decision making in each time period is:

$$F_i(x^i) = \min_u \left\{ G_i(u) + F_{i-1}(x^{i-1}) \right\}$$

where $F_i(x^i)$, $F_{i-1}(n^{i-1})$ = expected optimal returns for n through i^{th} and n through $(i-1)^{th}$ periods, respectively

$G_i(u)$ = expected total variable cost for i^{th} period

u = decision variables

x = stored power.

The basic equation is solved first for period n for a series of discrete values of x^1 (the power stored at the beginning of the period). The procedure is repeated for each period of interest, in the order n, n-1, n-2, ..., up to the current time period.

Solution to the problem yields a value to the water in the reservoir (stored power). If the value is lower than the purchase cost, then a sell decision is made; if the value is higher than the purchase cost, then a buy decision is made.

6. A VEHICLE SCHEDULING PROBLEM*

This problem is a complete departure from the rest of the scheduling problems discussed in this paper in the sense that none of the sophisticated Operations Research Scheduling techniques such as mathematical programming, queuing theory, and simulation were used in solving the problem. Also, no computer was used even for basic computations. A "common sense" analysis of the problem, backed up with traditional industrial engineering methods of time studies, yielded a solution to the problem resulting in 43% reduction in the operating costs.

When a tank truck arrives at a plant, it is driven from the plant terminal to the loading racks, waits until the load is completed and then returns to the terminal for departure to the customer. If the truck is held up over, say, two hours, then the driver should be paid at an hourly rate for as long as the truck is detained. The problem is to schedule the arrival to and departure from the loading racks of the tank trucks so that the detention time is minimized.

---

*A detailed discussion of the problem can be found in "How to Reduce Tank Truck Detention," J. R. Fell, Jr., presented at the 23rd Annual IE Conference for Management and Supervision, April 9-11, 1972.

The engineering study team set up to solve the problem first analyzed the volume of business at the loading racks which showed that about 62% of the detention cost was traceable to the two highest volume racks. An on-site study of these two racks further showed up several detention-causing delays such as waiting for the laboratory test of samples and waiting for the tractor. The study also indicated that travel times to and from the racks, and weigh-in and weigh-out times were very reasonable and did not appear to be a significant factor in causing detention.

The arrivals of the trucks were not random; they were scheduled. Also, there was excess rack capacity. However, as indicated above there were still problems causing delays. Armed with the knowledge from the problem analysis, a step-by-step procedure was followed to come up with a good scheduling procedure.

First, up-to-date rack layouts were drawn, showing locations of lines and their materials. This gave a picture of the rack specifications, enabling intelligent scheduling decisions to be made. Second, expected loading rates in minutes per thousand gallons were developed by time study. Third, procedures to reduce the delays identified in the rack analysis were developed. Finally, to effectively use the loading rates and rack specifications, a new scheduling method was developed in conjunction with the local distribution group. It was based on Short Interval Scheduling, a technique widely used in industry for controlling production in accordance with predetermined norms. It involved communications between the scheduler and the loading operation (via direct line) regarding the assignment of work and the status of work assigned, and between the scheduler and the foreman should the assignment termination significantly exceed the expected time. This technique allows the scheduler to monitor his planning activities effectively, and use the foreman to correct problems as they develop.

A loading priority was developed from information on the loading tickets received during the day. The rack was checked for limitations and physical configuration using the rack layouts and specifications which had been developed. Then the schedule was developed using the loading rates and adding in known delay factors. This schedule was plotted on the planning sheet, considering the priorities discussed earlier. Truck arrivals and filling times were plotted on the planning sheet. The scheduled plan was set up with the carriers, and any problem loads identified by the evening scheduler for investigation the next day.

On the following day, the day man would check out the identified problems revising the schedule and contacting the carrier if necessary. The scheduler would communicate with the loaders, foremen, and supervisors regarding progress or problems, record any major delays for management attention and keep the schedule current.

7. PROJECT SCHEDULING SYSTEMS

Two major project scheduling and control systems currently in use in Union Carbide are discussed here: (1) A Network Guidance and Updating System (ANGUS) for scheduling engineering and construction projects and (2) Maintenance Updating SCheduling and LEveling (MUSCLE) system for maintenance planning and scheduling. ANGUS is written in FORTRAN (Level H) language and, MUSCLE, in PL/I language. Both systems are currently run on an IBM 370/165 computer system.

The fundamental approach used in the analysis of project activity networks is the Precedence Diagramming Method (PDM). There are two basic elements in a PDM network: (1) activities represented by nodes and (2) precedence relationships between activities represented by the sequence lines connecting the nodes. Special relationships between activities are represented by lag factors indicating time relationships between beginning or ending of one activity to the beginning or ending of another activity.

The PDM has evolved from the application of the Critical Path Method (CPM) and the Project Evaluation and Review Technique (PERT). PDM includes some of the advantages of CPM and PERT, specifically excludes some major disadvantages of CPM and adds many new capabilities. The advantages of PDM over CPM and PERT are, for example, simplification by the elimination of necessity of events, dummy activities and other artificial devices; assignment of a single meaningful identifying label to an activity which enables direct correlation of activity numbers with cost accounts.

The appropriate calculations include early, free, and late start and finish dates for each activity in a PDM project network. The early dates represent the earliest plan for execution of the activity. The free dates represent the limits to which an activity may be delayed without interfering with the start or completion of any other activity. The late dates represent the amount of time any activity may be delayed without affecting the completion date, although some other activities in the network will be delayed in starting or completing. For critical activities, all three sets of dates are identical. Noncritical activities have float values which represent the difference between the limits defined above. The free float of an activity is equal to the working day value obtained when the early finish is subtracted from the free finish. The total float is equal to the value generated by the subtraction of the early finish from the latest finish.

When once a PDM network is analyzed, all the relevant information about the network can be stored on a magnetic tape or a direct access device so that, in the future, the same network can be updated without having to go through the calculations again. This provides the ability for the project control team to monitor the progress of individual projects without costly reruns of the PDM network analysis.

ANGUS permits the usage of weather or production factors which revise the span in which the activity occurs. The weather factors are calculated through the employment of rainfall and temperature probabilities. Production factors can be used to indicate vacations, the effect of training programs, the influences of multiple assignments, etc.

MUSCLES has resource scheduling capability. The basic concept in resource scheduling is to schedule the activity in such a way that it will satisfy the available resource requirements and at the same time satisfy the lag or technological constraint. There are cases, however, where the resource may be exceeded, but in no way should lag or technological constraints be violated.

The present algorithm, which is heuristic in nature, tries to schedule an activity between the early start and late start, and it will extend the actual start beyond the late start if necessary. Splitting is allowed when resource is not available on an activity.

The algorithm steps through time in one step at a time. During the first time period, it examines all the activity that can start at the time period, first considering the preceding lag relationship. If any of the activity violated the lag relationship, it will be pushed to the next time period. A priority is assigned to each activity as it enters the active list. The active list is sorted according to its priority in descending order. The first activity in the active list is scheduled first, and if no resource is available, it will be postponed to the next time period. This is done until the active list is exhausted. In the next time period, there will be activities that have started in the first period and continue on to the second time period. This is called the overlapped activity. Following the same way as the active list, the overlap list is created and subsequently the active list is created.

The overlap list is first scheduled and then the active list is scheduled. In scheduling the active list, resources can be borrowed from the overlap list by splitting.

8. CONCLUSIONS

The applications discussed above demonstrate the use of several scheduling techniques. Whenever optimization is required, the most commonly used approach is mathematical programming, particularly, linear programming. There are, however, situations where (1) what is required is not an optimization but the determination of the feasibility of a particular schedule or decision, (2) the system is too complicated to be formulated as a mathematical programming problem, or (3) the system has stochastic and/or dynamic characteristics. Under these circumstances, the most commonly used technique is simulation.

Sequencing problems of purely combinatorial nature are usually embedded in a much larger problem of production scheduling resulting in a mixed integer programming problem. Presently, the state of the art in integer and mixed integer programming does not offer any fast and economically feasible techniques for solving large problems. The branch and bound techniques are getting popular in some commercial MIP codes. However, for large problems, the search for global optimum is still expensive.

Dynamic programming is still very sparingly used for solving scheduling problems. A major difficulty in the way of using dynamic programming is the so called "Curse of Dimensionality." The problems usually involve more than one state variable. Also, several restrictions connecting the various stages further complicate the problem. Use of Lagrangian techniques and the method of successive approximations do not appear to appreciably reduce the dimensionality of the problem.

The network flow theory appears to be a promising field whose potential has not yet been fully utilized. There are currently some large scheduling problems which, when formulated as an LP problem, would exceed the capacity of the computer or whose computer solution would be expensive. However, even though these problems have network flows in the real world, their formulation as a network flow problem also gets complicated. There is a very interesting time-staged production scheduling problem which is yet unsolved. Some of the network characteristics of this problem which make it complicated are, for example, (1) multi-commodity flows, multiple terminals and multiple sources; (2) capacitated arcs, some having integrality requirements and minimum setup considerations, and (3) a user imposed restriction of desiring local optima over subsets of the network without violating the material balance relationship over the entire network.

ACKNOWLEDGMENTS

I would like to express my thanks to the people who provided some of the material contained in this report, gave me their comments and, finally, approved the material to be released for outside publication.

A SOLUTION TO A SPECIAL CLASS OF FLOW SHOP SCHEDULING PROBLEMS

Michael S. Salvador

Ernst & Ernst, Washington, D.C., District of Columbia, USA

This paper considers the most general type of "network" flow shop in which jobs pass through several stages, each of which is composed on one or more identical processors. Jobs are processed on any one of the processors at each stage in ascending order of stage numbers and the objective is minimization of makespan. The class of shops considered is characterized by prohibited in-process inventory and slightly restricted job ordering per processor. Originally designed for the scheduling of nylon polymerization, the algorithms developed in the paper have numerous applications, especially in the chemical processes and petrochemical production areas.

INTRODUCTION

Structurally, the problem being introduced in this research, when reduced to its most elementary and restrictive forms, resembles two common scheduling problems. One involves the scheduling of n immediately available jobs, each on one of m identical machines. The other, often called the simple flow shop scheduling problem, may be described as the sequencing of n immediately available jobs through each of m machines. A generalization of these two scheduling problems, namely a flow shop having one or more processors at each stage, is the topic of this research.

Many high volume production facilities consist of several independent simple flow shops. But, as is the case in almost all of these configurations, the nature of the processors at each level is such that they are effectively identical and hence interchangeable. The scheduling problem to which this work addresses itself deals with such a production system and is drawn from an actual application in the synthetic fibers industry. The solution technique has been generalized only to the extent of the assumptions gleaned from its original application; nevertheless, its applicability to other network-type flow shop scheduling problems is extensive.

THE BASIC ASSUMPTIONS

The following are the assumptions which will hold throughout. The first three serve to define the basic concept of a "network" flow shop. The remaining ones identify the particular scheduling problem to be solved.

1) Every job passes through each of $m \geq 2$ stages or levels in ascending sequence of level numbers.
2) The ith level $(i = 1,2...,m)$ is composed of n_i identical processors with equal capacity for processing any job, one at a time.
3) No job may be split on any level, i.e., the entire job must be processed on one and only one previously idle processor at each level.
4) Jobs may enter the shop in any sequence and their availability is unrestricted.
5) All processing times are known and deterministic for each job at each level*; $r_{i,j}$ will be the processing time of job j at level i.
6) An optimal schedule will be that schedule which minimizes the makespan, or total time required to process all jobs.
7) No in-process inventory is allowed between levels.
8) The immediately preceding job processed on any processor at any level must have commenced production in the shop (at level 1) at some time prior to the time when the job currently being processed on that processor commenced production in the shop.

A FORMAL STATEMENT OF THE PROBLEM

Consider the following definitions:

$P = \{p_1, p_2, \ldots, p_n\}$ is a permutation sequence of n jobs to be processed through a network flow shop. P will be called the job permutation sequence. The set of all job permutation sequences will be denoted $\tilde{P}$.

* Setup and teardown times of processors as well as transfer times of jobs between levels are assumed to be negligible. This is not included as a basic assumption since extensions of the solution algorithm so as to include these considerations is provided in the original research from which this abstract has been extracted.

$T(P) = \{t(p_1), t(p_2), \ldots, t(p_n)\}$ is a feasible sequence of relative ($t(p_1) \equiv 0$) times at which the respective jobs in P enter the shop at the first level; it follows that $t(p_1) \leq t(p_2) \leq \ldots \leq t(p_n)$. T(P) will be called the initiation time sequence. The set of all initiation time sequences will be denoted $\tilde{T}(P)$.

MS(P) is the makespan of the job permutation sequence P, given that each job enters the shop at the times specified by some initiation time sequence T(P).

It is assumed that, relative to some preset time base prior to the production of P, the availability time of each processor in the shop is known. The problem consists of finding optimal sequences P* and T*(P*) such that MS(P*) is a minimum. It can be shown that this will completely determine an optimal schedule.

The solution technique could assume three distinct phases:

Phase I: <u>Determination of a sequence of minimum initiation times</u>.

For any fixed P, a method for determining an initiation time sequence T*(P) associated with P which minimizes the makespan of P should be developed. It can be shown that a sufficient condition for T*(P) to minimize MS(P) is that $t^*(p_i) - t^*(p_{i-1}) \leq t(p_i) - t(p_{i-1})$ for $i=2,3,\ldots,n$ and for every $T(P) \varepsilon \tilde{T}(P)$. A technique for finding a T*(P) that satisfies this condition will be presented.

Phase II: <u>Determination of an optimal permutation sequence</u>.

Next, a method for finding a permutation sequence P* which induces minimum makespan, MS*(P*), must be developed. The branch-and-bound technique can be used to accomplish this, as will be seen.

Phase III: <u>Determination of the optimal schedule</u>.

Finally, using the results of Phases I and II, an optimal schedule can be easily extracted.

A SOLUTION TO THE NETWORK FLOW SHOP SCHEDULING PROBLEM

1. A Mathematical Model for the Network Flow Shop

As the terminology connotes, the problem structure being investigated here is easily represented by the mathematical notion of a directed network or directed linear graph. The model is constructed as follows:

i) Each processor in the shop and the bank of completed jobs are identified as nodes. Thus, there are a total of $n_1 + n_2 + \ldots + n_m + 1$ nodes in the network.

ii) An arc (x,y), i.e., "directed" from node x to node y, is an ordered pair with x being a node corresponding to a processor in level i and y being a node corresponding to a processor in level $i+1$ $(i = 1,2,\ldots,m)$. Level $m+1$ will be a fictitious level consisting of a single fictitious node to which all completed jobs pass from level m.

iii) The capacity or value $c(x,y)$ of an arc (x,y) is the minimum time (relative to a time base preset for all arcs) that must elapse before the processor corresponding to node x can commence to process a job and, immediately after completion, pass it to the presumably available processor corresponding to node y. This interpretation logically permits arc capacities to be negative numbers; negative arc capacities represent the duration of time a processor is idle.

As modeled, the directed network described above serves as both a graphical representation of the flow shop and a comprehensive status of the shop with regard to processor availabilities, job processing duration, magnitude of processor idling, etc. The arc capacities are, by construction, time dependent; the status of the shop is thus only dynamically represented to the degree to which the arc capacities are updated in time.

2. Determination of the Sequence of Minimum Initiation Times

It is necessarily assumed that, just prior to the initiation of the schedule, the status of every processor is known; status is defined in terms of the amount of time that must elapse before the processor can commence processing a job and then, upon completion of processing, transfer the job to a specific processor in the next level. In the context of the model defined in the previous section, this means that all arc capacities of all arcs in the network are known at some fixed point in time, after which the first job is to commence production as soon as possible.

Recall that the job permutation P is arbitrarily fixed; the intention is to define a methodology for determining $T^*(P)$. The procedure to be detailed below modifies the arc capacities of the network in such a way that the minimum time that must elapse before the next job can enter the shop without violating any problem constraints can be determined via definition of a suitable dynamic programming recursion equation.

Consider job p_k. Assume that at some point T in time all arc capacities have been defined reflecting all production activities prior to the initiation of job p_k at level 1. It can be shown that the minimum incremental amount of time beyond T that must elapse before the job p_k can enter level s is given by

$$\min_{(i,j)\varepsilon D(s,s+1)} \left[c(i,j) - \sum_{v=1}^{s-1} r_{v,p_k}\right] \qquad s = 2,3,\ldots,m,$$

where $D(s,s+1)$ is the Cartesan product of the set of all nodes in level s and the set of nodes in level $s+1$. Furthermore, assume that job p_k is to be processed along a "path" (a series of connected arcs) defined by the sequences $B(p_k) = \{b_1,b_2,\ldots b_m, b_{m+1}\}$, where b_i is the node representing the processor used in level i. (Recall that level m+1 contains only one fictitious node representing the bank of completed jobs.) Thus the absolute incremental amount of time that must elapse beyond T before job p_k can commence processing at level 1 and, at the same time, not encounter an unavailable processor at any level is given by

$$W = \max\{0,c'(b_1,b_2),c'(b_2,b_3),\ldots,c'(b_{m-1},b_m),c'(b_m,b_{m+1})\}$$

where the "modified" arc capacities, $c'(i,j)$, are computed as follows:

$$c'(i,j) = c(i,j), \qquad (i,j)\varepsilon D(1,2),$$

$$c'(i,j) = c(i,j) - \sum_{v=1}^{s-1} r_{v,p_k}, \qquad (i,j)\varepsilon D(s,s+1),\ s = 2,3,\ldots,m.$$

The problem of finding the minimum value of W and the optimal path $B^*(p_k)$ can be transformed into a dynamic programming formulation. In words, the problem reduces to finding that path over which the maximum modified arc capacity is a minimum.

Let

N_k = the set of all nodes in N which represent processors in level k $(k = 1,2,\ldots,m)$; N_{m+1} is a single element set containing the node representing the bank of completed jobs.

$f_n(x_n)$ = the value of the maximum modified arc capacity of a feasible path starting with processor x_n in level n $(n = 1,2,\ldots,m)$ and continuing to level $m+1$ (the collection of processed jobs) given that an optimal policy is employed up to level $m+1$.

Then utilizing the theory of Dynamic Programming, the minimum waiting time W^* can be found by solving the following set of recursion equations:

$$f_n(x_n) = \min_{y_n \varepsilon N_{n+1}} [\max c'(x_n,y_n), f_{n+1}(y_n)],$$

$$x_n \varepsilon N_n, \quad n = 1,2,\ldots,m-1,$$

$$f_m(x_m) = c'(x_m,y_{m+1}), \qquad x_n \ \varepsilon N_m, \ y_{m+1} \ N_{m+1}.$$

It follows that $W^* = \max\{0, \min_{x_1 \varepsilon N_1} [f_1(x_1)]\}$; the optimal path can be extracted from the stagewise solutions of the above recursion equations.

After W^* has been determined for p_k, the arcs which comprise the corresponding optimal path are updated to reflect the imposition of job p_k upon the system. Assume that the optimal path is given by the sequence $B^*(p_k) = \{b_1^*, b_2^*, \ldots, b_m^*, b_{m+1}^*\}$. Each processor represented in $B^*(p_k)$ cannot process subsequent jobs until it processes job p_k. Let $\bar{c}(x,y)$ be the current value of the capacity of arc (x,y), and $c(x,y)$ be the updated values. Then

$$c(b_i^*,y) = \sum_{v=1}^{i} r_{v,p_k}, \qquad i = 1,2,\ldots,m, \ y \ \varepsilon \ N_{i+1},$$

$$c(x,y) = \bar{c}(x,y) - W^*, \qquad x \notin B^*(p_k).$$

At this point, T is incremented by W^*, and the entire process described above is repeated for job p_{k+1} to determine its minimum initiation time. The same process continues until the minimum initiation time of job p_n has been determined. At that point, $T^*(P)$ is defined for the permutation sequence P.

3. Determination of the Optimal Job Permutation Sequence

To determine the optimal job permutation sequence P^*, the branch-and-bound technique can be used. To do this, however, a related subproblem must be solved. This subproblem involves finding lower bounds on the makespan of job permutations consisting of two contiguous partial permutations. The first of these partial permutations comprises a proper subset of the n jobs in a specific order; the second comprises the remaining jobs in an arbitrary order.

Let $\overline{P}_k$ represent a specific permutation and P_k an arbitrary sequence of k jobs. The notation $\overline{P}_k P_{n-k}$ then represents a permutation of the n jobs where the first k jobs are uniquely defined by $\overline{P}_k$ and the last n-k jobs by P_{n-k}. Here, $k \leq n - 2$, since neither the case $k = n - 1$ nor $k = n$ implicates inclusion of the latter of the two sequences, i.e., $\overline{P}_{n-1}P_1 \equiv \overline{P}_n$ and $\overline{P}_n P_0 \equiv P_n$.

Suppose k of n jobs $(1 \leq k \leq n - 2)$ have been scheduled to enter level 1 of a network flow shop according to the permutation $\overline{P}_k$. The task is to find a lower bound which is reasonably close to the minimum makespan achieved by the family of permutations $\overline{P}_k P_{n-k}$.

The simplest way to determine a valid lower bound on makespan is to imagine the conditions under which the n jobs can process through the shop soonest while, at the same time, not losing sight of practical considerations. For example, in this case, the most ideal situation for any level of production would be that the aggregated idle time of all the machines on that level be minimized; surely, then, that level would be operating at its "capacity", and the total of the n jobs would be processed at that level as quickly as possible. Practically speaking, however, it may not be possible for all jobs, even under optimal conditions, to reach, say, level m in time to make use of each processor as it becomes available there. This may be due in part to the particular constraints involved and the processing times of various jobs at other levels; it may more often be the result of a relatively small number of processors at some level upstream. It is clear that an optimal schedule in a network flow shop may require, at best, at least one level, the so-called "bottle-neck" of the shop, to operate at capacity. Logically, if minimum makespan were to be evaluated m times, sequentially assuming that each of the m levels was operating at its capacity, then the maximum of these m evaluations would serve as a lower bound on the makespan of the n jobs through the entire shop. An explicit albegraic expression for this lower bound, $LB(\overline{P}_k P_{n-k})$, is derived below.

Let p_k be the last job in $\overline{P}_k$. Then $t^*(p_k)$ is the time base to which the arc capacities refer when the lower bound is to be computed. It follows that the soonest level i will be totally idle is given by

$$A(i) = t^*(p_k) + \frac{\sum_{u \varepsilon N_i} \min_{v \varepsilon N_{i+1}} [c(u,v)] + \sum_{j \varepsilon P_{n-k}} r_{i,j}}{n_i},$$

where N_i is the set of processors in level i.

Of course, there is the possibility, in processing the jobs in $\overline{P}_k$, one or more of the processors could be active beyond the time required by the remaining processors to process the jobs in P_{n-k}. In this case, which is likely to occur as k increases, level i would be totally idle no sooner than

$$B(i) = t^*(p_k) + \max_{u \varepsilon N_i} \left\{ \min_{v \varepsilon N_{i+1}} [c(u,v)] \right\} .$$

Furthermore, the least amount of time required beyond A(i) or B(i) for all jobs to leave the shop is given by

$$R(i) = \begin{cases} \min_{j \varepsilon \overline{P}_k P_{n-k}} \left[\sum_{v=i+1}^{m} r_{v,j} \right] & \text{if } N(i) > 1 \\ \min_{j \varepsilon P_{n-k}} \left[\sum_{v=i+1} r_{v,j} \right] & \text{if } N(i) = 1 \end{cases}, \quad i=1,2,\ldots,m-1; R(m) \equiv 0.$$

In summary, then,

$$LB(\overline{P}_k P_{n-k}) = \max_{1 \leq i \leq m} [\max\{A(i),B(i)\}+R(i)].$$

Now, using this result, the branch-and-bound technique can be applied to find both the optimal permutation sequence P*, and the corresponding makespan MS*(P*).

4. Determination of the Optimal Schedule

An optimal schedule is derived by finding the job permutation sequence P* as described in the previous section. Each branching requires determination of the sequence of minimum initiation times, $T^*(\overline{P}_k)$, for the partial permutation $\overline{P}_k$ as the branch-and-bound algorithm progresses. At termination, P*, T*(P*), and MS*(P*) are known. Since no in-process inventory is allowed, P* and T*(P*) completely define the sequence of jobs entering every level of the shop and exiting from the shop. Processor assignments for each job $p_s^* \varepsilon P^*$, s=1,2,...,n, at each level are known by retaining $B^*(p_s^*)$ when $T^*(P^*)$ is evaluated.

CONCLUDING REMARKS

It should be pointed out that extensions to the solution algorithm above have been developed so as to consider positive processor setup and teardown times, restricted transfer lines between contiguous levels, restricted job/processor assignments, stochastic processing times, and arrival and due dates.

The algorithm has been successfully applied to the production of nylon through a 7-level polymerization process, substantially increasing the throughput of the plant and requiring approximately 18 minutes of computer time on an IBM 1800 process control computer to schedule roughly 60 batches of a variety of seven polymers. More recently the algorithm in its native state was coded and executed on an IBM 370-145; as an example of the results, a 6-job, 15 processor, 6-level problem required 59.49 seconds of computer time.

An approximating technique is now being researched which by-passes the branch-and-bound methodology with a substantial reduction in computational requirements; a near optimal permutation sequence is constructed by sequentially scheduling jobs that can be started soonest.

TWO RECENT DEVELOPMENTS IN SCHEDULING APPLICATIONS

Harry W. Steinhoff, Jr.

Hughes Aircraft Company

The application of sequence scheduling rules at a manufacturing facility is discussed. Three sequencing rules are contrasted - minimum slack time per operation, minimum start date with compression, and critical ratio.

The large scale data processing production environment is discussed. When computer failures occur there is often need for a fast recovery procedure which has decision rules to reschedule network jobs and minimize the lateness of production job delivery commitments. The problem is structured as a resource-constrained project network problem with multiple subnetworks and multiple resource requirements. Some potential methods for rapid solution are suggested for investigation.

Introduction

A wide gap is still evident between scheduling theory and its applications, judging from the topics of many of the papers presented at this Symposium. The gap, however, appears in most instances to be concerned with one of scale rather than one of emphasis.

Scheduling theory continues to struggle with problems of scheduling the job shop containing one or two machines, or n identical machines, or scheduling the single facility with multiproducts. In contrast, the applications aspect of industry often has the task of scheduling shops exceeding 80 jobs and 40 machines.[1] For these medium and large scale job shops the complete enumeration of all schedule combinations tends to be infeasible. Hopefully, scheduling theory can further contribute and provide additional insight into sequencing algorithms and scheduling techniques that will improve the scheduling function in these shops.

The area of resource-constrained project scheduling faces a similar dilemma of problem size and scheduling problems of this type "have proved to be frustrating to deal with, and no general solution techniques of practical value have been developed."[2] The scheduling is compounded when resources unexpectedly become more constrained and projects (jobs) must be rescheduled expeditiously to meet project completion times.

The large scale data processing production facility encounters this type of situation. Many large complex networks of jobs are processed and often concurrently. When computer failures occur there is need for a <u>fast recovery</u> procedure which has decision rules to reschedule network jobs and minimize the lateness of production delivery commitments. Very few rules are designed with that objective in mind. This problem can be structured as a resource-constrained project network problem with multiple subnetworks and multiple resource requirements.

On this note, I would like to present one application of a sequence scheduling algorithm in use in the large scale job shop, and one situation--the large scale data processing production facility--where an improved scheduling algorithm(s) is needed for rescheduling network jobs.

(1) Sequence Scheduling Algorithms for the Job Shop

The El Segundo Manufacturing Division of Hughes Aircraft Company maintains a medium- to large-size fabrication department. Besides machining, it includes other manufacturing processes such as sheet metal assembly, manufacture of printed circuit boards, and wave-guide assembly. It is strictly a job shop. The fabrication department handles approximately 6,000 orders at any one time, of which some 2,000 are active in the shop. The production control system used by the shop provides a daily listing of all orders at each work center in the shop in the sequence in which the orders are to be processed. Recent developments in sequence scheduling are presented here and two lesser known sequencing rules are highlighted. The mathematical expressions for determining order (same as job) sequence values for each of the three sequencing rules are presented in the Appendix.

Sequencing Rules at El Segundo Manufacturing Division

In a job shop one of the important elements of scheduling is the sequence in which orders are selected for processing at each work center. The El Segundo Manufacturing Division became active in this area in the early 1960's. At that time a simulation process was developed whereby actual operating data contained in the existing data processing system was used as input to an IBM 7090 computer for simulation of the El Segundo fabrication shop. A study was made of six dispatch rules using the El Segundo simulation process and using selected criteria (i.e., number of orders completed, percent of orders completed late). When the results were weighed giving completion criteria the most consideration, the study indicated that the Minimum Slack Time per Operation (MINSOP) dispatch rule achieved the best performance for the fabrication shop; the Minimum Planned Start Date per Operation (MINSD) dispatch rule in use in the shop at that time gave the lowest performance.[3]

The MINSOP rule assigns to each job in the work center a MINSOP value determined by subtracting the remaining processing time for the job from the total time remaining (due date minus present time) and dividing the result by the number of remaining operations. The jobs are then sequenced by ascending MINSOP value and assigned for processing in this sequence, the job with the smallest MINSOP value first.

The MINSD rule assigns to each job in the work center a planned start date for that operation determined by subtracting in reverse routing sequence the processing time and expected queue time (move plus wait) for each of the job's remaining operations from the job's due date. The jobs are then sequenced by ascending start date and assigned for processing in this sequence, the job with the earliest start date first.

In 1964 a simulation process was developed at the El Segundo Manufacturing Division in which actual shop operating data was used as input to an IBM 1410-1301 computer system for predicting daily order flow in the El Segundo fabrication shop.[4] This system was implemented in 1964 using the MINSOP dispatch rule.[5]

During the late 1960's the volume of work in the El Segundo fabrication shop decreased significantly. The decision was made to review the fabrication control system with the goal of improving the performance of the fabrication shop as well as economizing on data processing costs. Certain criteria were established as goals. The primary objectives were:

1. No "hot" orders on the shop floor. (i.e., no different

classes of orders or different priorities.)

2. A meaningful interpretation of the sequence scheduling number. (i.e., a date, as opposed to a number or ratio which is then reinterpreted.)
3. No expeditors should be required on the shop floor except for exceptional circumstances.
4. If a date were used the sequence scheduling number displayed for the job at a work center should reflect a date either in the present or future, but not in the past. (i.e., it is difficult to assign a job to a person today and tell him that according to the schedule he should have started the job last week.)

The result of this analysis was a decision to change the dispatch rule from MINSOP with two classes of orders, "hot" followed by regular, to an operational start date dispatch rule with slack compression (MINSDC) and one class of regular orders.

The MINSDC rule is identical to the MINSD rule except that the initial planned start date of a job for that operation is recalculated when that date is earlier than the current date. The expected queue time is successively compressed by ten percent increments of the original queue time and the start date recalculated each time until a start date is found equal to or exceeding the current date. The maximum compression is to ten percent of the original time and start dates still occurring in the past are retained. The compression factor is used as a second level sequence within jobs with equal start dates, maximum compression first.

The MINSDC dispatch rule is the rule currently in use at the El Segundo Manufacturing Division and Tucson Manufacturing Division of

Hughes Aircraft Company.

Evaluation of Sequence Scheduling Rules

How effective is this sequencing rule of earliest start date per operation using compression of the slack or queue time? An analysis was reported in 1971 regarding slack-time priority rules.[6] Three rules and their relative merits were discussed in the study -- MINSOP, MINSDC (called slack compression plan in the study) and Critical Ratio (C.R.).

The C.R. rule assigns to each job in the work center a C.R. value determined by dividing the total time remaining (due date minus current date) for the job by the sum of the processing times and expected queue times for each of the job's remaining operations. The jobs are then sequenced by ascending C.R. value and assigned for processing in this sequence, the job with the smallest C.R. value first.

The authors who reviewed these three scheduling rules indicated that the Critical Ratio rule is a useful refinement over MINSOP when lead times (processing time plus expected queue time) are known. Their comments on MINSDC were that while it has a major "human" benefit working with more meaningful numbers than C.R., its two readily apparent limitations using Work Center start dates are: 1) no tie-breaking capability except the compression ratio, and 2) the number of operations remaining is not recognized in the priority calculations.[7] Some observations are thus in order.

The MINSOP rule considers the number of potential job delays by looking at the number of remaining operations and ignores expected queue lengths. How important are queue lengths?

The C.R. rule emphasizes total processing time remaining. How important is the number of remaining operations? Note also that if processing times are significantly less than expected queue times and can be neglected, then the resulting C.R. Rule is the same as the compression ratio of MINSDC.

The MINSDC rule emphasizes available slack time versus expected queue time. Note that when no compression is required (the desired criterion) the MINSDC rule is the same as MINSD. But what if five to ten percent of the jobs in the shop are consistently behind schedule; does schedule compression improve the shop performance over other scheduling rules?

Which of these rules is preferred according to schedule theory and under what conditions? It may be that simulation is needed to provide the answer. The MINSOP rule has received considerable attention in the literature but the other two rules are not well known and have not been reported in any of the major simulation reported to date. It appears that further investigation is needed.

(2) Scheduling Jobs in a Large-Scale Data Processing System

The computer has been used extensively for solving management science problems, but little has been done to solve problems involving the computer as part of a large scale data processing system. One such system involves the blending of the scientific community of users with the commercial community of users. The resulting job mix in the data processing system consists of unscheduled jobs intermingled with production jobs which run on a scheduled repetitive basis and may be independent or highly interactive with one another. Resources are limited, hence, a basic objective is to schedule jobs using resources most efficiently consistent with meeting production schedules. By structuring interactive jobs as subnetworks which combine to form project networks the situation can be interpreted as one of finding solutions (optimal or otherwise) to the resource-constrained project network problem under conditions of multiple project networks and with multiple resource requirements per job (or activity).

The problem is complex as evidenced by the literature. Davis and Heidorn[8] have looked at a single project network with the so-called "multi-source" problem in which more than one resource type may be required by every job in the project and the availability of each resource is limited. Their comment was that "Except for trivial examples, the multi-resource case has so far resisted all attempts at analytical formulation in a manner conducive to rapid computation."[9] The authors then describe a procedure for determining the optimal (minimum duration) to problems involving requirements of several resource types per job.

In this paper the focus is on clarifying and formally defining this scheduling problem for the large scale data processing system. The goal is to develop a methodology which will in a timely manner answer such questions as:

1. When the computer goes down for x hours, how should production jobs be rescheduled (or perhaps canceled) in order to recover lost time and meet deadline schedules of the production jobs?
2. Are the data processing resources adequate for future needs?

Problem Situation

The large scale data processing system is designed to process scientific and business programs in the Batch, Remote Batch, and Time Sharing modes of computer operation as well as performing scheduled business processing such as payroll, financial systems, administrative and manufacturing applications. The services provided by Data Processing may be grouped conveniently into Scheduled Production Jobs and Unscheduled Production Jobs. There are an estimated 300 Production Jobs to be processed daily with an estimated 10 activities per job.

To process these jobs two multiprogrammed computer systems are available, each with its own memory and central processor but having shared or dedicated access to disc storage units, card readers, card punches, printers, and tape drives. The adjective multiprogrammed is meant to imply that two or more jobs simultaneously occupy the main memory of the computer and share resources such as the central processor, machine operators, and input/output devices. Once a job is in core it stays there until it is finished (batch).

Scheduled production jobs are characterized by operations which may or may not have predetermined completion times and which have operation processing times that fluctuate according to the amount of data input. Individual jobs run on a scheduled repetitive basis and may be independent or highly interactive with many other jobs forming several interconnected networks. Unscheduled jobs tend to be random in nature but the time of their inputs to data processing tend to reflect the working patterns of the organization.

Structuring the Problem

Assume as a first approximation that the category of Unscheduled Jobs generates statistically determinate demand curves over time. This category can then be treated as a single Scheduled Production Job with known resource demands over time which cannot be rescheduled to other times.

Figure 1 is a flow diagram depicting the basic sequence of operations (activities) of a Scheduled Production Job from start to finish. Individual jobs may not require all these activities.

The following conditions[10] prevail at any one point in time:

1. A given planning horizon of periods not necessarily of the same length denoted by t $(t = 1, 2, \ldots, T)$.
2. A set of I jobs must be processed during time period T. The ith job consists of n_i tasks numbered 1 to n_i. The time to perform each task is a known integer represented by p_{ij} for the jth task of the ith job. Task ij may have a specified start time b_{ij} and/or finish time f_{ij}.
3. A set of K different resources is available. R_{kt} is the amount of the kth resource available in time period t. r_{ijk} is the amount of the kth resource required by task ij during its processing. For example, in the data processing problem the computer task requires specified amounts of core, the central processor unit, and perhaps other resources.
4. As a first approximation no preemption of task performance is allowed. Once task ij is started it must be processed until completed in p_{ij} time units. This condition is tantamount to requiring

   $$f_{ij} = t_{ij} + p_{ij},\ j = 1, \ldots, n_i,\ i = 1, \ldots, I \quad .$$

5. The start times of the tasks on a given job are constrained by a cycle-free precedence network of the CPM-PERT type. This network is determined by a set of constraints for each job of the form

   $$t_{ij} \geq f_{i\ell} = t_{i\ell} + p_{i\ell}\ ,\ \ell \varepsilon P(ij)$$

   where P(ij) is the predecessor set for task ij and contains all tasks of job i which must be completed before ij is started.

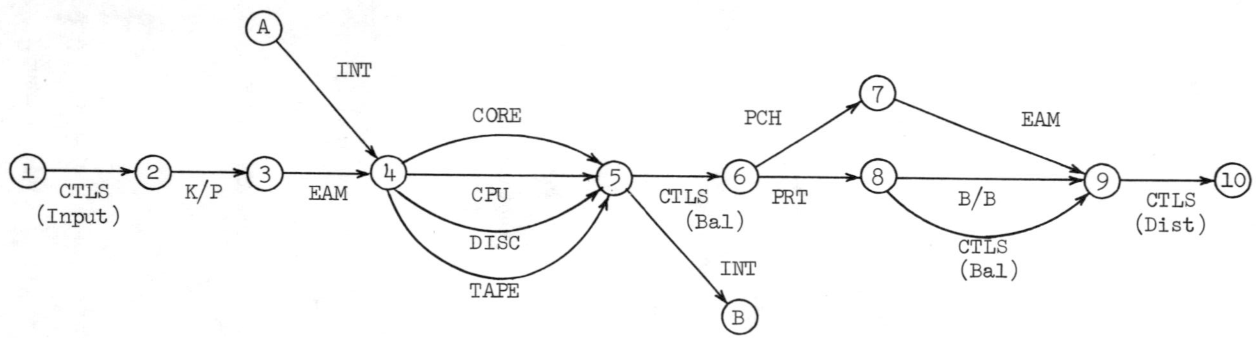

Resource Key

B/B - Bursting & Binding Operations
CORE - Computer Memory
CPU - Central Processor Unit
CTLS - Controls
DISC - Disc Packs
EAM - Electrical Accounting Machinery
INT - Interface with other Job Networks
K/P - Key Punch
PCH - Card Punch
PRT - Printers
TAPE - Tape Drives

Notes

1. Every job has one and only one CPU activity. Each of the other activities may or may not exist for a given job.
2. Node A denotes an interface node to a precedent job activity of a different job.
3. Node B denotes an interface node to a subsequent job activity of a different job.
4. Bal = Balancing; Dist = Distribution.

FIGURE 1

Fundamental Job Network for a Scheduled Production Job

6. A set of R network jobs consists of mutually exclusive subsets of the set of I jobs. The rth network job consists of m_r jobs numbered 1 to m_r.

7. The start times of the tasks on a given network job are constrained by a cycle-free precedence network of the CPM-PERT type. This network is determined by a set of constraints for each network job of the form

$$t_{rj} > f_{r\ell} = t_{r\ell} + p_{r\ell} \ , \ \ell \varepsilon P(rj)$$

where $P(rj)$ is the predecessor set for task rj of the rth network job and contains all tasks of network job r which must be completed before rj is started.

8. Define x_{kt} as the vector of resources used in period t with associated unit vector cost c_{kt}. Total resource cost C is

$$C = \sum_{t=1}^{T} c_{kt} \cdot x_{kt}$$

9. We are required to find values, if they exist or are not otherwise specified, for b_{ij}, $j = 1, \ldots, n_i$, $i = 1, \ldots, I$ which satisfy conditions 1-8 and which minimize C. If a feasible solution does not exist, let V be the subset of R network jobs with specified finish times f_{rj} and the subset of I jobs with specified finish times f_{ij} which are not in the set R. The objective is to select from the set V the minimum number of jobs which, when eliminated from the set of I jobs, generates a feasible solution for the remaining I jobs or exhausts the set V. Note that when a job is selected from set V and eliminated from the set of I jobs, the resources it used become available for redistribution and assignment to the remaining I jobs.

Solution Approaches

A rapid solution to the problem structured above is desired to answer the question of how to reschedule (and perhaps cancel) jobs when the computer goes down for x hours. By rapid is meant within a few minutes. Several approaches seem worthy of investigation. Three are mentioned here.

One generalized approach is to structure the initial set of I jobs as a rather large CPM-PERT type network -- given adequate computer processing capability, solve that problem if solvable or if not, then solve it for the unconstrained case, select the job to be canceled, and repeat the procedure. This process seems intuitively lengthy if many jobs or networks must be selected.

A second approach is to first reduce, if possible, the initial set of I jobs before using a generalized approach. One way of doing this is to find and eliminate from the set of I jobs all jobs which are automatically late, given the time the computer again becomes available and assuming unconstrained resources. Another way is to categorize jobs with deadline schedules into critical and noncritical (postponable) jobs, and then eliminate the noncritical jobs immediately from the set of I jobs. Either way the resulting set of I jobs may not be substantially reduced.

The third approach is to utilize a zero-one (0-1) linear programming formulation of scheduling problems such as reported by Pritsker, Watters, and Wolfe[11] which can accommodate a wide range of conditions. Their model considers only the job level and the project level; expansion of the model to include operations or tasks for individual jobs complicates the notational problem and perhaps expands the computational time of solution beyond acceptable limits.

As mentioned previously, analytical formulation of the multi-resource case with limited resources has not been conducive to rapid computation. It is hoped that presenting this scheduling problem will focus attention on this area of scheduling research and produce meaningful results in the near future.

Appendix

Job sequence values for the three dispatching rules discussed can be mathematically defined in the following manner. Let:

j = job number

i = operation number for job j ($i = 1, 2, \ldots$)

c = current operation number to be processed for job j

N_j = total number of operations for job j

P_{ji} = processing time for operation i for job j

Q_{ji} = expected queue time (move and wait) for operation i for job j

S_{jc} = computed sequence value of job j at operation c

D_j = due date of job j

C = current date

The sequence value of a job using the MINSOP rule is

$$S_{jc} = \left(D_j - C - \sum_{i=c}^{N_j} P_{ji}\right) / \; (N_j - c + 1)$$

The sequence value of a job using the Critical Ratio rule is

$$S_{jc} = (D_j - C) \; / \left(\sum_{i=c}^{N_j} P_{ji} + \sum_{i=c}^{N_j} Q_{ji}\right)$$

The sequence value of a job using the MINSDC rule is

$$S_{jc} = D_j - \sum_{i=c}^{N_j} P_{ji} - k_{jc}\left(\sum_{i=c}^{N_j} Q_{ji}\right) \qquad \text{where}$$

$k_{jc} = 1.0, .9, .8, \ldots, .1$ as determined by z_{jc} where

$$k_{jc} = \begin{cases} 1.0 & z_{jc} \geq 10 \\ .1 & z_{jc} < 2 \\ .1 z_{jc} & \text{otherwise} \end{cases}$$

Appendix

$$\text{for} \quad z_{jc} = 10 \left(D_j - C - \sum_{i=c}^{N_j} P_{ji} \right) / \sum_{i=c}^{N_j} Q_{ji} \quad ,$$

z_{jc} integer.

Footnotes

1. Panwalkar, S. S. Dudek, R. A., and Smith, M. L., "Sequencing Research and the Industrial Scheduling Problem" (paper presented at Symposium on Theory of Scheduling and Its Applications, North Carolina State University, Raleigh, N. C., May 15-17, 1972), 3.

2. Bennington, G. E., and McGinnis, L. F., "A Critique of Project Planning Under Constrained Resources" (paper presented at Symposium on Theory of Scheduling and Its Applications, North Carolina State University, Raleigh, N. C., May 15-17, 1972), 1.

3. LeGrande, E.,"The Development of a Factory Simulation System Using Actual Operating Data", Management Technology III No. 1 (May 1963), 1.

4. Steinhoff, H. W., Jr., "Daily System for Sequencing Orders in a Large Scale Job Shop", 6th Annual ORSA/TIMS Joint Western Regional Meeting, Orcas Island Washington, April, 1964. Also in Buffa, E. (ed.) Readings in Production and Operations Management, Wiley, 1966. 155-166.

5. Bulkin, G. H., Colley, J. L., and Steinhoff, H. W., Jr., "Load Forecasting, Priority Sequencing and Simulation in a Job Shop Control System", Management Science XIII (October 1966), B-29 - B-51.

6. Putnam, A. O., Everdell, R., Dorman, G. H., Cronan, R. R., and Lindgren, L. H., "Updating Critical Ratio and Slack-Time Priority Scheduling Rules", Production Inventory Management XII No. 4 (Fourth Quarter 1971), 51-73.

7. Ibid., pp. 64-65.

8. Davis, E. W., and Heidorn, G. E., "An Algorithm for Optimal Project Scheduling Under Multiple Resource Constraints", Management Science XVII (April 1971), B-803 - B-816.

9. Ibid., p. B-803.

10. Fisher, M. L., "Optimal Solution of Scheduling Problems Using Lagrange Multipliers: Part II" (paper presented at Symposium on Theory of Scheduling and Its Applications, North Carolina State University, Raleigh, N. C., May 15-17, 1972), 2-3. The six conditions described by Fisher and his nomenclature have been adopted as a reference and then modified and expanded as necessary to describe the data processing scheduling problem.

11. Pritsker, A. Allen B., Watters, L. J., and Wolfe, P. M., "Multiproject Scheduling with Limited Resources: A Zero-One Programming Approach", Management Science XVI (September 1969), 93-108.

TOWARD A MAN-MACHINE INTERACTIVE SYSTEM FOR PROJECT SCHEDULING

J. D. Wiest

Rice University, Houston, Texas, USA

A preliminary man-machine interactive system for project scheduling is described in which computer graphics facilities are combined with a heuristic program for allocating resources to activities in a project network. The user can quickly generate feasible schedules and graphically display the results on a cathode ray tube. By means of a light pen and tablet, he can adjust resource levels, change scheduling parameters or modify program heuristics, and readily determine the effects of these changes on the project schedule. The combination of man's intuitive skills and the computer's computational powers enhances the effectiveness of both in searching for good project schedules.

Introduction: Problems in Using Network Techniques

Network techniques, since their introduction in the late 1950's, have undoubtedly helped many managers in the planning, scheduling and monitoring of projects. Over the years of their use, shortcomings and weaknesses of the methods became evident, were widely discussed, and sometimes were alleviated. Many man-years of effort were expended in trying to solve these problems and in other ways to extend the usefulness of the techniques.

Some of the problems were conceptual in nature (see, for example, 8, 13, 21), but mostly they were pragmatic. The probabilistic feature of PERT and the time-cost trade-off capability of CPM were largely abandoned because of difficulties in securing reliable data necessary for their use. Another problem was the models' early focus on time scheduling only. Eventually this focus was widened to include resource scheduling (4, 16, 24), and many heuristic scheduling programs were developed which took into account limitations

This study was performed under a research grant from the National Space and Aeronautics Administration. The author is also grateful to the Computer Science Department, University of Utah, for making available the use of its graphics laboratory facilities, and to the College of Business at Utah for financial support in the use of these facilities.

An earlier version of this paper was first presented at the Second Symposium/Seminar of the Project Management Institute in St. Louis, Mo.

on the availability of needed resources. While these models have increased in power and the range of their application, a number of basic problems remain which somewhat limit the managerial usefulness of these techniques:

1) Projects differ considerably one from another, and the heuristics of one resource scheduling program may work well on one project but not so well on another. With most heuristic programs, it is inconvenient to try various heuristics on a given project because of difficulties in changing the program, problems of computer turn-around time and evaluating the printed output, and so on.

2) Another computer-related difficulty that has bothered managers from the inception of network-based techniques is the "problem of the paper flood." The mere size of some projects invites voluminous output, and ingenious programmers have found almost unlimited ways to "massage" the data and expand the output from a PERT or CPM run. While much of this information may be useful, especially at the operating level where men and machines are actually assigned to jobs, for the manager who is concerned with overall planning and control, a thick stack of printed output from a single computer run can be bewildering and frustrating. The endless tables of data are virtually impossible to comprehend and interpret for purposes of managerial decision making, without considerable condensation and analysis. And even if the data were greatly reduced, printed tabular output is not an ideal form for human comprehension. Most programs are not well-engineered for managerial use.

3) A further deterrent of most network-type computer programs is the awkwardness of managerial interaction. Once a manager has finally waded through the output, if he has any "what-if" type questions (What if we combine these two operations? What if we hire some more engineers, or rent another bulldozer? What if we delay the start of this job?), the answers usually require repunching of data cards, resubmission of the program (with turn-around requiring hours at best and sometimes a day or more), and rewading through reems of output. Somewhere in the process, much of the potential usefulness

of network techniques for true managerial decision-making gets lost. Most managers lack the time or patience to use the techniques in answering the important kinds of questions a manager can (and should) ask. The techniques as presently implemented were not designed with a man-machine interactive capability in mind. Hence the limited amount of manager-computer dialogue.

Most of the above difficulties are essentially problems of communication between man and computer. Such problems are not unique to project management, however; and a good deal of attention has been paid to these problems in other contexts. Time sharing computer systems have drastically reduced turn-around times, and developments in computer graphics--both hardware and software systems--have increasingly facilitated input-output processes.

It seemed to me that the potential existed for greatly extending the usefulness of project management techniques by combining such techniques with newly developed computer technology. In particular I was interested in determining if computer graphics equipment and associated software could be married to heuristic project scheduling programs, using a time-shared computer system, in such a way as to achieve a measure of manager-machine interactive capability. The results of that investigation, which has really just begun, are the basis for this interim report.

Computer Graphics

The term "computer graphics" is widely used in computer literature but only vaguely defined. It is generally interpreted to mean the communication between man and computer by means of lines, curves, dots, or other graphic symbols. Some writers have used "graphics" in direct opposition to "alphanumerics." Perhaps this serves to emphasize the essential feature of computer graphic systems--the extending of information flows to include data in graphical forms--but it remains true that in most computer graphics applications, data communicated includes both graphical and alphanumeric forms.

Computer graphics installations vary widely in equipment and capabilities, but a typical system would likely include a cathode ray tube (CRT), a light pen and tablet and/or a joy stick, an alphanumeric keyboard and function keys, plus the usual input/output devices ordinarily available (punched card readers, line printer, etc.). The CRT is basic hardware for graphical output. Appropriately driven by a computer, it can display pictures made up of lines, curves, dots, and alphanumeric characters. The light pen, a stylus-photocell arrangement, can be used to locate positions on the CRT, to be interpreted as a call for a subprogram or as an input of pictorial data for display on the CRT. Along with a tablet such as the RAND Tablet (5) and associated software, the light pen provides a convenient method of inputting graphs, diagrams, flow charts, alphanumeric characters, and other data that can be drawn or sketched by the user. Non-graphic communication, in or out, can be accomplished by means of the alphanumeric keyboard, and the function keyboard provides a convenient means of directing the interaction process by specifying certain functions to be performed and displays to be mounted on the CRT. Generally, the CRT and other gear are driven by a small computer, which in turn connects on line with a larger computer which performs the bulk of computations required.

An essential feature of computer graphics systems is a time-sharing or program-interrupt capability of the central computer. Conversational-style communication between man and computer, whether graphical or digital, would be impossible without rapid response from the central computer; and such response requires on-line access to the computer and the capability of almost instantaneous processing, whether by interrupting a currently running program, multiprocessing, or some other means.

The "2+2=5" Effect

Ease of communication is what largely determines the degree of "synergism" achieved in man-machine interactive systems (18). It is not just that man and computer are used where they have the greatest relative advantage, but that the combination of the two achieves results which neither could accomplish alone. Man's intuitive skills and ability to solve problems heuristically, when combined with a computer's great computational power, can yield a problem-solving system which is better than the sum of its parts. Computer graphics can enhance that synergistic effect, by greatly improving communications between the two.

Perhaps the most widely publicised application of computer graphics has been in the area of computer-aided design (18), following the pioneering work of Sutherland in developing hardware and software systems for drawing diagrams (20). Other uses have ranged from debugging computer programs to check-out procedures for space vehicles (6). Applications in the area of management decision-making have been fairly recent, beginning with real time graphical report generators (17) and continuing with more interactive systems for scheduling maintenance (6) and job shop operations (9). The potential of computer graphic systems for decision making and the extent of its power for aiding managers are just beginning to be perceived (10, 15).

Development of Graphic SPAR

For a number of years I have been interested in the problem of large project scheduling with limited resources (22, 23). As part of my research on this topic, I developed a computer model capable of scheduling single or multiple projects within specified resource limits and other contraints commonly encountered by a project manager. Called SPAR*, the model is based on a collection of heuristics or rules which determine how available resources are to be allocated to activities requiring the resources. By changing certain parameters and heuristics in the model, one can simulate project scheduling under varying constraints and scheduling rules.

To develop interactive capabilities for SPAR, I used the University of

*For "Scheduling Program for Allocating Resources." A more detailed description of the model is contained in (24).

Utah's Graphics Lab facilities,* which consisted of an IDI display scope and light pen, a Sylvania tablet, and a model 35 teletype, which combined to form an interactive console with keyboard and display. A PDP-8 computer acted as a controller for the graphics equipment and served as a communications link between that equipment and the main computer, a Univac 1108, to which it was connected on-line. Also within the laboratory was an on-line Univac 1004 reader/printer and 1108 console monitor which permitted the user to read in, execute and obtain printed results of his program while in the Graphics Lab.

To improve the usefulness of a project scheduling program like SPAR, and to increase a project manager's interaction with the computer, it seemed to me that a man-machine system would have to provide at least the following capabilities:

1) Easy input of scheduling program and project data.
2) Minimum turn-around time--ideally immediate response.
3) At the user's option, graphical output of certain data, for faster comprehension and interpretation.
4) Easy and rapid modification of program parameters and/or project data.
5) Hard copy of selected output data.

Fortunately, programming for the graphics system was sufficiently developed and documented so that I could write my own programs in a FORTRAN-like language the Department had developed, with a little help from an experienced user of the system. A simplified flow chart for Graphic SPAR is shown in Figure 1.

---

*The Computer Science Department at the University of Utah, whose facilities I used, has been engaged for several years in an extensive program of research and development in computer graphics. While its own research program has been dedicated to computing principles and techniques per se, as contrasted with applications, it has encouraged others within various departments of the University to explore the use of computer graphics in their own fields of study. My research was sponsored by the College of Business, which was interested in exploring possible uses of computer graphics in management.

Figure 1

Simplified Flow Chart for Graphic SPAR

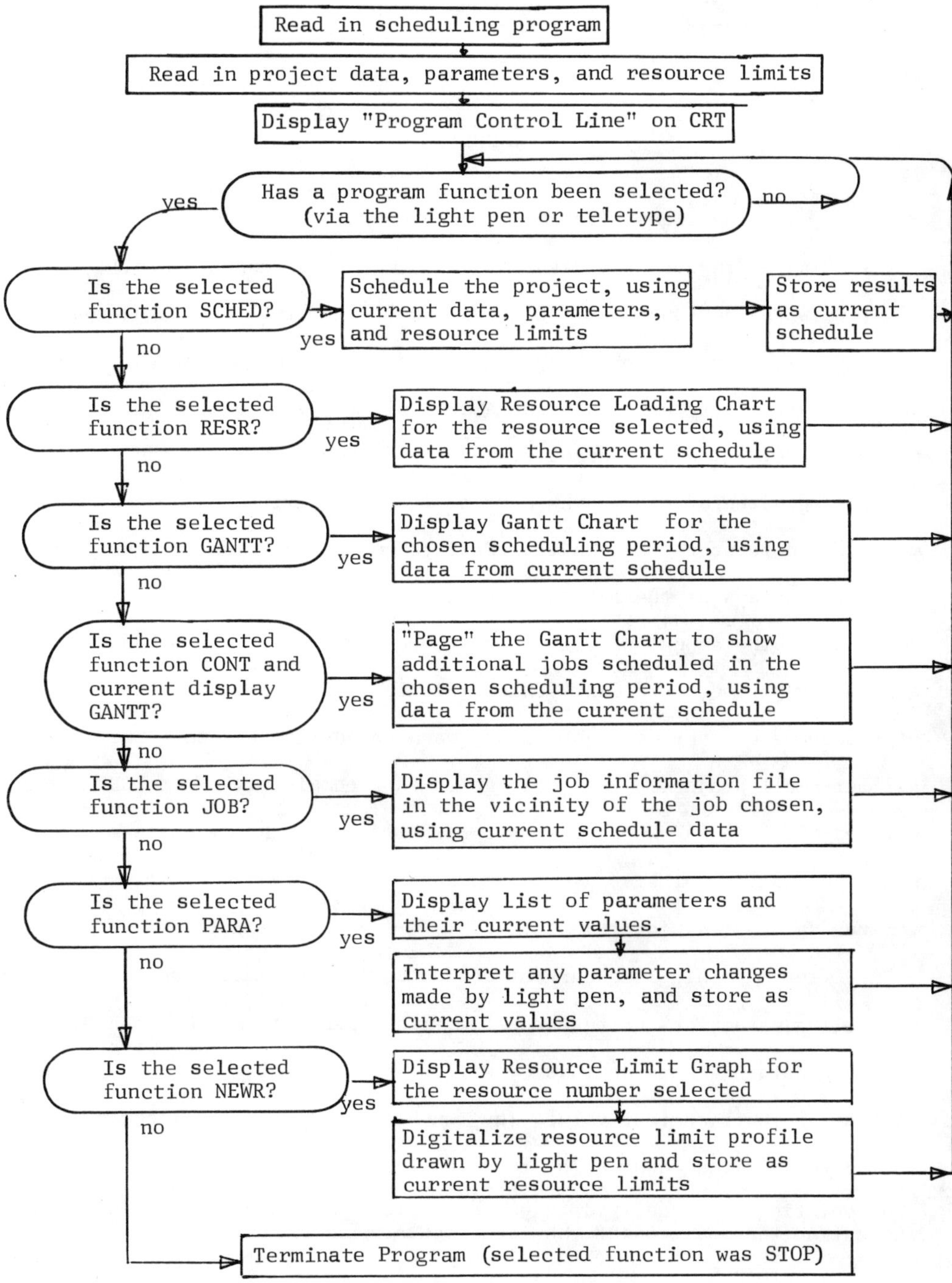

Operation of the Program

Briefly, the program functions as follows: After the program and a project data deck are read into the computer, a "Program Control Line" appears at the bottom of the CRT (Figure 2). Names of several control functions are permanently displayed which the user (manager) can select at will, by pointing the light pen at a particular word.* The functions are of three types:

1) Action functions ("Schedule the project," or "Stop")

2) Display functions ("Display resource loading chart," etc.)

3) Data change functions ("Change resource limits," or "Change parameters")

The user might start by changing some initial parameters or resource data, but typically would ask for a schedule to be generated (by pointing the pen at SCHED). The main computer is called upon to perform the appropriate scheduling calculations, using current resource limits and scheduling parameters, and to store the result. Next the user would ask for one or more graphic displays of the results--perhaps a Resource Loading Chart (Figure 3), which shows the amount of a given resource required by the schedule in each scheduling period, or a Gantt Chart (Figure 4), which displays the amount of resources used by each job during each scheduling period. The latter, for all but small projects, is too large to show completely on the CRT and is divided into "pages" which the user can "flip through" by means of the light pen. The user might also ask to see a section of the job list, showing various data pertaining to each job.

After examining the current schedule, the user may wish to change some scheduling parameters or resource limits and generate another schedule with this new data. He can do this by calling for a display of some thirteen parameters and their current values (by touching PARA with the light pen) and making any desired changes (by touching the parameter to be changed, followed

*The pointing is actually done on the tablet, but the words and light dot indicating the position of the light pen are on the CRT.

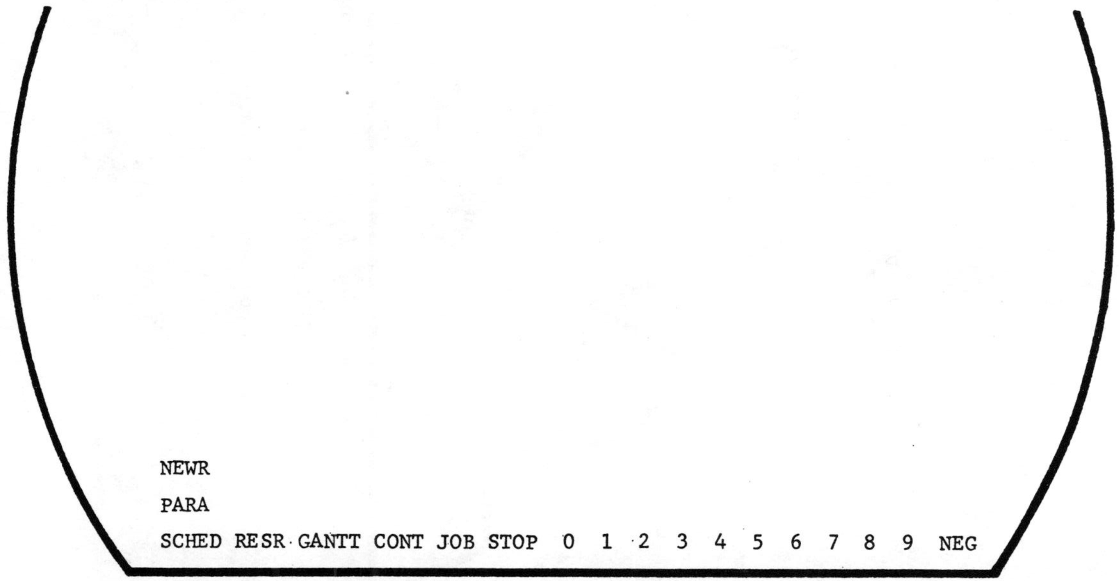

FIGURE 2: Program Control Line displayed at bottom of Cathode Ray Tube

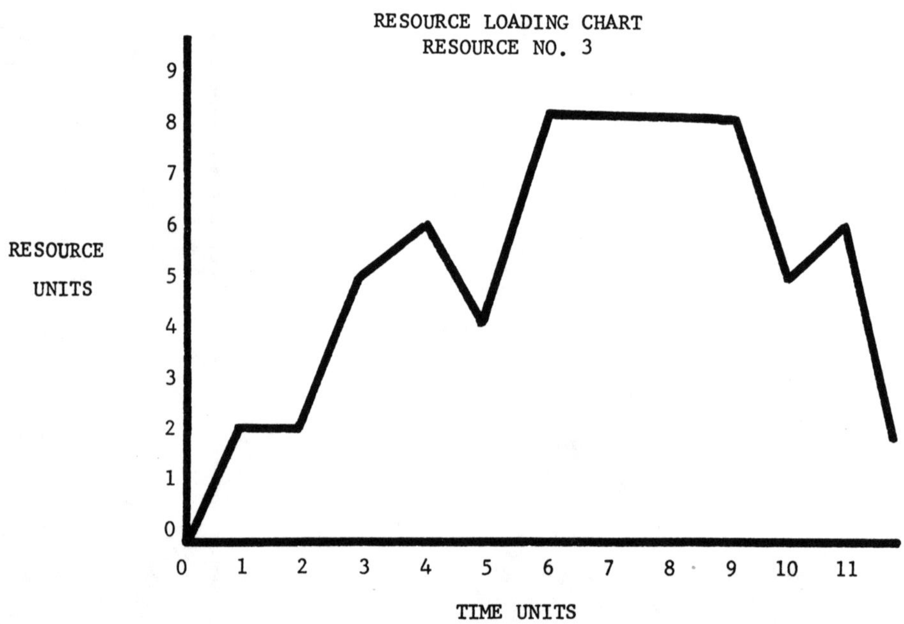

FIGURE 3: Graphic display of Resource Loading Chart

GANTT CHART
PERIODS 1 THROUGH 20

ACTIVITY	1	2	3	4	5	6	7	8	9	10	11	12	13	14	15	16	17	18	19	20
1	2	2	2	2	2															
3		6	8	8	4															
4				5	5	5	5	5	5	5	5	5	5	5						
7			2	2	2	2	2	2	2	2	1									
9						10	10	10	10	10	10	10	8	8	8	8	8	8	8	
14										9	9	9	7	5	5	5				
15								12	12	12	12	12								
17								8	8	8	8	8								
PERIOD:	1	2	3	4	5	6	7	8	9	10	11	12	13	14	15	16	17	18	19	20

FIGURE 4: Graphic display of Gantt Chart

by the number desired as the new value--again with the light pen). Or he can change resource limits for any desired resource by calling for the Resource Limit Graph (RESR) and sketching a new resource-time profile on the graph with the light pen (Figure 5). The program automatically converts the continuous profile to a stepped-profile of the nearest discrete units of resource. Then the user calls for a new schedule.

The process is repeated as often as the user wishes to do so. He can try different program heuristics by changing the scheduling parameters, or he can investigate the effects of altering resource limits, as a manager might affect by hiring additional resources, sub-contracting, shifting existing resources from one use to another, etc. In this way he can generate a variety of schedules, simulating the scheduling environment (resource limits and scheduling rules), and pick the schedule which best satisfies whatever criteria he is using for a "good schedule."

The whole cycle from schedule to new schedule may take a minute (or several minutes, depending on how long the graphs are examined and how extensive are the changes being made). By contrast, without the graphics capabilities, the process may require hours or days, while the user hand plots the schedule data, repunches input data cards, and waits for normal computer turnaround. In 30 minutes at the graphics console, I found that I could accomplish more than could be done in two or three weeks by conventional methods. The improvement is not just a matter or reduced turnaround time, but of improved user efficiency. The user can maintain a train of thought and a sense of direction better if his experiments occur over a span of minutes rather than weeks. Start-up costs, always incurred when one returns to a pursuit after extended interruptions, are eliminated. The user's attention is constantly maintained while at the graphics console.

The importance of experimenting with different heuristic scheduling rules and various constraint patterns should not be overlooked. The complex interaction of heuristics in a program like SPAR, applied to a complex

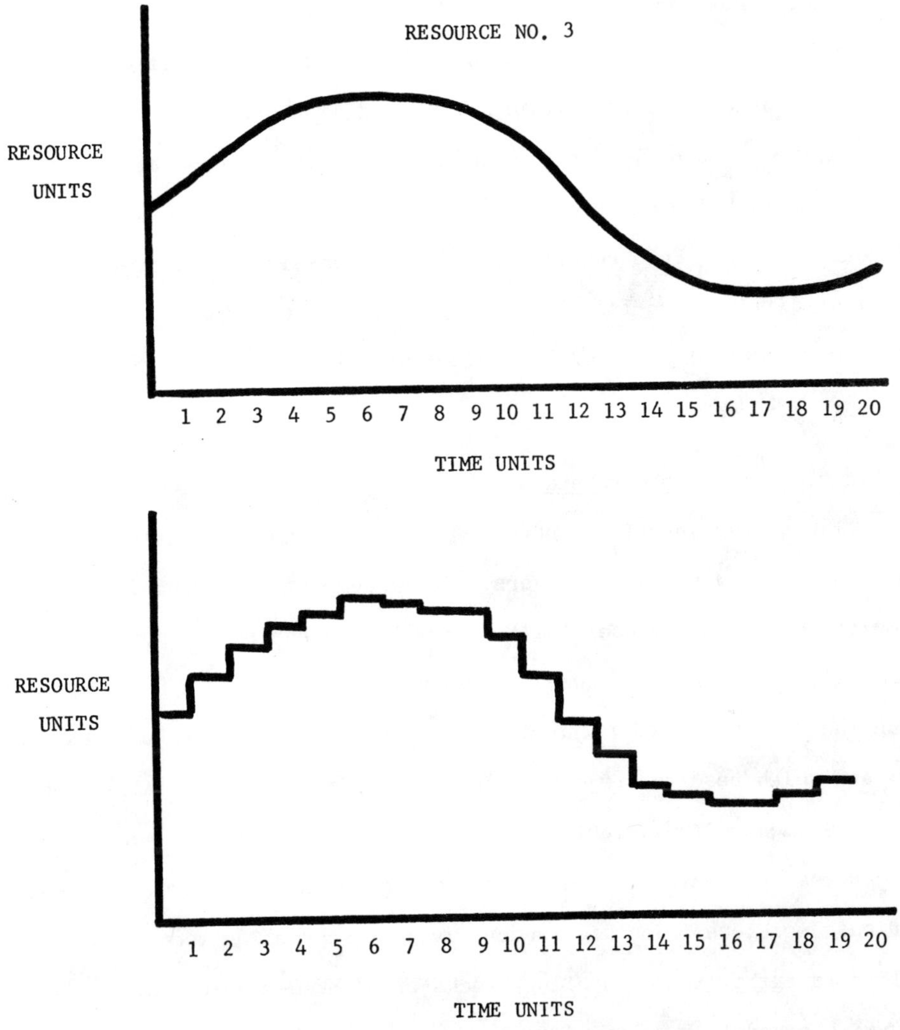

FIGURE 5: Resource Limit Graph, as drawn by light pen (above) and as approximated in discrete units by computer (below).

structure of activities in a project network, can sometimes produce anomalous results. In one case, for example, I picked out what appeared to be a bottleneck resource and successively increased its limit in a series of trials. The project schedule, instead of becoming shorter, grew successively longer and longer! Not until the resource level was 50% above its original level did the schedule length stop increasing, and the resource level had to be doubled before the schedule length returned to what it had been originally. Then varying the resource limits downward, I found that they could be reduced several units below their original level without adversely affecting the schedule length. Such results are almost impossible to anticipate *a priori*, and most difficult to find by conventional scheduling procedures. But the problem was detected and explored in just a few minutes at the graphics console.

Implications for Project Managers

The system described above is preliminary and obviously not ready for immediate use by project managers, but it demonstrates clearly, I believe, some of the potential usefulness of computer graphics for project management. Primarily it can help managers to manage. No one is in a better position than the manager to know what decisions can be made, where changes are possible, and which results are desirable. Operating at the graphics console, he can experiment with variables that he knows are within his power to change. He can test his intuition, use his experience, and simulate the results of various decisions. In this manner, project scheduling techniques become useful tools rather than stumbling blocks of stacked-high output. By experimenting with scheduling constraints and parameters, with immediate feedback in the form of easily comprehended graphic displays, he will likely find a better project schedule than would otherwise be possible. Such a manager-machine interactive system would combine the best of his heuristic skills and experience with the computational power of a large computer, increasing the effectiveness of both.

Future Work

While the present graphic SPAR program has served to demonstrate the potential benefits from interactive project scheduling, there are several modifications and additions I would like to make that should increase its usefulness. For example, I would like to enable the user to do the following:

1) Display the network itself on the CRT, or some selected part of it.
2) Add or delete activities from the network.
3) Change the logic of the network, by changing predecessor relationships.
4) Modify activity data, such as resource usage rates allowed.
5) "Back up" in a schedule and reassign one or more jobs, letting the computer make whatever subsequent modifications are necessary in the schedule.
6) Supply his own criteria function, by which the computer would evaluate each schedule produced.

Some of these program changes could readily be made (e.g. 4); others would require greater effort. All of them would add to the program's flexibility and should enhance a manager's ability to effectively plan and schedule. Much more experience needs to be gained, however, to determine the nature and essential features of a system that would best facilitate manager-machine interaction. (1, 2)

Operating at a graphics console will be a totally new experience for most managers, and the process will require the development of certain skills and procedures. As opposed to the usual situation in which the number of possible schedules he can explore is very limited, he will be faced with almost unlimited possibilities, the prospect of which may overwhelm him. He must develop guidelines to make his explorations at the graphics console reasonably efficient. Whether a set of generally useful guidelines can be developed and given to him beforehand, or whether he must evolve these as part of the process of gaining skill in using the system is another question which must be researched.

Presumably, scheduling heuristics which can be recognized and clearly described can be programmed for the computer. The *man* in the man-machine system is useful for providing heuristics that are too complicated or too difficult to generalize to put in a computer program, and for supplying *ad hoc* rules and procedures. When he explores the effect of various resource levels, for example, he (as a manager) should have some notion of the limits within which such changes should be made, to keep them consistent with reality; but he may have to develop some rules of his own for how best to explore for better solutions within this feasible range of resource limits.

Perhaps, after observing a user evolve such procedures, we might generalize them and incorporate them in the program, further increasing the efficiency of the user's search methods; but others may be too "*ad hoc*," too dependent upon a specific project or project environment to be programmed, and must be left to the user to devise as needed.

Many of the above remarks are conjectural, of course, but they follow from questions raised by the possibility of a man-machine interactive scheduling system. While we are just beginning to glimpse the potential power that computer technology may provide for managerial decision making, we are also beginning to see the considerable amount of research that must be done to find efficient ways of using that technology.

REFERENCES

1. Bell, T. E., "Graphical Analysis Procedures for Simulation and Scheduling," Ph.D. Dissertation, University of California, Los Angeles, 1968.

2. __________, "Computer Graphics for Simulation Problem-Solving," The RAND Corporation, RM-6112-PR, December 1969.

3. Chesler, L. and R. Turn, "The Application of On-Line Graphical Techniques for Programming and Operating a 'Moving Network' Monitoring Display," The RAND Corporation, RM-5183-PR, January, 1967.

4. Davis, E. W., "Resource Allocation in Project Network Models - A Survey," Journal of Industrial Engineering, Vol. 17 (April 1966).

5. Davis, M. R. and T. O. Ellis, "The RAND Tablet: A Man-Machine Graphical Communication Device," The RAND Corporation, RM-4122-ARPA, August 1964.

6. Drezner, S. M. and R. L. Van Horn, "Design Considerations for CAMCOS - A Computer-Assisted Maintenance Planning and Control System," The RAND Corporation, RM-t255-PR, July, 1967.

7. Elmaghraby, S. E., "The Determination of Optimal Activity Duration in Project Scheduling," Journal of Industrial Engineering, 19, 1 (January, 1968), pp. 48-51.

8. Fulkerson, D. R., "Expected Critical Path Lengths in PERT Networks," Operations Research, 10, 6 (Nov.-Dec., 1962), pp. 808-817.

9. Jones, C. H., "Job Shop Scheduling Problem: Notes on Operating Computer Aided Decisions for Scheduling," unpublished instructions to operators, Harvard Business School, 1967.

10. __________, "At Last: Real Computer Power for Decision Makers," Harvard Business Review 48, 5 (October-November, 1970), pp. 75-89.

11. Lamberson, L. R. and R. R. Hocking, "Optimum Time-Compression in Project Scheduling," Management Science, 16, 10 (June, 1970), pp. 597-606.

12. Licklider, J. C. R., "Graphic Inputs - A Survey of Techniques," Computer Graphics, Thompson Book Company, Washington, D.C. (1967).

13. Mac Crimmon, K. R. and C. A. Ryavec, "An Analytical Study of the PERT Assumptions," Operations Research, 12, 1 (January-February, 1964), pp. 16 -37.

14. Meyer, W. L. and L. R. Shaffer, "Extensions of the Critical Path Method Through the Application of Integer Programming," Department of Civil Engineering, University of Illinois (July, 1963).

15. Miller, I. M., "Computer Graphics for Decision Making," Harvard Business Review 47, 6 (November-December, 1969), pp. 121-132.

16. Moshman, J., J. Johnson and M. Larsen, "RAMPS: A Technique for Resource Allocation and Multiproject Scheduling," Proceedings of the 1963 Spring Joint Computer Conference.

17. O'Lone, R., "Cathode Ray Display Speeds Scram Data," Aviation Week and Space Technology, March 13, 1967, pp. 74-79.

18. Prince, M. D., S. H. Chasen, et. al., "Computer Aided Design," Lockheed-Georgia Quarterly, (Summer 1965).

19. Pritsker, A. A. B., L. J. Walters, and P. M. Wolfe, "Multiproject Scheduling with Limited Resources: A Zero-One Programming Approach," Management Science, Vol. 16, No. 1 (September 1969), pp. 93-108.

20. Sutherland, I. E., "Sketchpad: A Man-Machine Graphical Communication System," Lincoln Laboratory, M.I.T., Technical Report 296, January 1963.

21. Welsh, D. J. A., "Errors Introduced by a PERT Assumption," Operations Research, 13, 1 (January-February, 1965) pp. 141-143.

22. Wiest, J. D., "The Scheduling of Large Projects with Limited Resources," Ph.D. thesis, Carnegie Institute of Technology (1963) (pp. 23-27).

23. _________, "Some Properties of Schedules for Large Projects with Limited Resources," Operations Research, Vol. 12 (May-June 1964).

24. _________, "A Heuristic Model for Scheduling Large Projects with Limited Resources," Management Science, Vol. 13, No. 6 (February 1967).

EFFICIENT SOLUTION PROCEDURES FOR CERTAIN SCHEDULING AND SEQUENCING PROBLEMS

L. G. Mitten and C. A. Tsou

The University of British Columbia, Vancouver, British Columiba, Canada

Introduction

The solution of scheduling and sequencing problems has in most cases proved to involve serious difficulties. Direct solution methods are available only for problems of very special structure -- e.g., see [7] -- and computational experience with recursive procedures has generally been quite disappointing. Dynmaic programming has been of very limited utility (e.g., see [4] and the comments of [8] on [2]). Branch and bound methods, although frequently proposed (e.g., see [1] and [6]), have generally not been efficient enough to provide a practical solution procedure.

The results presented in the present paper grew out of investigations aimed at developing solutions for practical project scheduling problems of the type discussed in [2] and [8] but the methods are applicable to a considerably broader range of scheduling and sequencing problems. Basically, our method consists of an implicit enumeration procedure employing a combination of branch and bound and sequence perturbation conditions. The branch and bound portion of the algorithm is itself quite efficient, combining very good bounds with a very good (usually optimal) initial sequence. The sequence perturbation condition is used both to eliminate sets of sequences and to decompose sets of sequences into independent subsets, thus effecting very large reductions in the number of possibilities which need to be considered.

In subsequent sections, we will define the class of problems to be considered, develop the solution methods, and present some computational experience.

Statement of the Problem

Consider a finite set S containing N (> 1) elements, let P_σ be the set of all permutations of the elements of any subset $\sigma \subset S$, and let $P = \bigcup P_\sigma$, with the union taken over the set of all subsets of S. For any permutation $p \in P$, the unordered set S_p consists of precisely those elements contained in p, and we take $\bar{S}_p = S - S_p$. By way of illustration, S might represent a set of N projects and a permutation $p = (p_1, p_2, \ldots, p_n) \in P$ might represent the order in which a subset $S_p = \{p_1, \ldots, p_n\} \subset S$ of n projects are to be carried out. If $p = (p_1, \ldots, p_n) \in P$, then for any $0 \leq m \leq n$ the permutation $p^m = (p_1, \ldots, p_m)$ will consist of the first m elements of p, with p^0 the null permutation. If p and q are two disjoint permutations from P (i.e., if $S_p \cap S_q = \emptyset$), then (p,q) is the permutation consisting of the permutation p followed by the permutation q. In particular, if $p = (p_1, \ldots, p_n) \in P$ and $x \in S$ but $x \notin S_p$, then $(p,x) = (p_1, \ldots, p_n, x)$.

Associated with each element $x \in S$ there are two given real constants, c_x and d_x with $0 < c_x < \infty$ and $0 \leq d_x < \infty$, and for any permutation $p \in P$ we define $D_p = \sum_{x \in S_p} d_x$. For any real constant $0 \leq D_o < \infty$, the cost or penalty associated with a permutation $p = (p_1, \ldots, p_n) \in P$ is given by the objective function

$$C(p,D_o) = \sum_{m=1}^{n} c_{p_m} f(D_o + D_{p^{m-1}}) \tag{1}$$

where $f(\xi) \geq 0$ is a real valued bounded function and $D_{p^0} = 0$. We note that $C[(p,q),D_o] = C(p,D_o) + C(q,D_o + D_p)$.

For an interpretation of the above, in a project scheduling problem we might have the following identifications: d_x = the duration of project x; c_x = the net cost of project x, discounted to the beginning of the project; $f(\xi)$ = present worth of a unit expenditure ξ time periods hence; D_o = the number of time periods from the present to the start of the first project; and $C(p,D_o)$ = the present (discounted) net cost of carrying out the projects sequentially in the order p, with the first project starting D_o time periods hence.

For given D_o, the minimum cost is given by $C^*(D_o) = \min_{p\varepsilon P_S} C(p,D_o)$ and the set of optimal solutions is $P^*(D_o) = \{p \varepsilon P_S | C(p,D_o) = C^*(D_o)\}$. The objective is to find a permutation $p^* \varepsilon P^*(D_o)$.

General Solution Method

To find an optimal solution we employ a tree search implicit enumeration scheme. Elimination of branches is accomplished through the use of two types of conditions; bounds and element interchanges. The bounds used are simple and quite good, but by themselves do not provide an elimination procedure which is efficient enough to solve practical problems. The interchange conditions are very efficient at eliminating non-optimal permutations, but cannot by themselves guarantee convergence to an optimal solution. Used together, the bounding and interchange operations assure both efficiency and convergence.

The branching operation is the standard one employed in tree search procedures (see [1] or [6]). Two bounds are used -- an upper bound $B(D_o)$ on the minimum cost $C^*(D_o)$, and a lower bound $b_\sigma(D_o)$ on the value of $C(p,D_o)$ for any permutation p of the elements of a set σ.

We may now state the bound elimination condition. Let p be a given partial permutation and let $\sigma = S - S_p \neq \emptyset$ and let q be any permutation of the elements of σ. Then

$$(2) \qquad (p,q) \notin P^*(D_o) \text{ if } C(p,D_o) + b_\sigma(D_o + D_p) > B(D_o)$$

The justification for (2) is obtained by noting that $C[(p,q),D_o] = C(p,D_o) + C(q,D_o + D_p) \geq C(p,D_o) + b_\sigma(D_o + D_p) > B(D_o) \geq C^*(D_o)$.

The interchange conditions are based on the effects of interchanging the order of two elements in a permutation.* More specifically, let x and y be two distinct elements of S and let q and r be any two disjoint permutations not containing x or y (either q or r or both may be null permutations). If p = (q,x,y,r) then p' = (q,y,x,r) is the permutation obtained by interchanging the order of the elements x and y in p. A direct calculation using (1) shows that

$$(3) \qquad C(p,D_o) - C(p',D_o) = c_y[f(D_o + D_q + d_x) - f(D_o + D_q)] - c_x[f(D_o + D_q + d_y) - f(D_o + D_q)].$$

Now taking $R_w(D) = [f(D + d_w) - f(D)]/c_w$ for any $w \in S$ and $0 \leq D < \infty$, it can be seen from (3) that $R_x(D_o + D_q) \leq R_y(D_o + D_q)$ is equivalent to $C(p,D_o) \leq C(p',D_o)$. Thus, we have obtained the following basic interchange condition for p to be an optimal permutation:

---

* This type of condition has been used previously to provide direct solutions for sequencing problems of very special forms (e.g., see [5] and [7]).

(4) $$p \in P^*(D_o) \text{ only if } R_x(D_o + D_q) \leq R_y(D_o + D_q).$$

If $p \in P_S$ and if condition (4) holds for every adjacent pair of elements in p, then p is termed a locally optimal permutation.

The necessary condition (4), or the contra-positive thereof, will be used to develop several schemes for eliminating non-optimal permutations, and it will also be used to calculate a sharp upper bound $B(D_o)$.

Bounds

Since $C(p,D_o) \geq C^*(D_o)$ for all $p \in P_S$, any permutation $p \in P_S$ provides an upper bound $B(D_o) = C(p,D_o)$ on the minimum cost $C^*(D_o)$. A particularly good bound is provided by a locally optimal permutation $t = (t_1, \ldots, t_N)$ constructed as follows: chose t_1 such that $R_{t_1}(D_o) \leq R_x(D_o)$ for all $x \in S$; choose t_2 such that $R_{t_2}(D_o + d_{t_1}) \leq R_x(D_o + d_{t_1})$ for all $x \in S - \{t_1\}$; etc.

Example: Suppose $S = \{\alpha, \beta, \gamma, \delta\}$ and $c_\alpha = 10$, $c_\beta = 10$, $c_\gamma = 20$, $c_\delta = 5$, $d_\alpha = 2$, $d_\beta = 1$, $d_\gamma = 3$, $d_\delta = 2$, $D_o = 0$, and $f(\xi) = \xi^2$. Then $R_\alpha(0) = 0.4$, $R_\beta(0) = 0.1$, $R_\gamma(0) = 0.45$, and $R_\delta(0) = 0.8$ and so $t_1 = \beta$ since $R_\beta(0)$ is the smallest value. Next, $D_o + d_\beta = 1$ and $R_\alpha(1) = 0.8$, $R_\gamma(1) = 0.75$, and $R_\delta(1) = 1.6$ so that $t_2 = \gamma$. Finally, $D_o + d_\beta + d_\gamma = 4$ and $R_\alpha(4) = 2.0$ and $R_\delta(4) = 4.0$ so that $t_3 = \alpha$, leaving $\delta = t_4$. Thus, the locally optimal permutation is $t = (\beta, \gamma, \alpha, \delta)$ and the corresponding upper bound is $B(0) = 360$.

In our experience, the locally optimal sequence t constructed by the above procedure will usually be optimal, or very close to it (in all the problems we have solved, including the example above, it has turned out to be optimal in every single case). The bound $B(D_o) = C(t,D_o)$ is easy to calculate and will generally be quite good at eliminating non-optimal sequences.

Turning to the lower bound on the cost of a permutation, we first note that $C(p,D_o) \leq C[(p,q),D_o]$, so that $C(p,D_o)$ always provides a lower bound on the cost of all sequences containing p as an initial segment. When $f(\xi)$ is a monotone function (which it usually is in practice), we may obtain a sharper lower bound by direct application of an inequality from [3]. Let σ be any non-empty proper subset of S and let P_σ be the set of all permutations of the elements of σ. Now suppose that $f(\xi)$ is non-decreasing and form the two permutations $r = (r_1, \ldots, r_n) \in P_\sigma$ and $s = (s_1, \ldots, s_n) \in P_\sigma$ defined by $c_{r_1} \geq c_{r_2} \geq \ldots \geq c_{r_n}$ and $d_{s_1} \leq d_{s_2} \leq \ldots \leq d_{s_n}$; if $f(\xi)$ is non-increasing, reverse the order of the elements in both r and s. The lower bound is given by $b_\sigma(D_o) = \sum_{i=1}^{n} c_{r_i} f(D_o + D_{s^{i-1}})$ and we are assured that $b_\sigma(D_o) \leq C(p,D_o)$ for all $p \in P_\sigma$.

Example: The data from the example above may be used to illustrate the calculation of $b_\sigma(D_o)$. Suppose that we wish to obtain a lower bound on the cost of all permutations whose first element is δ. As a preliminary step we note that $C[(\delta,p),D_o] = C(\delta,D_o) + C(p,D_o + d_\delta) = 0 + C(p,2)$ and so the required bound will be given by $b_\sigma(2)$ since $b_\sigma(2) \leq C(p,2)$ for any permutation p of the elements of $\sigma = S - \{\delta\} = \{\alpha, \beta, \gamma\}$. Next, the permutations r and s are formed and we obtain $r = (\gamma, \alpha, \beta)$ or $r = (\gamma, \beta, \alpha)$ since $c_\gamma = 20 > c_\alpha = 10 = c_\beta$, and $s = (\beta, \alpha, \gamma)$ since $d_\beta = 1 < d_\alpha = 2 < d_\gamma = 3$. Finally, the bound is calculated as $b_\sigma(2) = c_\gamma f(2) + c_\alpha f(2 + d_\beta) + c_\beta f(2 + d_\beta + d_\alpha) = 20(2)^2 + 10(3)^2 + 10(5)^2 = 420$. Therefore, any permutation with δ as the first element will have a cost of at least 420. We may note that this lower bound is greater than $B(0) = 360$, and so no permutation with δ as the first element could be optimal.

Element Interchange

We now show how condition (4) can be used as the basis for several powerful methods for eliminating partial permutations from consideration. The first scheme provides a means for determining that particular elements cannot be the next ones added to a partial permutation. Let $p = (p_1, \ldots, p_n)$ be a partial permutation containing $n < N$ elements. Let z be an element of $\bar{S}_p$ and define the set $\sigma_z = \{w \in \bar{S}_p | R_z(D_o + D_p) \leq R_w(D_o + D_p)\}$ and let $A_z = \sum_{w\epsilon\sigma_z} d_w - \min_{w\epsilon\sigma_z} d_w$.

(5) If $\sigma_z \neq \bar{S}_p$ and if $R_x(D) < R_y(D)$ for all $x \in \bar{S}_p - \sigma_z$ and all $y \in \sigma_z$ and all $D_o + D_p \leq D \leq D_o + D_p + A_z$, then $(p,z,q) \notin P^*(D_o)$ for any permutation q of the elements of the set $\bar{S}_p - \{z\}$.

We establish (5) by contradiction. Suppose that the hypothesis of (5) is satisfied, but that $(p,z,q) \in P^*(D_o)$. To be specific, suppose that $q = (q_1, \ldots, q_k)$ and that σ_z contains m + 1 elements. Now, if $q_i \in \sigma_z$ for all $1 \leq i \leq m$, then $q_m \in \sigma_z$ and $q_{m+1} \notin \sigma_z$ and so $R_{q_{m+1}}(D) < R_{q_m}(D)$ by the hypothesis of (5) with $D = D_o + D_p + d_z + D_{q^{m-1}}$, noting that $D \leq D_o + D_p + A_z$. But $R_{q_{m+1}}(d) < R_{q_m}(D)$ violates condition (4) and so $(p,z,q) \notin P^*(D_o)$, contrary to assumption. On the other hand, if $q_i \notin \sigma_z$ for some $1 \leq i \leq m$, then by the hypothesis of (5) there will exist an adjacent pair of elements in (z,q^m) violating condition (4) and thus $(p,z,q) \notin P^*(D_o)$ contrary to assumption.

When $\sigma_z = \{z\}$ in (5) then $A_z = 0$ and we have the following much simpler condition.

(6) If $z \in \bar{S}_p$ and if $R_w(D_o + D_p) < R_z(D_o + D_p)$ for all $w \in \bar{S}_p - \{z\}$, then $(p,z,q) \notin P^*(D_o)$ for any permutation q of the elements of $\bar{S}_p - \{z\}$.

Our illustration of the application of conditions (5) and (6) will make use of the fact that in any specific problem, D can take on only a finite number of values. Also, for any D, conditions (5) and (6) depend only on the relative values (rather than the actual numerical values) of the $R_x(D)$. Thus, for each possible value of D we may calculate the $R_x(D)$ value of each $x \in S$ and form a permutation $p(D) = [p_1(D), \ldots, p_N(D)]$ such that $R_{p_1(D)}(D) \leq \ldots \leq R_{p_N(D)}(D)$. A listing of the permutations p(D) for all possible values of D is called a _preference table_ -- see Table 1 for an example with $S = \{\eta, \lambda, \mu, \rho\}$. Each row of the preference table lists the elements of S in order of ascending $R_x(D)$ values, from lowest (left-most) to highest (right-most).

Table 1

Preference Table for Four Projects

D	$p_1(d)$	$p_2(D)$	$p_3(D)$	$p_4(D)$
0	ρ	λ	η	μ
1	λ	ρ	μ	η
2	ρ	λ	μ	η
3	λ	ρ	η	μ
4	λ	ρ	μ	η
5	ρ	λ	η	μ
6	λ	ρ	η	μ
7	ρ	λ	μ	η
8	ρ	λ	μ	η

<u>Example</u>: Suppose that $d_\eta = 3$, $d_\lambda = d_\mu = 2$, $d_\rho = 1$, $D_o = 0$, and that Table 1 gives the preference orders for the various values of D. For convenience, assume that all inequalities among the $R_x(D)$ values are strict. Starting with $D = D_o = 0$ and no elements selected (i.e., p = the null permutation in (5) and (6)), take $z = \mu$. Then from (6) we see that μ cannot be the first element in an optimal permutation, since $R_\mu(D_o)$ is larger than the other $R_x(D_o)$ values. Next, try $z = \eta$ so that in (5) $\sigma_z = \{\eta, \mu\}$ and $A_\mu = d_\eta + d_\mu - d_\mu = 3$; the hypothesis of (5) is satisfied since $R_x(D) < R_y(D)$ for all $x \in S - \sigma_z = \{\lambda, \rho\}$ and $y \in \sigma_z$ and $0 \leq D \leq 3$. Thus, η cannot be the first element in an optimal permutation. Next, try $z = \lambda$ so that $\sigma_z = \{\eta, \lambda, \mu\}$ and $A_\lambda = d_\eta + d_\lambda + d_\mu - d_\mu = 5$; the hypothesis of (5) is not satisfied since, for example, $R_\lambda(3) < R_\rho(3)$ with $\rho \in S - \sigma_z = \{\rho\}$ and $\lambda \in \sigma_z$ and $0 \leq D = 3 \leq 5$. Thus, λ might be the first element in an optimal permutation. We cannot chose $z = \rho$, since then $\sigma_z = \bar{S}_p = S$ and so the hypothesis (5) is not satisfied.

We conclude, then, that neither η nor μ could be the first element in an optimal permutation, but that λ or ρ might be. To extend the example one step further, suppose that we choose ρ as the first element of the permutation so that $p = (\rho)$ and $\bar{S}_p = S - S_p = \{\eta, \lambda, \mu\}$ and $D_o + D_p = 1$. Then by (6), η cannot be the second element, since $R_w(1) < R_\eta(1)$ for all $x \in \bar{S}_p - \{\eta\} = \{\lambda, \mu\}$. Next, try $z = \mu$ so that $\sigma_\mu = \{\eta, \mu\}$ and $A_\mu = d_\eta + d_\mu - d_\mu = 3$; the hypothesis of (5) is satisfied since $R_\lambda(D) < R_y(D)$ for all $y \in \sigma_\mu$ and $1 \leq D \leq 4$, and thus μ cannot be the second element. This leaves only λ as a second element. Thus, if ρ is first, then λ must be second, and another application of (6) at D = 3 shows that the third element must be η, leaving μ as the last element. (Note that $(\rho, \lambda, \eta, \mu)$ is not necessarily the optimal permutation since we assumed ρ as the first element; permutations starting with λ would have to be checked also.)

Proceeding now to the second exclusion condition, under certain circumstances it is possible to decompose the problem of finding an optimal permutation into the problem of finding two independent sub-permutations. Such a decomposition can dramatically reduce the difficulty of the problem. For example, if S contains 14 elements, then the total number of permutations which must be considered (either explicitly or implicitly) by the enumeration scheme is about 2.4×10^{11}, while if S can be decomposed into two sets of seven elements each, the total number of permutations which must be considered is about 2.7×10^4 ; thus, the original problem is about 10 million times as large as the decomposed problem.

Specifically, let $p \in P$ be a partial permutation, let σ be a non-empty subset of $\bar{S}_p$, and let $A_\sigma = \sum_{w\in\sigma} d_w - \min_{w\in\sigma} d_s$. Suppose that σ is partitioned into two disjoint non-empty subsets, σ' and σ'', and let $D' = \sum_{w\in\sigma'} d_w$. Finally, suppose that $q^* \in P_{\sigma'}$ and $C(q^*,D_o + D_p) \leq C(q, D_o + D_p)$ for all $q \in P_{\sigma'}$, and that $r^* \in P_{\sigma''}$ and $C(r^*, D_o + D_p + D') \leq D(r, D_o + D_p + D')$ for all $r \in P_{\sigma''}$. Then

(7) If $R_x(D) < R_y(D)$ for all $x \in \sigma'$ and all $y \in \sigma''$ and all $D_o + D_p \leq D \leq D_o + D_p + A_\sigma$, then $C[(p,q^*,r^*),D_o] \leq C[(p,s),D_o]$ for all $s \in P$.

To establish (7), let $Q = \{(q,r) | q \in P_{\sigma'}$ and $r \in P_{\sigma''}\}$. Now if $s^* \in P_\sigma$ and $C(s^*,D_o + D_p \leq C(s,D_o + D_p)$ for all $s \in P_\sigma$, then $s^* \in Q$, for otherwise there would exist adjacent elements in s^* violating the necessary condition (4).

Then we may write $\min_{s\varepsilon P_q} C[(p,s),D_o] = C(p,D_o) + \min_{q\varepsilon P_{\sigma'};r\varepsilon P_{\sigma''}}$ $C[(q,r),D_o + D_p] = C(p,D_o) + \min_{q\varepsilon P_{\sigma'}} C(q,D_o + D_p) + \min_{r\varepsilon P_{\sigma''}} C(r,D_o + D_p + D')$ $= C(p,D_o) + C(q^*,D_o + D_p) + C(r^*,D_o + D_p + D') = C[(p,q^*,r^*),D_o]$ as claimed.

Example: The "decomposition" result (7) can be illustrated with the Preference Table 1. Take p as the null permutation so that $\sigma = S$ and $A_\sigma = 7$. It is seen that the hypothesis of (7) is satisfied by taking $\sigma' = \{\lambda, \rho\}$ and $\sigma'' = \{\eta, \mu\}$ for then $R_x(D) < R_y(D)$ for all $x \varepsilon \sigma'$ and $y \varepsilon \sigma''$ and $0 \leq D \leq 7$. Thus, the four projects are partitioned into two sets of two projects each with the elements of σ' occupying the last two positions. In fact, the optimal permutation is easily found. Applying (6) to σ' it is seen that λ cannot be the first element of q^* and so ρ is first. Then applying (6) to σ'' with $p = (\rho, \lambda)$ and $D_o + D_p = 3$, we see that μ cannot be the first element of r^* and so η is first. Thus, the optimal permutation is $(q^*,r^*) = (\rho, \lambda, \eta, \mu)$. It might be noted that the locally optimal permutation is easily constructed from the preference table (and, in the case of Table 1, it is again the optimal permutation).

It is possible to derive further exclusion conditions from the necessary condition (4), but there is some question as to whether the additional programming and computational time are justified by the resulting gain in efficiency of permutation elimination.

Solution Procedure

The following outline presents the basic logic of the recursive solution procedure. At the outset a locally optimal permutation p^* is formed and the initial upper bound $B_1 = C(p^*,D_o)$ is computed. Initially, the set of candidate solutions is $K = \{p^*\}$, and the initial list L_1 of partial solutions contains only the null permutation. At step k the operations are as follows:

(a) Selection - Select a partial permutation p from L_k. Delete p from L_k and go to (b).

(b) Decomposition - If the set $\bar{S}_p$ cannot be decomposed, go to (c). If $\bar{S}_p$ can be decomposed, find the optimal sub-permutations and combine them with p to form a sequence q; go to (d) with $\Pi_k = \{q\}$.

(c) Branch - Form the set $\Pi_k = \{r \in P | r = (p,w) \text{ and } w \in \bar{S}_p\}$; go to (d).

(d) Complete permutation check - If $r \in \Pi_k$ contains less than N elements, go to (e). Otherwise, add the elements of Π_k to the list K of candidate solutions and then set $\Pi_k = \emptyset$. Let $B_k = \min_{w \in K} C(w,D_o)$ and delete from K any permutation r for which $C(r,D_o) > B_k$; go to (g).

(e) Interchange eliminations - Delete from Π_k any elements satisfying the hypotheses of (5) or (6); go to (f).

(f) Bound elimination - Compute lower bounds on all $r \in \Pi_k$ and delete from Π_k all permutations with lower bounds greater than B_k; go to (g).

(g) Update and stop check - Add the sequences remaining in Π_k to I_k to obtain L_{k+1}. If $L_{k+1} = \emptyset$, stop with $p^*(D_o) = K$; otherwise, return to (a).

Computational Experience

The preceding algorithm has yielded very good results in solving a number of project scheduling problems of the type described below. There is given a set S of N electric power generation projects with d_x the output capacity and c_x the net cost of project $x \in S$, the latter evaluated at the time of initiation of the project. It is desired to determine the order in which the various projects should be installed so as to minimize the net present cost of the projects while meeting the projected demand rate at all times. The demand rate (above present generating capacity) at time $\tau \geq 0$ is given by a monotone non-decreasing function $h(\tau)$, and the inverse function $g(D) = h^{-1}(D)$ gives the time at which the demand has reached a rate of D. Then the present value of a unit expenditure at the time that the demand reaches a rate D is given by $f(D) = (1 + i)^{-g(D)}$, with i the annual discount rate. Thus, the present value of the net discounted cost of installing the projects sequentially in the order $p = (p_1, \ldots, p_N)$ is given by the objective function (1), assuming that project p_{n+1} is installed at time $g(D_{p_n})$.

To test the solution procedure, we have solved a number of problems. In cases i and iii the demand function was given by $h(\tau) = 28.28\tau^2$ for $0 \leq \tau \leq 10.3$ and $h(\tau) = 10^4\sqrt{(\tau - 9.2)/12.5}$ for $\tau > 10.3$ with results as shown in Table 2. Case ii lists the results for 13 runs using a variety of demand functions and values for the c_x and d_x. Case iv gives the results for two problems using real demand functions and cost data. In each case, the number of permutations in P_S is the total number of possible orders in which the N projects could be run, and the number of permutations in P is the total number which must be considered (either explicitly or implicitly) by the solution procedure. The computer program was written in FORTRAN and executed on an IBM 360/67 computer.

Table 2

Computational Results

Case	Number of Projects	Number of Runs	Number of Permutations			Average CPU Time, secs.
			In P_S	In P	Explicitly Considered	
i	10	1	$3.6\text{x}10^6$	$4.0\text{x}10^6$	142	25
ii	10	13	"	"		13.6
iii	15	1	$1.3\text{x}10^{12}$	$1.4\text{x}10^{12}$	118	28
iv	18	2	$7.4\text{x}10^{15}$	$7.8\text{x}10^{15}$		25
v	10	1	$3.6\text{x}10^6$	$4.0\text{x}10^6$	>151200*	>226

*Exceeded storage capacity and run terminated.

It is seen from Table 2 that the number of permutations explicitly examined by the solution procedure in cases i to iv is small and the computation times are quite reasonable. For comparison, we have solved the problem in case i by a straight branch and bound procedure using the conventional bounds (rather than the special bounds given in this paper); the computer run had to be stopped after 226 seconds (over 151,200 permutations considered) without reaching the optimal solution. The improvement obtained by using the tighter bounds and the additional elimination conditions (5), (6), and (7) is certainly impressive!

Summary

For problems with an objective function of the form (1), the solution procedure presented in this paper provides an efficient method for finding an optimal permutation. The combination of good bounds and the several element interchange conditions is very effective in eliminating most of the possible permutations from explicit considerations. Computational experience with exercise problems and with real problems demonstrates the practical feasibility of the method.

Acknowledgements

The authors wish to thank Professor Dennis Russell of our institution for having brought the project scheduling problems to our attention. We also express appreciation to Dr. Eric Denardo for helpful suggestions on notation.

REFERENCES

1. Agin, M., "Optimum Seeking with Branch and Bound", Mgt.Sc., Vol. 13 (1966), B176-B185.

2. Butcher, W. S., Y.Y. Haimes, and W.A. Hall, "Dynamic Programming for the Optimal Sequencing of Water Supply Projects", Water Resources Research, Vol. 5 (1969), pp. 1196-1204.

3. Hardy, G. H., J. E. Littlewood, and G. Bolya, Inequalities, Cambridge University Press, 1934.

4. Held, M. and R. M. Karp, "A Dynamic Programming Approach to Sequencing Problems", Jour. Sec. Ind. and Appl. Math., Vol. 10, No. 1 (March,1962), pp. 196-210.

5. Johnson, S.M., "Optimal Two and Three Stage Production Schedules with Setup Times Included", Nav. Res. Log. Quart., Vol. 1, No. 1 (March, 1954), pp. 61-68.

6. Lawler, E. L. and D. E. Wood, "Branch and Bound Methods: A Survey", Opns.Res., Vol. 14 (1966), pp. 669-719.

7. Rau, J. G., "Minimizing a Function of Permutations of n Integers", Opns.Res., Vol. 19, No. 1 (Jan.-Feb., 1971), pp. 237-240.

8. Erlenketter, D., "The Sequencing of Expansion Projects", Working Paper No. 166, Western Management Science Institute, UCLA, (Nov., 1970).

ON THE SET REPRESENTATION AND SET COVERING PROBLEMS*

K. G. Murty

The University of Michigan, Ann Arbor, Michigan, USA

June 1972

Give a finite set $\Gamma = \{1,2,\ldots,n\}$ and a class $C = \{c_1, c_2 \ldots c_m\}$ of *nonempty* subsets of Γ, a subset $E \subset \Gamma$ is said to *represent the class C* if

$$E \cap c_i \neq \phi \text{ for all } c_i \in C$$

The *minimum cardinality set representation problem for C* is the problem of finding a minimum cardinality subset of Γ representing the class C.

Given a class $\boldsymbol{\Theta} = \{\xi_j\}$ of subsets of some nonempty finite set ϑ, the problem of determining the minimum cardinality subclass $\theta \subset \boldsymbol{\Theta}$ such that

$$\bigcup_{\xi_j \in \theta} \xi_j = \bigcup_{\xi_j \in \boldsymbol{\Theta}} \xi_j$$

is know as the *minimum cardinality set covering problem for* $\boldsymbol{\Theta}$.

The set representation and the set covering problems are equivalent. Corresponding to each problem of one type, we can construct a problem of the other type among which a one to one correspondence exists between the respective sets of optimum solutions.

Various results on representing subsets are discussed. An algorithm is developed for solving the set representation problem using only set theoretic operations. The algorithm is inductive on m, the number of subsets in the class C.

Algorithms for solving the travelling salesman problem and the problem of finding all the complementary feasible bases for a linear complementarity problem, have recently been developed which require the solving of a sequence of set representation problems [8,10]. In such applications the class C for the associated set representation problem is augmented with additional subsets in each stage. The inductive algorithm for the set representation problem developed here, will be very suitable in such applications.

---

* This research has been supported by the National Science Foundation under Contract no. GK-27872 with the University of Michigan

1. Introduction

1.1 The Set Representation Problem

Given a finite set $\Gamma = \{1,..,n\}$ and a class of nonempty subsets of Γ, $C = \{c_1,....,c_m\}$, $E \subset \Gamma$ is said to *represent* C or is a *represent* of C if $E \cap c_i \neq \phi$ for all $c_i \varepsilon C$.

Obviously, Γ represents C, and hence, the class of represents for C is nonempty. A minimum cardinality represent for C will be called an MCR for C. $F \subset \Gamma$ is said to be a *Minimal represent* (or an MR, in abbreviation) for C if F represents C but no proper subset of F does. Obviously every MCR for C is an MR for C but the converse may not be true.

s(C) = class of all MCR's for C

δ(C) = cardinality of an MCR for C.

S(C) = class of all MR's for C.

The problem of finding δ(C) and a subset in s(C) is the *minimum cardinality set representation problem for C*. We will just refer to it as the *set representation problem* for C.

1.2 The Set Covering Problem

Let $\vartheta = \{1,...,n_1\}$ and $\boldsymbol{\Theta} = \{\xi_1,..., \xi_{m_1}\}$ be a class of subsets of ϑ. The subclass $\theta \subset \boldsymbol{\Theta}$ is said to cover the class $\boldsymbol{\Theta}$ if

$$\bigcup_{\xi_j \varepsilon \theta} \xi_j = \bigcup_{\xi_j \varepsilon \boldsymbol{\Theta}} \xi_j$$

The *minimum cardinality set covering problem for* Θ is the problem of finding the minimum cardinality subclass of Θ covering it. We will just refer to it as the *set covering problem for* Θ.

Consider the set representation problem for C discussed in 1.1. Let

$$\vartheta^* = \{1,\ldots,m\} \quad \text{where } m \text{ is the number of subsets in C.} \tag{1}$$

and for j = 1 to n let

$$\xi_j^* = \left\{ i: \ 1 \leq i \leq m, \text{ is such that } j \in c_i \right\} \tag{2}$$

The fact that each subset in C is nonempty implies that

$$\bigcup_{j=1}^{n} \xi_j^* = \vartheta^* \tag{3}$$

Let $$\Theta^* = \{\xi_1^*,\ldots,\xi_n^*\} \tag{4}$$

From (1), (2), (3), (4) it is evident that $\theta^* \subset \Theta^*$ is a minimum cardinality cover for Θ^* iff

$$E = \{j: \ j \text{ such that } \xi_j^* \ \epsilon\, \theta^* \}$$

is an M C R for C. Conversely if E is an M C R for C then

$$\theta^* = \{ \xi_j^* : \ j \in E\}$$

is a minimum cardinality cover for Θ^*.

Thus the set representation problem for C is equivalent to the set covering problem for θ^* when θ^* is defined as in (2), (4). In a similar manner, given a set covering problem, a set representation problem equivalent to it can be constructed.

1.3 The set covering problem is discussed extensively in literature. I encountered this problem in developing an algorithm for a fundamental problem in linear inequalities [8] which includes the traveling salesman problem as a special case. It is also useful in an algorithm for obtaining all the complementary feasible bases for a linear complementarity problem [10].

In these applications one is required to solve a sequence of set representation problems, in which the class C grows as the algorithm progresses. To be useful in such applications, we develop here an algorithm for solving the set representation problem which proceeds inductively on m, the number of subsets in the class C. The algorithm uses only set theoretic operations.

Applications of the set representation problem in system representation etc. are obvious. Various other applications are discussed in [2,3,4,7].

2. Applications

The set representation problem finds various applications in scheduling and sequencing and other areas. Some of them are discussed here.

2.1 Delivery problem: (Balinski & Quandt [13])

A warehouse has to deliver material to m clients at different destinations in a region by a carrier. The carrier can combine a number of clients to be delivered together to form a route. A client may receive delivery via a number of different routes. A list of all the feasible routes is made. Suppose the list contains n routes, called 1,2,...,n. Let

$$c_i = \{j: \text{ route } j \text{ contains } i\text{-th client}\} .$$

for i = 1 to m. Then the minimum cardinality subset of routes representing the class $C = \{c_1,\dots,c_m\}$ provides the minimum number of routes to service all the clients.

2.2 Location of fire hydrants

Given a network of streets (arcs in the network) and traffic centers (nodes in the network), to find a subset of nodes for locating the fire hydrants so that each street contains at least one fire hydrant. This is the problem of covering the arcs of the network by a subset of the nodes.

Suppose the network has n nodes and m arcs. Let

$$c_i = \{j: j \text{ is a node on arc } i\} \quad \text{for } i=1 \text{ to } m.$$

Then an MCR for $C = \{c_1,\ldots,c_m\}$ provides the optimal location for fire hydrants.

2.3 Information retrieval (Day [14])

Let $F_1,\ldots,F_n$ be files on which information is stored. There are m requests for information. Let

$$c_i = \{j: \text{ i-th request can be fulfilled by searching } F_j\}$$

for i=1 to m. Then an MCR for $C = \{c_1,\ldots,c_m\}$ provides the minimal cardinality subset of files to be searched to satisfy all the requests for information.

2.4 Airline crew scheduling: (Arabeyre et. al. [16])

A *flight segment* is either a flight or a portion of a flight between two stations. A *trip* is a combination of one or more flight segments into duty periods for the crew, creating as many good days of work as possible which meet the given restrictions (like the restriction that the first flight segment of a trip must depart from a station which is the same as the return station of the last flight segment. Also if there is more than a day's work for a crew member in a trip, then the trip must include enough layover periods, etc.)

The various flight segments to be run by the airline can be combined into various possible trips. Suppose there are m flight segments and we have n possible trips. Let

$$c_i = \{j: \text{ j-th trip contains the i-th flight segment}\}$$

for i=1 to m. Then any MCR to $C = \{c_1,\ldots,c_m\}$ provides a minimum cardinality subset of trips which covers all the flight segments.

2.5 Assembly line balancing problem: (Salvesan [15])

A set of elemental tasks i=1 to m have to be performed. The problem is to group these tasks into collections of elemental tasks called *assembly stations*. An operator or a group of operators will be assigned to perform the tasks in each assembly station. There may be a precedence relationships among the elemental tasks. Also there is a *cycle time constraint* which requires that the time required to perform the tasks in each assembly station is less than or equal to the cycle time.

A list of possible assembly stations is made by looking at the precedence network and the estimated times required to perform the elemental tasks. Let n be the total number of assembly stations listed. Let

$$c_i = \{j: \begin{array}{l}\text{j-th assembly station contains the} \\ \text{i-th elemental task}\end{array}\}$$

for i=1 to m. Then an MCR for $C = \{c_1,\ldots,c_m\}$ provides a minimum cardinality set of assembly stations for carrying out all the elemental tasks. Having obtained this set, modifications can be made in the assembly stations so that each elemental task appears in only one assembly station.

2.6 The traveling salesman problem (Murty [8])

Given a system of linear equations in nonnegative variables

$$\begin{aligned} Ax &= b \\ x &\geqq 0 \end{aligned} \tag{5}$$

and a specified subset of variables $x_1,\ldots,x_r$ which correspond to a linearly independent subset of column vectors in (5), the *special basis problem* is the problem of finding a feasible basis for (5) which contains all the special variables as basic variables or concluding that no such basis exists. In [8] it has been shown that the traveling salesman problem is a special case of this problem.

An algorithm for solving the special basis problem is the following. Let $A_{.j}$ denote the j-th column vector of the matrix A. Suppose A is of order mxn and let $\Gamma = \{1,\ldots,n\}$. If $E \subset \Gamma$ let A_E denote the submatrix of A whose column vectors are $A_{.j}$ for $j \varepsilon E$. Define a class C, of subsets of Γ, which will be augmented with additional subsets of Γ in each stage of the algorithm until termination.
Initially let

$$c_i = \{i: \ x_i \text{ is a special variable}\}$$

for i = 1 to r and $C = \{c_1,\ldots,c_r\}$ initially. In a general stage of the algorithm find an MCR, E, for the class C at this stage. If the cardinality of E is $\geqq$ m+1, then terminate with the conclusion that there is no feasible basis containing all the special variables as basic variables.

If the cardinality of E is $\leqq$ m and the column vectors of E form a linearly dependent set then find the subset $c_p = \{j: \ A_{.j}$ is not in the linear span of $A_E\}$ and include c_p in the class C and go to the next stage.

If the column vectors in A_E form a linearly independent set, then complete A_E to a basis, B, for (5). If $\bar{b} = B^{-1}b \geqq 0$, then B is a desired special basis and we terminate. On the other hand if $\bar{b} = B^{-1}b \not\geqq 0$, then for each i, such that $\bar{b}_i < 0$, define a set,

$$c_{p_i} = \{j: \bar{a}_{ij} < 0\}$$

where $(\bar{a}_{ij}) = \bar{A} = B^{-1}A.$

Include all such subsets c_{p_i} in the class C and go to the next stage.

In a finite number of steps, this algorithm will solve the special basis problem. Here we have to solve a sequence of set representation problems, in which the class C grows after each stage.

2.7 Enumerating All the Complementary Feasible Bases Corresponding to a Linear Complementarity Problem

Given a square n x n matrix M and a n x 1 column vector q, the linear complementarity problem (M,q) is to find a $w \in R^n$, $z \in R^n$ satisfying

$$\begin{aligned} w - Mz &= q \\ w \geqq 0,\ z &\geqq 0 \end{aligned} \qquad (6)$$

$$\text{and } w^T z = 0 \qquad (7)$$

In this problem let I be the identity matrix of order n x n. A *Complementary basis* is a square submatrix B of $(I \vdots -M)$ of order n such that the j-th column vector in it, $B_{.j}$, is either $I_{.j}$ or $-M_{.j}$. A Complementary basis, B, is a *feasible complementary basis* to this problem if it is a feasible basis to (6), i.e., if $B^{-1}q \geqq 0$. The problem of finding all the feasible Complementary bases has important applications in solving the general nonvex quadratic programming problem [18], in finding all the equilibrium strategies of a bimatrix game [19], etc.

An algorithm for this problem discussed in [10] based on an approach similar to that in 2.6 is the following. Let $\Gamma = \{w_1, \ldots, w_n; z_1, \ldots, z_n\}$. Let A be the n x 2n matrix $(I : -M)$. For j = 1 to n, $A_{.j} = I_{.j}$ and this is the column vector in (6) *associated with* the variable w_j. For $j = n + 1$ to 2n, $A_{.j} = -M_{.j}$ and this is the column vector in (6) *associated with* the variable z_j. If $E \subset \Gamma$ let A_E be the matrix whose column vectors are those of A which are associated with the variables in E.

As before, define a class C, of subsets of Γ. Initially,

$$c_r = \{w_r, z_r\} \qquad \text{for } r = 1 \text{ to } n$$

and $C = \{c_1, \ldots, c_n\}$.

In a general stage of the algorithm, find an MCR, E, for the class C at this stage. If the cardinality of E is $\geqq n + 1$, the algorithm terminates.

If the cardinality of E is n, and if A_E is singular, then find the subset c_p = subset of variables in Γ such that the column vectors of A associated with them are not in the linear span of column vectors of A_E. Include c_p in the class C and go to the next stage.

If A_E is nonsingular and if $A_E^{-1}q \geqq 0$, then by rearranging the columns in proper order, if necessary, A_E becomes a feasible complementary basis. Store this basis in a *list of feasible complementary bases.* Then add the complement of E to the class C and go back to the next stage.

If $\bar{b} = A_E^{-1}q \not\geq 0$, then for each i such that $\bar{b}_i < 0$, define the subset of Γ

$$c_{p_i} = \left\{ \begin{array}{lcl} \text{variable in (6)} & : & j \text{ such that } \bar{a}_{ij} < 0 \\ \text{associated with } A_{.j} & & \text{for } j = 1 \text{ to } 2n \end{array} \right\}$$

where $(\bar{a}_{ij}) = \bar{A} = A_E^{-1}A$. Include all such subsets c_{p_i} in the class C and go to the next stage.

After a finite number of steps the cardinality of an MCR for C becomes $\geq n + 1$. When this happens all the feasible complementary bases are contained in the list generated during the algorithm.

2.8 Other Applications

Given $\Gamma = \{1,\ldots,n\}$ and the class $C = \{c_1,\ldots,c_m\}$ of nonempty subsets of Γ let

$$A_{i.} = (a_{i1},\ldots,a_{in})$$

where $a_{ij} = 1$ if $j \varepsilon c_i$ or 0 otherwise, and let A be the matrix consisting of $A_{i.}$ as rows. Then clearly the problem of finding an MCR for C is equivalent to solving the integer program

$$\begin{array}{lll} \min & \sum_j x_j & \\ \text{subject to} & A_{i.}\ x & \geq 1 \text{ for all } i=1 \text{ to } m \\ & x_j & = 0 \text{ or } 1 \text{ for all } j. \end{array}$$

In this problem if we replace the objective function by $\sum_j w_j x_j$ where the w_j are given positive weights, then we get the weighted set representation problem. The weighted set representation problems finds a vast number of applications in sequencing, scheduling and other areas, but we will restrict our attention to a discussion of the set representation problem here.

3. Results

Some of the following results are obtained directly from the definitions or discussed in [4, 6, 7, 12].
We use the symbol "$\setminus$" to indicate set theoretic difference. Thus if E and F are two sets, then $E \setminus F$ is the set of elements of E which are not in F.

For our applications here, the order in which elements of a set are listed in the set is unimportant. Hence, if $E_1, \ldots, E_t$ are disjoint sets, we will use the term *cartesian product* $E_1 \times \ldots \times E_t$ to refer to the class of sets, each set in which has cardinality t and contains one element from each of E_r for r=1 to t.

3.1 Given $\delta(C)$, no subset of Γ of cardinality $\leq \delta(C)-1$ can represent C.

3.2 If E represents C and $F \supset E$ then F represents C.

3.3 If $\hat{C} \subset C$ and E represents C then E represents $\hat{C}$. Thus $\hat{C} \subset C$ implies $\delta(C) \geq \delta(\hat{C})$.

3.4 If the singleton set $\{j\} \varepsilon C$, then every represent for C must contain the element j. Also if $j \notin c_i$ for all $c_i \varepsilon C$, then j is not contained in any MCR for C.

3.5 If E represents C then obviously $E \supset F$ for some $F \varepsilon S(C)$. Thus given S(C) all represents for C can easily be obtained.

3.6 s(C) is the sub-family of S(C) containing all the minimum cardinality subsets in S(C).

3.7 If $C = \{c_1,\ldots,c_m\}$ and $c_1,\ldots,c_m$ are all mutally disjoint then every set in the cartesian product $c_1 \times c_2 \times\ldots\times c_m$ is an MCR for C.

Thus in this case

$$s(C) = c_1 \times c_2 \times\ldots\times c_m \tag{5}$$

and

$$\delta(C) = m. \tag{6}$$

Also in this case $s(C) = S(C)$ (7)

Conversely if $\delta(C) = m$ = cardinality of the class C then the subsets in the class C must be mutally disjoint and (5) and (7) hold.

3.8 Proposition: If $E_1,\ldots,E_r$ are disjoint subsets of Γ of cardinalities $\delta_1,\ldots,\delta_r$ such that $\delta_1 \leqq \delta_2 \leqq \ldots\leqq \delta_r$ and $C = E_1 \times E_2 \times\ldots\times E_r$ = the cartesian product of $E_1,\ldots,E_r$ then $\delta(C) = \delta_1$ and $s(C) = \{E_i : E_i$ has cardinality $\delta_1\}$ and $S(C) = \{E_1,\ldots,E_r\}$

Proof: $E \subset \Gamma$ be such that $E_i \setminus E \neq \emptyset$ for each i =1 to r, where "$\setminus$" indicates set theoretic difference. Pick $j_i \varepsilon E_i \setminus E$ for each i = 1 to r. Then $\{j_1,\ldots,j_r\} \in C$ and $E \cap \{j_1,\ldots,j_r\} = \emptyset$, and hence, E does not represent C. Thus every represent for C must contain some E_i as a subset. Also each E_i for i=1 to r is obviously a represent for C. Hence $S(C) = \{E_1,\ldots,E_r\}$, and hence, the minimum cardinality subsets among $E_1,\ldots,E_r$ are the MCR's for C.

3.9 Let $C(\bar{j}) = \{c_i : c_i \in C \text{ and } j \notin c_i\}$ for $j \in \Gamma$. Also for any $F \subset \Gamma$ let $C(\bar{F}) = \{c_i : c_i \in C \text{ and } c_i \cap F = \phi\}$.

3.10 Suppose E is an MCR for C and $j \in E$. If $E \setminus \{j\} \neq \phi$ then, $C(\bar{j}) \neq \phi$ and $E \setminus \{j\}$ is an MCR for $C(\bar{j})$.

3.11 Theorem: Pick a $j \in \Gamma$ and fix it.

(i) E is a M C R for $C(\bar{j})$ and $E \cap c_i \neq \phi$ for all $c_i \in C \setminus C(\bar{j})$ then E is an M C R for C and j cannot be contained in any M C R for C. In this case

$$s(C) = \{E: \ E \in s[C(\bar{j})] \text{ and } E \cap c_i \neq \phi \text{ for all } c_i \notin C(\bar{j})\} \quad (9)$$

(ii) On the other hand if $E \in s[C(\bar{j})]$ implies that $E \cap c_i = \phi$ for some $c_i \in C \setminus C(\bar{j})$ then $\delta(C) = 1 + \delta[C(\bar{j})]$, and any subset of the form

$$F = E \cup \{j\} \quad \text{for some } E \in s[C(\bar{j})] \quad (10)$$

is an M C R for C. Every M C R for C which contains j is of the form F in (10).

Proof

(i) Under that hypothesis obviously E represents C and by 3.3 it is an M C R for C. Also let $F \subset \Gamma$ be an M C R for C and suppose F contains j. Let $G = F \setminus \{j\}$. Since F represents C, G must represent $C(\bar{j})$. By the above result we have

$$\text{cardinality of } F = \delta\ (C) = \delta[C(\bar{j})]$$

But cardinality of G = cardinality of F-1 which implies by 3.1 that G cannot represent $C(\bar{j})$, leading to a contradition.

(ii) By 3.3 every subset of Γ which represents C must represent its subclass $C(\bar{j})$. However, by the hypothesis all subsets of Γ of cardinality $\delta\ [C(\bar{j})]$ which represent $C(\bar{j})$ do not represent C. Hence, $\delta(C) > \delta[C(\bar{j})]$. Thus, if we can find a subset of Γ of cardinality $1+\delta[C(\bar{j})]$ which represents C, that must be an M C R for C.

Since every subset in $C \setminus C(\bar{j})$ contains j, it is clear that subsets of the form F in (10) represents C. Each of these sets is of cardinality $1 + \delta[C(\bar{j})]$, and hence, all of them are M C R's for C.

By 2.9, we also see that every M C R for C which contains j must be of the form defined in (10).

3.12 Corollary: Either the union of all the subsets in s(C) is Γ or there exists a $j \in \Gamma$ such that $s(C) \subset s[C(\bar{j})]$ and s(C) is obtained as in (9).

3.13 Corollary: Either there exists a $j \in \Gamma$ such that s(C) is obtained as in (9) in one step, or s(C) is the class containing all subsets of the form (10) as j varies over

$$\bigcup_{c_i \in C} c_i$$

3.14 If F is a subset of some MCR for C then $F \cup G$ is an MCR for C for every $G \in s[C(\bar{F})]$. Conversely if E is an MCR for C and F is a proper subset of E then $C(\bar{F}) \neq \emptyset$ and $E \setminus F$ is an MCR for $C(\bar{F})$.

Similarly if F is a subset of some MR for C then $F \cup G$ is an MR for C for all $G \in S[C(\bar{F})]$. And if F, G are disjoint and $F \cup G \in S(C)$, then $G \in S[C(\bar{F})]$.

4. Hakimi's method for finding all the represents [17]:

If the class C contains two subsets c_{i_1} and c_{i_2} such that $c_{i_1} \subset c_{i_2}$, then obviously the subset c_{i_2} can be deleted from the class C without changing the class of represents. We will assume that this has been done and hence, that no element in the class C is a subset of another element in C.

Associate the variable x_j with j. If $E \subset \Gamma$ let $x_E = \sum_{j \varepsilon E} x_j$. The *representing function* for the class C is

$$f(x_1,\ldots,x_n) = x_{c_1} \times x_{c_2} \times \ldots \times x_{c_m} \qquad (11)$$

where, in multiplying the terms on the right hand side, remember to change the exponent of any variable x_j in any term to 1 whenever it becomes > 1 and also change the coefficient of every term to 1. When it is obtained in this manner, each term in $f(x_1,\ldots,x_n)$ is the product of all the variables x_j associated with j in some represent for C and conversely the product of all the variables x_j associated with any MR for C appears as a term in $f(x_1,\ldots,x_n)$.

5. Algorithm

5.1 Let S_r (C) be the class of all MR for C which have cardinality r. Then $S(C) = \bigcup_{r=1}^{n} S_r$ (C)
Obviously for any $r < \delta$ (C), S_r (C) = ϕ and $S_{\delta(C)}$ (C) = s(C).

5.2 Let $C' = \{c_1, \ldots, c_{m-1}\}$ and $C = \{c_1, \ldots, c_{m-1}, c_m\}$. If $F \varepsilon S_r(C')$ and $F \cap c_m \neq \phi$, then $F \varepsilon S_r(C)$. Also if $F \varepsilon S_{r-1}(C')$ and $F \cap c_m = \phi$ then for $j \varepsilon c_m$, $F \cup \{j\} \varepsilon S_r(C)$ if no subset of $F \cup \{j\}$ is contained in any $S_t(C)$ for $t < r$. These results also follow from those of Lawler [6].

5.3 For each r, if $S_r(C')$ is nonempty, it can be expressed in a compact form as the union of several cartesian products. If an additional subset c_m is augmented to the class C' making it the class C, then $S_r(C)$ for r=1 to n can be obtained by using 5.2 starting from $r = \delta(C')$.

If $E_1, \ldots, E_r$ are mutually disjoint subsets of Γ and if the Cartesian product $E_1 \times \ldots \times E_r$ is a subset of $S_r(C')$, then every subset in $E_1 \times \ldots \times E_r$ which has a nonempty intersection with c_m is in $S_r(C)$. The class of all subsets of $E_1 \times \ldots \times E_r$ which have a nonempty intersection with c_m can easily be identified using only set theoretic operations on the sets $E_1, \ldots, E_r, c_m$, and this class itself can be expressed as a union of several Cartesian products all of which are in $S_r(C)$. Also the remaining subsets which form the class of all subsets in the Cartesian product $E_1 \times \ldots \times E_r$ which have an empty intersection with c_m can be expressed as a union of several Cartesian products by simple set theoretic operations on the sets $E_1, \ldots, E_r, c_m$.

If $F_1, \ldots, F_r$ are mutually disjoint subsets of Γ and $F_1 \times \ldots \times F_r$ is a subset of $S_r(C')$ and if each $F_1, \ldots, F_r$ has an empty intersection with c_m, then every subset in $F_1 \times \ldots \times F_r$ has an empty intersection with c_m. Then consider the Cartesian product, $F_1 \times \ldots \times F_r \times c_m$, each subset in which is a represent for the class C. If each $S_t(C)$ for $t \leqq r$ has already been expressed as a union of several Cartesian products, then by comparing those Cartesian products with $F_1 \times \ldots \times F_r \times c_m$, we can easily obtain the class of all subsets in $F_1 \times \ldots \times F_r \times c_m$ which do not contain some subset of $S_t(C)$ for $t \leqq r$ as a subset; and represent this class itself as a union of several Cartesian products. All these Cartesian products belong to $S_{r+1}(C)$.

Expressing each $S_r(C')$ as the union of several cartesian products makes it very convenient to carry out these operations.

5.4 If for each $F \in S_{\delta(C')}$ we have $F \cap C_m = \phi$, then $\delta(C) = 1+\delta(C')$ and $s(C) = S_{\delta(C)}(C) = \left\{F \cup \{j\} : F \in S_{\delta(C')},\ j \in c_m\right\}$.

5.5 To solve a sequence of set representation problems where the class C gets augmented with additional subsets at each stage, it is necessary to obtain

$S(C) = (S_1(C), \ldots\ldots, S_n(C))$ at each stage and update it whenever any additional subsets are augmented to C. Even though the number of subsets in S(C) tends to be large, representing each $S_r(C)$ as the union of several cartesian products makes it possible to store it in a compact form and perform operations on them efficiently.

6.1 <u>Numerical example</u>

Consider problem in which $c_1 = \{1,2,3,4,5,6\}$, $c_2 = \{1,2,7,8,9\}$, $c_3 = \{2,3,7,10\}$, $c_4 = \{3,4,8,11,12\}$. Let $C^1 = \{c_1\}$, $C^2 = \{c_1,c_2\}$,

$C^3 = \{c_1, c_2, c_3\}$, $C^4 = \{c_1, c_2, c_3, c_4\}$.

Considering C^1 it is clear that:

$$s(C^1) = S(C^1) = \{1,2,3,4,5,6\}.$$

Considering C^2 and by applying 5.2 to 5.4 it is clear that

$$s(C^2) = S_1 (C^2) = \{1,2\}$$

and $$S_2 (C^2) = \{3,4,5,6\} \times \{7,8,9\}.$$

Using this and considering C^3 we conclude that

$$S_1(C^3) = \{2\},\quad S_2(C^3) = \{1\}, \times \{3,7,10\}, \{3\} \times \{7,8,9\}, \{4,5,6\} \times \{7\}\}.$$

and $S_3(C^3) = \{\{4,5,6\} \times \{8,9\} \times \{10\}\}$.

Now considering C^4 and using the same arguments we get:

$$S_2(C^4) = \{\{2\} \times \{3,4,8,11,12\} , \{1\} \times \{3\} , \{3\} \times \{7,8,9\} , \{4\} \times \{7\}\}.$$

$$S_3(C^4) = \{\{1\} \times \{7, 10\} \times \{ 8,11,12\} , \{1\} \times \{10\} \times \{4\}, \{5,6\} \times \{7\} \times \{8,11,12\}, \{4\} \times \{8,9\} \times \{10\}, \{5,6\} \times \{8\} \times \{10\}$$

and $S_4(C^4) = \{\{5,6\} \times \{9\} \times \{10\} \times \{11,12\}\}$.

Hence, $s(C) = s(C^4) = S_2(C^4)$.

REFERENCES

[1] Balas, E. and M. W. Padberg. "On the Set Covering Problem". W.P. #74-69-7, Management Science Research Group, Carnegie-Mellon University, Feb., 1970.

[2] Balinski, M. L. "Integer Programming: Methods, Uses, Computation", Proceedings of the Princeton Symposium on Mathematical Programming, 1970.

[3] Edmonds, J. "Covers and Packings in a Family of Sets", Bulletin American Mathematical Society, 68, pp. 494-99, 1962.

[4] Garfinkel, R. S. "Set Covering A Survey", Graduate School of Management, University of Rochester, July 1970.

[5] Glover, F. "A Note on Extreme Point Solutions and a Paper by Lemke, Salkin and Spielberg", Operations Research, 19, No. 4, July-Aug., 1971, 1023-1025.

[6] Lawler, E. L. "Covering Problems, Duality Relations and a New Method of Solution", SIAM, Applied Mathematics, 14, No. 5. Sept., 1966, pp. 1115-1132.

[7] Lemke, C. E., H. M. Salkin and K. Spielberg, "Set Covering by Single Branch Enuemration with Linear Programming Subproblems", Operations Research, 19, No. 4, July-Aug., 1971, 998-1022.

[8] Murty, Katta G. "A Fundamental Problem in Linear Inequalities with Applications to the Traveling Salesman Problem". Mathemathical Programming . 2, No. 3 June 1972, pp. 296-308.

[9] Murty, Katta G. "Adjacency on Convex Polyhedra", SIAM Review, 13, No. 3, pp. 377-386, July, 1971.

[10] Murty, Katta G. "An Algorithm for Generating all the Complementary Feasible Bases for a Linear Complementarity Problem", Department of Industrial and Operations Engineering, The University of Michigan, 1971.

[11] Roth, R. "Computer Solutions to Minimum Cover Problems", Operations Research, 17, pp. 455-465, 1969.

[12] Salkin, H. M., and R. D. Koncal. "A Pseudoual All-Integer Algorithm for the Set Covering Problem", Technical Memorandum 204, Operations Research Department, Case Western Reserve University, Nov., 1970.

[13] Balinski, M. L. and R. Quandt. "On an Integer Program for a Delivery Problem", Operations Reserach, 13, 300-304, 1964.

[14] Day, R. H. "On Optimal Extracting from a multiple File Data Storage System: An Application of Integer Programming", Operations Research, 13, No. 3, 482-494, 1965.

[15] Salveson, M. E. "The Assembly Line Balancing Problem", Journal of Industrial Engineering, 6, No. 3, 18-25, 1955.

References

[16] Arabeyre, J. P., J. Fearnley, F. C., Steiger, W. Teather. "The Airline Crew Scheduling Problem" A Survey" Transportation Science, 3, 140-163, May 1969.

[17] Hakimi, S. L. "Optimum Distribution of Switching Centers in a Communication Network and Some Related Graph Theoretic Problems", Operations Research, 13, No. 3, 462-475, 1965.

[18] Keller, E. L., "Quadratic Optimization and Linear Complementarity," Ph.D. dissertation, Dept. of Mathematics, University of Michigan, 1969.

[19] Kuhn, H. W., "An Algorithm for Equilibrium Points in Bimatrix Games," Proc. Nat. Acad. Sci., USA, Vol. 47, pp. 1657-1662, 1961.

DISCUSSION OF "ON THE SET REPRESENTATION AND THE SET COVERING PROBLEM"

Eugene L. Lawler

University of California at Berkeley, Berkeley, California, USA

Murty points out that the set representation problem and the set covering problems are equivalent. A very easy way to see this is to construct a bipartite graph for each problem. For the representation problem, the nodes in one part are identified with elements of the set Γ and nodes in the other part with subsets in the class C. There exists an arc (i,j) iff element i belongs to subset c_j. For the covering problem, the nodes in one part are identified with subsets ξ_i in the class θ and nodes in the other part with elements of the underlying set. There exists an arc (i,j) iff element j belongs to subset ξ_i. In both cases, a feasible solution (i.e., a represent or a cover) consists of a subset S of nodes in the first-mentioned part such that each node in the other part is adjacent to at least one node in S. Minimal solutions and minimum cardinality solutions have obvious interpretations.

The set covering problem was the subject of intensive study by switching theorists and computer logic designers at least twenty years ago. The method for generating all minimal covers (or represents) which Murty attributes to Hakimi in Section 4 of his paper was well known to these earlier investigators. I attributed this method to S. R. Petrick [1] in my 1966 paper (Murty's reference [6]). Notice that if the variables x_j are considered to be Boolean variables, then the procedure simply involves multiplying out a product-of-sums Boolean expression to obtain a sum-of-products expression. (With Boolean variables, it is unnecessary to introduce ad hoc rules concerning exponents and coefficients.)

Murty's contribution is to notice that the Petrick ("Hakimi") method may be improved by means of a compact representation for sets of minimal covers (or sums-of-products) generated at intermediate steps of the computation. The suggested representation is in terms of unions of cartesian products.

(Possibly a better term might be "unordered" cartesian products, since the elements of these products are unordered sets, contrary to the usual definition.)

Murty's improvement is likely to be effective only if a significant number of the sets entering into the cartesian products contain more than a single element. If this is not the case, there will not be a substantial compactification of data. Moreover, unless a great deal of care is given to details of programming, any savings occasioned by data compactification may be more than offset by added complications in set theoretic operations, compared with the unadorned Petrick procedure. It is a non-trivial task to compare two cartesian products to determine if one contains a superset of a set contained in the other. The efficient implementation of such tasks is crucial to the success of Murty's proposal.

I would speculate that Murty's idea might be helpful for problems in which the number of elements in Γ is large compared with the number of subsets in C, and not to be worthwhile when the reverse is the case. It appears that, indeed, this is the type of problem Murty is most interested in solving.

It should also be noted that the Petrick method can be made more efficient by employing a different nesting of iteration loops than that used in the paper. Murty first finds $S_1(C^1)$, then $S_1(C^2)$, $S_2(C^2)$, then $S_1(C^3)$, $S_2(C^3)$, $S_3(C^3)$,etc. Instead one can first find $S_1(C^1)$, $S_1(C^2)$, $S_1(C^3)$,..., until finally, for some $q<m$, $S_1(C^q) = \emptyset$. Continue this way until, for some r, $S_r(C^m) \neq \emptyset$. All minimum cardinality solutions are contained in $S_r(C^m)$. It is unnecessary to generate $S_{r+1}(C^{r+1})$, etc. Thus, in the example at the end of Murty's paper, it is unnecessary to find $S_3(C^3)$, $S_3(C^4)$, and $S_4(C^4)$.

The order of iteration chosen by Murty appears to have been motivated by the applications he had in mind. That is, the solution procedure would be applied to a covering problem in which the sets $c_1, c_2,\ldots,c_m$ are generated one at a time and $s(C^1)$, $s(C^2)$,...,$s(C^m)$ are successively required. However, even in this case, a change in order of iteration may be worthwhile.

The Petrick-Murty method may prove useful for a restricted class of set covering problems, although it is hard to see how it could be as generally satisfactory as a well-designed branch-and-bound procedure. Computational experience with the traveling salesman problem and the linear complementarity problem, as described in Section 2.6 and 2.7 would certainly be most welcome.

It is unlikely that anyone will ever discover a procedure for the set covering problem which is efficient in the sense that the number of computational steps required is a polynomial function of $|\Gamma|$ and $|C|$. Recent results of Karp [2] show that the set covering problem is in the same equivalence class, with respect to polynomial complexity, as a number of other classic heartbreakers. For example, the problem of finding a Hamiltonian tour in a graph, computing the chromatic number of a graph, minimal covering of arcs by nodes, etc. A polynomial-bounded algorithm for any one of these problems, and many others, could be made over into a polynomial-bounded algorithm for any other problem in this equivalence class. I cannot prove it, but I feel sure that no such algorithm will ever be found.

[1] Petrick, S. R., "A Direct Determination of the Irredundant Forms of a Boolean Function from the Set of Prime Implicants", AFCRC-TR-56-110, Air Force Cambridge Research Center (1956).

[2] Karp, R. M., "Reducibility Among Combinatorial Problems", Proceedings of IBM Symposium on Complexity of Computer Computations, to be published by Plenum Press, N. Y. (1972).

SELECTED COMMENTS CONCERNING OPTIMIZATION TECHNIQUES FOR FUNCTIONS OF PERMUTATIONS

John G. Rau
ULTRASYSTEMS, Inc.
Newport Beach, California, USA

In this paper is provided a brief overview of the status of techniques in optimizing functions of permutations both with and without constraints. Problems of this type are readily found in the literature on sequencing and scheduling as well as in sequential search and testing. The kinds of techniques that are applicable range from dynamic programming to random search and simple pairwise switching of coordinates. In general, necessary and sufficient conditions for local optima do not exist, however the known conditions for special cases are presented.

I. INTRODUCTION

The general problem we wish to address in this paper is the following:

(1) Minimize $F(i_1,\ldots, i_n)$

where

(2) $(i_1,\ldots, i_n)$ is a permutation of the integers $1,2,\ldots,n$.

Since there are n! permutations of these integers, for n=20 we have 20!= 2,432,902,008,176,640,000, which means that one evaluation per microsecond would require nearly 80,000 years to try each of the 20! permutations. Consequently, more efficient schemes than that of trial-and-error must be found.

Special cases of this problem abound, both with and without constraints and/or precedence relationships, in the literature. In particular, this type of problem arises frequently in the sequencing of jobs on machines [1, 4, 8, 9, 11, 15, 17, 23], the selection of research tasks to be performed in a research and development program [5], the sequential search for an object of a particular type in a given collection of objects [1, 6, 7, 13, 20], the sequential testing of a device in a quality control process [2, 19, 20], and in the classical Traveling Salesman Problem [14, 24].

The unconstrained problem stated above can be easily generalized to one with constraints in the following way:

(3) Minimize $F(i_1,\ldots, i_n)$

subject to

(4) $H_j(i_1,\ldots, i_n) \leq C_j$ for $j=1,\ldots, m$

(5) $(i_1,\ldots, i_n)$ is a permutation of the integers $1,2,\ldots,n$.

This type of problem arises in sequencing when there exists a prescribed ordering between certain subsets of jobs or in the case of sequential testing when tests must be performed in a prescribed order.

In this paper we present a selected review of the state-of-the-art in optimization theory for functions of permutations and discuss the known results that are pertinent to both of the previously stated problems.

II. TYPICAL OBJECTIVE FUNCTIONS

The following examples serve to illustrate the types of objective functions that arise in practice.

Example 1 (see References [1], [6], [7], [13] and [23])

$$F(i_1,\ldots, i_n) = \sum_{j=1}^{n} a_{i_j} \sum_{k=1}^{j} b_{i_k} \tag{6}$$

(a) Consider a system consisting of n independent components with the property that, whenever one component fails, the system fails. When the system fails, then the ith component is known to have an a priori probability a_i of being the component that caused system failure. If each component can be tested one at a time, requiring b_i time units for testing the ith component, then the problem is to determine the order in which the components should be tested in order to minimize the expected time until the failed component is found. For the test sequence $(i_1,\ldots, i_n)$, the expected time until determination of the failed component is given by $F(i_1,\ldots, i_n)$.

(b) Suppose we are given a set of jobs $J_1, J_2,\ldots, J_n$ which are to be executed successively on a single machine. Any given job J_i is assumed to consume b_i time units on the machine. With job J_i is associated a cost per unit time factor a_i such that if job J_i is completed at time t then the cost associated with completing it at that time is $a_i t$. If the machine is to be constantly in use and no job is interrupted before completion, then the problem is to find the order in which the jobs should be performed on the machine so as to minimize the total cost associated with completion of each job; that is, if the job sequence is $(i_1,\ldots, i_n)$, then the cost of completion for Job J_{i_j} is given by $a_{i_j} \sum_{k=1}^{j} b_{i_k}$ and the total cost for all jobs is given by $F(i_1,\ldots, i_n)$. Equivalently, this objective function can be interpreted as the weighted sum of the job completion times.

Example 2 (see References [2], [5] and [19])

$$F(i_1,\ldots, i_n) = \sum_{j=1}^{n} a_{i_j} \prod_{k=1}^{j-1} b_{i_k} \tag{7}$$

(a) A research project is characterized by a series of n independent tasks each with a high degree of uncertainty. The project is successful if, and only if, all the tasks are successful. With task i is associated a cost a_i of performing the task and a probability b_i that it is successful. The problem is to determine the order in which the tasks should be performed in order to minimize the expected total cost of the project. If the tasks are performed in the order $(i_1,\ldots, i_n)$, then the total expected cost is given by $F(i_1,\ldots, i_n)$.

(b) An item is subjected to a series of n tests --- say, hardness, tensile strength, weight, length, size, etc. Associated with the ith test are two constants: a_i= the cost per item of conducting the ith test, and b_i= the probability of accepting the item on the ith test. The tests are assumed to be independent in the sense that they may be run in any order whatsoever, and the constants a_i and b_i are taken to be independent of the order in which the tests are run. For a given choice of the sequence of tests, the rule is established that an item is subjected to each test in the sequence in turn as long as the tests accept the item; if the item is rejected by a test, no further tests in the sequence are performed. The objective is to determine the test sequence that minimizes the total expected cost. For the test sequence $(i_1, \ldots, i_n)$ this cost is given by $F(i_1, \ldots, i_n)$.

Example 3 (see Reference [11])

(8) $$F(i_1, \ldots, i_n) = \sum_{j=1}^{n} \max \left\{0, \sum_{k=1}^{j} a_{i_k} - b_{i_j}\right\}$$

(a) Consider the problem of sequencing a set of jobs $J_1, J_2, \ldots, J_n$ on a single machine. Job J_i has a processing time a_i and a due date b_i that are given. Setup times are neglected and are assumed to be independent of the order in which the jobs are processed. If the jobs are processed in the order $(i_1, \ldots, i_n)$, then the tardiness of job J_{i_j} is given by

$T_{i_j} = \max \left\{0, \sum_{k=1}^{j} a_{i_k} - b_{i_j}\right\}$. For the objective of determining the sequence that minimizes the total tardiness, the objective function is given by $F(i_1, \ldots, i_n)$.

Example 4 (see Reference [17])

(9) $$F(i_1, \ldots, i_n) = \max_{k=1,\ldots,n} \left\{\sum_{j=1}^{k} a_{i_j} - \sum_{j=1}^{k-1} b_{i_j}\right\}$$

(a) Consider a two-stage production process that requires n items to be processed sequentially on two machines, A and B. At most one item can be on a machine at a given time and the ordering of the items is the same on both machines. The objective is to determine the ordering of items that minimizes the total elapsed time for the process. Since the total elapsed time is equal to the sum of the total time machine B is in use (a constant) and the total time machine B is idle, an equivalent problem is to minimize the total idle time of machine B.

If we define

a_i = the setup time plus work time of the ith item on machine A
b_i = the setup time plus work time of the ith item on machine B
X_i = the inactive period of time for machine B immediately before the ith item comes onto machine B

then, for the sequence (1,2,...,n) and referring to the illustrative time scale below,

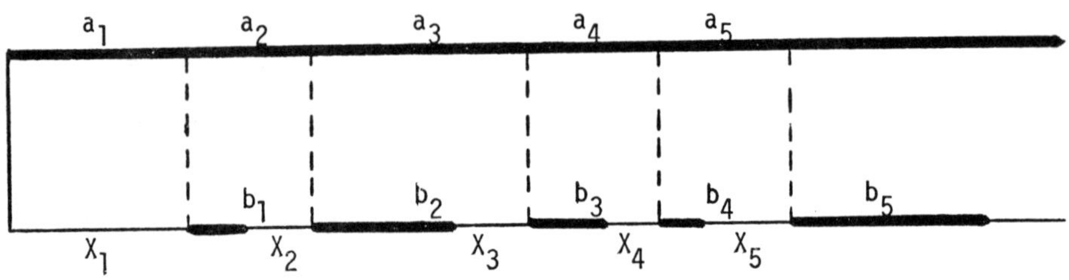

we have

(10) $$X_1 = a_1$$

(11) $$X_2 = \max \left| a_1 + a_2 - b_1 - X_1, 0 \right|$$

(12) $$X_1 + X_2 = \max \left| a_1 + a_2 - b_1, a_1 \right|$$

(13) $$X_3 = \max \left| a_1 + a_2 + a_3 - b_1 - b_2 - X_1 - X_2, 0 \right|$$

(14) $$X_1 + X_2 + X_3 = \max \left| a_1 + a_2 + a_3 - b_1 - b_2, X_1 + X_2 \right|$$

(15) $$= \max \left| a_1 + a_2 + a_3 - b_1 - b_2, a_1 + a_2 - b_1, a_1 \right|$$

and, in general, for all n items

(16) $$\text{Total idle time of machine B} = \sum_{i=1}^{n} X_i = \max_{k=1,\ldots,n} \left\{ \sum_{i=1}^{k} a_i - \sum_{i=1}^{k-1} b_i \right\} .$$

For the arbitrary sequence of items $(i_1, \ldots, i_n)$ the total idle time is given by $F(i_1, \ldots, i_n)$.

Example 5 (see Reference [22])

(17) $$F(i_1,\dots,i_n) = a_{i_1}a_{i_2} + a_{i_3}a_{i_4} + \dots + a_{i_{n-1}}a_{i_n}$$

where n is even.

(a) In the room-and-piller method for extracting coal from underground seams, each section of the mining operation consists of a continuous miner which has assigned to it two shuttle cars that make periodic trips from the continuous miner to the conveyor belt. If both shuttle cars in a section are operating, the section effectiveness is 1; if only one car is operating its effectiveness is δ, where $1/2 < \delta < 1$; otherwise the section effectiveness is 0. If there are n possible shuttle cars, where a_i is the probability of failure of the ith car, then the probability of both shuttle cars, say i and j, failing in a section is given by the product $a_i\,a_j$, assuming independent events. The objective is to pair the shuttle cars so as to maximize the expected effectiveness of the entire mining operation. If they are paired according to the scheme $\{i_1, i_2\}$, $\{i_3, i_4\},\dots, \{i_{n-1}, i_n\}$, then the expected effectiveness is given by

(18) $$E(i_1,\dots,i_n) = (1-a_{i_1})(1-a_{i_2}) + \delta[a_{i_1}(1-a_{i_2}) + a_{i_2}(1-a_{i_1})] + \dots + (1-a_{i_{n-1}})(1-a_{i_n}) + \delta[a_{i_{n-1}}(1-a_{i_n}) + a_{i_n}(1-a_{i_{n-1}})]$$

(19) $$= n + (\delta-1)\sum_{j=1}^{n} a_{i_j} - (2\delta-1)(a_{i_1}a_{i_2} + \dots + a_{i_{n-1}}a_{i_n})$$

To maximize $E(i_1,\dots,i_n)$ is equivalent to minimizing $F(i_1,\dots,i_n)$ as given by (17), since $\delta > 1/2$.

Example 6 (see References [14] and [24])

(20) $$F(i_1,\dots,i_n) = a_{i_1i_2} + a_{i_2i_3} + \dots + a_{i_ni_1}$$

(a) The classical Traveling Salesman Problem consists of finding the tour thru each of n cities, passing thru each city only once and returning to the starting point, that results in the minimum total cost (time, money or any other type of penalty). If the cities in the tour are ordered according to the n-tuple $(i_1,\ldots, i_n)$ and a_{ij} denotes the cost of traveling from city i to city j, then the total cost of this tour is given by $F(i_1,\ldots, i_n)$. This particular problem has been formulated (see [24]) as an integer linear programming problem with $n^2 - n + 2$ linear constraints and $n^2 + n - 1$ integer valued variables.

III. STATE-OF-THE-ART IN TECHNIQUES

The basic problem is to find necessary and sufficient conditions for a function $F(i_1,\ldots, i_n)$ to have an extremum at $(i_1^*,\ldots, i_n^*)$ where $(i_1,\ldots, i_n)$ is a permutation of the integers 1,2,...,n. If F were a function of continuous variables or even discrete variables, then the conditions are well known. When one examines the subset of Euclidean n-space that consists of points whose coordinates are a permutation of the first n positive integers, then no analogous conditions are known to exist. These results are briefly summarized in Table 1, below, for the case of minimizing the objective function.

Table 1 Necessary and Sufficient Conditions for a Minimum

Variables	Function	Necessary Condition	Sufficient Condition
Continuous	$F(x)$	$\frac{dF}{dx} = 0$	$\frac{d^2F}{dx^2} > 0$
	$F(x_1,\ldots, x_n)$	$\frac{\partial F}{\partial x_i} = 0$ for each i	Hessian has positive principal minors
Discrete	$F(n)$	$F(n_0-1) > F(n_0)$ $F(n_0+1) > F(n_0)$	$\Delta^2F(n) > 0$
Permutations	$F(i_1,\ldots, i_n)$	?	?

Partial results do exist in the form of somewhat specialized sufficiency conditions. The most significant of these is due to W. E. Smith [23] and can be stated as follows:

Smith's Condition A sufficient condition that $F(\bar{p}) \leq F(p)$ for all permutations p of n objects is that:

(A) there is a real-valued function g of ordered pairs of elements such that if $p = (p_1, p_2, \ldots, p_n)$ is any permutation and p' the permutation obtained from p by the interchange of the ith and $(i+1)^{st}$ elements, then $F(p) \leq F(p')$ if $g(p_i, p_{i+1}) \leq g(p_{i+1}, p_i)$ and

(B) $\bar{p}$ is such that the ith object precedes the jth object if $g(i, j) \leq g(j, i)$.

In the course of the proof of this theorem evolved the classical Switch-and-Check technique (also called the Contiguous-Binary-Switching-Rule, see [10]) which is based on comparing the value of the objective function at two permutations which differ by two adjacent terms being interchanged. This comparison is made by examining the difference in the two function values. For example, consider the difference

$$F(1,2,\ldots,j+1,j,\ldots,n) - F(1,2,\ldots,j,j+1,\ldots,n).$$

If this yields a simple condition for j to precede j+1, such as a negative difference, or vice-versa, then by a pairwise comparison of terms, or a series of transpositions of adjacent terms, one reduces the initial permutation to an optimal one. In Smith's Condition the criterion for deciding which term precedes another depends on the function g. The remarkable property asserted by this theorem is that if such interchange is not profitable from the point of view of g, then it is not profitable from the point of view of F.

Unfortunately, Smith's Condition does not give a method for obtaining the function g, which we would expect, of course, to depend on the objective function F. Upon interchanging the ith and $(i+1)^{st}$ elements of a permutation, the value of F changes by a certain amount. Now, if this change can be attributed to quantities attached to just the ith and $(i+1)^{st}$ terms, say $F(p) = A + Bg(p_i, p_{i+1})$ and $F(p') = A + Bg(p_{i+1}, p_i)$, then one may be able to eliminate constant values from both values of the function F to obtain $F(p) - F(p') = B(g(p_i, p_{i+1}) - g(p_{i+1}, p_i))$ and then examine the difference in the g values. This is the essence of the Switch-and-Check technique.

Two recent attempts to characterize the objective functions for which Smith's Condition is applicable can be found in References [9] and [20]. These are, respectively, as follows:

(1) (Elmaghraby) Let Q by any permutation of n jobs and let F(Q) be a bounded function defined on Q. Then there exists a function g satisfying Smith's Condition if F(Q) can be represented as

(21) $$F(Q) = g(i,j) \circ \phi \{Q - (i,j); g(i,j)\}$$

where the symbol "o" denotes a binary operation which preserves inequalities on the real line, $\{Q - (i,j)\}$ denotes the sequence Q less the pair i and j, in that order, and ϕ is a monotone function of g(i,j).

and

(2) (Rau) If f is a positive function of one integer variable and g_i (for i=1,..., n) is a function of i integer variables such that the following two conditions are satisfied:

(A) $g_i(x_1,\ldots, x_{i-1}, x_i) = g_i(y_1,\ldots, y_{i-1}, x_i)$ for all $i(1 \leq i \leq n)$ and all permutations $(y_1,\ldots, y_{i-1})$ of the integers $x_1,\ldots, x_{i-1}$, where $x_1,\ldots, x_i$ are distinct integers selected from the set consisting of 1, 2,..., n;

(B) there exist a real-valued function G and positive functions $G_1,\ldots, G_{n-1}$ (with G_1=1) such that for $1 \leq i \leq n-1$ and for all sets of distinct integers $\{x_1,\ldots, x_{i-1}\}$ and $\{y_1,\ldots, y_i\}$ selected from the integers 1, 2,..., n with the property that $\{x_1,\ldots, x_{i-1}\} \subset \{y_1,\ldots, y_i\}$,
$$g_i(x_1,\ldots, x_i) - g_{i+1}(y_1,\ldots, y_i, x_i) = G(y_j)G_i(x_1,\ldots, x_{i-1})$$
where $y_j \neq x_k$ for $1 \leq k \leq i-1$,

then the minimum value of

(22) $$F(i_1,\ldots, i_n) = \sum_{j=1}^{n} f(i_j)g_j(i_1,\ldots, i_j)$$

is obtained at $(i_1^*,\ldots, i_n^*)$ where

(23) $G(i_1^*)/f(i_1^*) \geq G(i_2^*)/f(i_2^*) \geq \ldots \geq G(i_n^*)/f(i_n^*)$.

In the derivation of (23), it is shown that the g function is defined by

(24) $g(i_j, i_{j+1}) = f(i_j)g_j(i_1,\ldots, i_{j-1}, i_j) + f(i_{j+1})g_{j+1}(i_1,\ldots, i_j, i_{j+1})$

and

(25) $g(i_{j+1}, i_j) = f(i_{j+1})g_j(i_1,\ldots, i_{j-1}, i_{j+1}) + f(i_j)g_{j+1}(i_1,\ldots, i_{j+1}, i_j)$

as a result, we obtain the forms

(26) $F(i_1,\ldots, i_j, i_{j+1},\ldots, i_n) = A + g(i_j, i_{j+1})$

and

(27) $F(i_1,\ldots, i_{j+1}, i_j,\ldots, i_n) = A + g(i_{j+1}, i_j)$

Using condition (B),

(28) $F(i_1,\ldots, i_j, i_{j+1},\ldots, i_n) - F(i_1,\ldots, i_{j+1}, i_j,\ldots, i_n)$
$= [f(i_j)G(i_{j+1}) - f(i_{j+1})G(i_j)]G_j(i_1,\ldots, i_{j-1})$,

from which (23) follows.

IV. ILLUSTRATION OF THE SOLUTION APPROACH

To illustrate the use of the Switch-and-Check technique, referring to Example 1 and equation (6), we obtain

(29) $F(1, 2,\ldots, j, j+1,\ldots, n) - F(1, 2,\ldots, j+1, j,\ldots, n) =$
$a_{j+1}b_j - a_jb_{j+1}$

from which it follows that the optimal permutation $(i_1^*,\ldots, i_n^*)$ must satisfy the condition

(30) $a_{i_1^*}/b_{i_1^*} \geq a_{i_2^*}/b_{i_2^*} \geq \ldots \geq a_{i_n^*}/b_{i_n^*}$

Similarly, referring to Example 2 and equation (7), we obtain

(31) $F(1, 2,\ldots, j, j+1,\ldots, n) - F(1, 2,\ldots, j+1, j,\ldots, n) =$
$$\left(\prod_{i=1}^{j-1} b_i\right)\left((1-b_{j+1})a_j - (1-b_j)a_{j+1}\right)$$

from which it follows that the optimal permutation $(i_1^*,\ldots, i_n^*)$ must satisfy the condition

(32) $(1-b_{i_1^*})/a_{i_1^*} \geq (1-b_{i_2^*})/a_{i_2^*} \geq \ldots \geq (1-b_{i_n^*})/a_{i_n^*}$

It can also be verified via the Switch-and-Check technique that the permutation, say $(i_1, \ldots, i_n)$, which minimizes the objective function of equation (17) is determined by the following rule, assuming the a_i's are ordered by decreasing values:

(1) Set $i_1 = i$ for which a_i is the largest;
(2) Set $i_2 = i$ for which a_i is the smallest;
(3) Set $i_3 = i$ for which a_i is the second largest;
(4) Set $i_4 = i$ for which a_i is the second smallest;
etc.

In the case of illustration (b) in Example 1, if the cost of completing job J_i at time t is given by the arbitrary cost function $c_i(t)$ for $i = 1, 2, \ldots, n$, then the total cost incurred when the jobs are completed in the order $J_{i_1}, J_{i_2}, \ldots, J_{i_n}$ is given by

$$(33)\qquad F(i_1, \ldots, i_n) = \sum_{k=1}^{n} c_{i_k}\left(\sum_{j=1}^{k} b_{i_j}\right)$$

To find the permutation that minimizes $F(i_1, \ldots, i_n)$ by using the Switch-and-Check technique, in general cannot be accomplished without specifying the appropriate form of the cost functions. However, using the Principle of Optimality (see [1]), this problem can be formulated via the dynamic programming technique (see[15]). The approach employed here is to let $S = \{i_1, i_2, \ldots, i_{n(S)}\}$ be a subset of $\{1, 2, \ldots, n\}$, and denote by T_S the quantity $\sum_{j \in S} b_j$. Let C(S) be the minimum cost incurred in executing the jobs $J_{i_1}, J_{i_2}, \ldots, J_{i_{n(S)}}$, in any order whatsoever, in the interval $[0, T_S]$. Then

(a) if $n(S) = 1$,

$$(34)\qquad C(\{j\}) = c_j(b_j) \quad \text{for any } j;$$

(b) if $n(S) > 1$,

$$(35)\qquad C(S) = \min_{j \in S} \left\{C(S - \{j\}) + c_j(T_S)\right\}$$

Condition (b) follows from the observation that in an optimal order of execution for the jobs with indices in S some job, call it J_j, must be executed last and the remaining jobs (those with indices in $S-\{j\}$) must be executed optimally in the interval $[0,T_{S-\{j\}}]$; in which case, the total cost incurred by such an ordering will be $C(S-\{j\}) + c_j(T_S)$. Taking the minimum over all choices of j, we obtain equation (35). Furthermore, it follows that an ordering $(i_1,\ldots, i_n)$ is optimum if, and only if,

(36) $$C(\{i_1,\ldots, i_j\}) = C(\{i_1,\ldots, i_{j-1}\}) + c_{i_j}\left(T_{\{i_1,\ldots, i_j\}}\right)$$

for $j = 2, 3,\ldots, n$.

The procedure for finding the optimum solution to the dynamic programming formulation of this job ordering problem is to proceed in two phases. In the first phase, the quantities C(S), for all subsets S of $\{1, 2,\ldots, n\}$, are computed recursively from equations (34) and (35). In the second phase, the condition given by equation (36) is used to generate an optimum ordering $(i_1,\ldots, i_n)$, where i_n is obtained first and then, successively, $i_{n-1}, i_{n-2},\ldots, i_1$.

The Traveling Salesman Problem, presented in Example 6, can also (see [16]) be formulated by means of the Principle of Optimality. In particular, suppose the starting city is specified as 1 and so the problem becomes one of finding the permutation $(1, i_2, i_3,\ldots, i_n)$ of the integers $1, 2,\ldots, n$ that minimizes the quantity $a_{1i_2} + a_{i_2i_3} +\ldots+ a_{i_n1}$. In this case, for a subset $S = \{i_1, i_2,\ldots,i_{n(S)}\}$ of $\{2, 3,\ldots, n\}$ and arbitrary $j\in S$, let us define C(S,j) as the minimum cost of starting from city 1 and visiting all cities in the set S, terminating at city j. Then

(a) if $n(S) = 1$,

(37) $$C(\{j\}, j) = a_{1j} \quad \text{for any } j;$$

(b) if $n(S) > 1$,

(38) $$C(S, j) = \min_{i\in S-\{j\}} \left\{C(S-\{j\}, i) + a_{ij}\right\}.$$

In order to verify (b), we observe that in visiting the cities in S and terminating at city j, some city i must immediately precede city j. Thus, assuming that the other cities are visited in an optimum order, the cost incurred is $C(S-\{j\}, i) + a_{ij}$. Therefore, taking the minimum over all possible immediate predecessors of city j, we obtain condition (b).

Finally, if $\mathcal{C}$ denotes the minimum cost of a complete tour of all n cities, starting and returning at city 1, then

(39) $$\mathcal{C} = \min_{j=2,\ldots,n} \left\{ C(\{2,\ldots,n\}, j) + a_{j1} \right\}$$

It follows that a permutation $(1, i_2, i_3, \ldots, i_n)$ is optimum if, and only if,

(40) $$\mathcal{C} = C(\{2,\ldots, n\}, i_n) + a_{i_n 1}$$

and

(41) $$C(\{i_2,\ldots, i_{j+1}\}, i_{j+1}) = C(\{i_2,\ldots, i_j\}, i_j) + a_{i_j i_{j+1}}$$

for $j=2,3,\ldots, n-1$.

Here again a two-phase computation is used to obtain an optimum solution. In the first phase, the quantities $C(S,j)$ are computed recursively from (37) and (38). In the second phase, equation (41) is used to determine i_n and then, successively, $i_{n-1}, i_{n-2},\ldots, i_2$.

V. SEARCH THEORY APPROACH

An approach that can be used to solve optimization problems of the type addressed here is that of developing a search procedure based on the concept of neighborhoods of points p that are permutations of the integers 1, 2,..., n. The results presented in this section can basically be found in Reference [21].

For convenience, let $\mathcal{P}$ denote the set of all permutations of the integers 1, 2,..., n, and suppose F is a real-valued function defined on $\mathcal{P}$. We assume the objective is to minimize F on $\mathcal{P}$.

<u>DEFINITION 1</u> A tree structure on $\mathcal{P}$ is a pair $(\mathcal{N}, s)$ where

I. $\mathcal{N}$ is a collection of subsets $\{\mathcal{N}(p) : p \in \mathcal{P}\}$ such that

(a) for each $p \in \mathcal{P}$ there is precisely one $\mathcal{N}(p) \in \mathcal{N}$,

(b) for each $p \in \mathcal{P}$, $p \in \mathcal{N}(p)$;

II. s is a function defined on $\mathcal{P}$ satisfying

(a) $s(p) \in \mathcal{N}(p)$

(b) $F(s(p)) \leq F(p)$

(c) $F(s(p)) = F(p) \Rightarrow s(p) = p$

(d) $s(p) = p \Rightarrow F(p') \geq F(p)$ for all $p' \in \mathcal{N}(p)$.

The collection $\mathcal{N}$ is called a neighborhood structure on $\mathcal{P}$, $\mathcal{N}(p)$ being the neighborhood of p; the function s is called a successor function on $\mathcal{P}$.

A tree structure amounts to a monotone non-increasing (local) search procedure; that is, the successor function associates with each point a point in its neighborhood at which the function is at most as large. It is possible to define the successor function without reference to neighborhoods, however, the neighborhood structure permits us to describe the particular local search pattern in a convenient way.

To illustrate the concept of a tree structure, let $p = (i_1, \ldots, i_n) \in \mathcal{P}$, then possible choices for a neighborhood $\mathcal{N}(p)$ are as follows:

(a) $\mathcal{N}(p) = \{(i_2, i_1, i_3, \ldots, i_n), (i_2, i_3, i_1, i_4, \ldots, i_n), \ldots, (i_2, i_3, \ldots, i_n, i_1), (i_1, i_2, \ldots, i_n)\}$

(b) $\mathcal{N}(p) = \{(i_3, i_1, i_2, i_4, \ldots, i_n), (i_3, i_4, i_1, i_2, i_5, \ldots, i_n), \ldots, (i_3, i_4, \ldots, i_n, i_1, i_2), (i_1, i_2, \ldots, i_n)\}$

(c) $\mathcal{N}(p) = \{(i_2, i_1, i_3, \ldots, i_n), (i_1, i_3, i_2, \ldots, i_n), \ldots, (i_1, i_2, \ldots, i_{n-2}, i_n, i_{n-1}), (i_1, i_2, \ldots, i_n)\}$,

i.e., successive pairs of coordinates are interchanged. We observe that each $\mathcal{N}(p)$ is ordered and so let us define p', to be the first element in $\mathcal{N}(p)$ for which $F(p') < F(p)$. Then define $s(p) = p'$. If there is no such p', then define $s(p) = p$. It is trivially verified that in each case $(\mathcal{N}, s)$ is a tree structure according to Definition 1. We further observe that in case (c) the neighborhood structure is essentially that which one considers when employing the Switch-and-Check technique.

For n=3 the neighborhood structures defined by (a) - (c) can be illustrated as follows:

(1) p = (1,2,3)

Case	$\mathcal{N}(p)$
(a)	{(2,1,3), (2,3,1), (1,2,3)}
(b)	{(3,1,2), (1,2,3)}
(c)	{(2,1,3), (1,3,2), (1,2,3)}

(2) p = (2,1,3)

Case	$\mathcal{N}(p)$
(a)	{(1,2,3), (1,3,2), (2,1,3)}
(b)	{(3,2,1), (2,1,3)}
(c)	{(1,2,3), (2,3,1), (2,1,3)}

(3) p = (1,3,2)

Case	$\mathcal{N}(p)$
(a)	{(3,1,2), (3,2,1), (1,3,2)}
(b)	{(2,1,3), (1,3,2)}
(c)	{(3,1,2), (1,2,3), (1,3,2)}

(4) p = (3,2,1)

Case	$\mathcal{N}(p)$
(a)	{(2,3,1), (2,1,3), (3,2,1)}
(b)	{(1,3,2), (3,2,1)}
(c)	{(2,3,1), (3,1,2), (3,2,1)}

(5) p = (2,3,1)

Case	$\mathcal{N}(p)$
(a)	{(3,2,1), (3,1,2), (2,3,1)}
(b)	{(1,2,3), (2,3,1)}
(c)	{(3,2,1), (2,1,3), (2,3,1)}

(6) p = (3,1,2)

Case	$\mathcal{N}(p)$
(a)	{(1,3,2), (1,2,3), (3,1,2)}
(b)	{(2,3,1), (3,1,2)}
(c)	{(1,3,2), (3,2,1), (3,1,2)}

DEFINITION 2

(a) If $s(p) = p$, then the point $p \in \mathcal{P}$ is called a locally minimal point with respect to $(\mathcal{N}, s)$ and $F(p)$ a local minimum.

(b) The nth successor of p, denoted by $s^n(p)$, is given recursively by $s^n(p) = s(s^{n-1}(p))$ for $n = 2,3,\ldots$.

(c) A point $p^* \in \mathcal{P}$ such that $F(p^*) \leq F(p)$ for all $p \in \mathcal{P}$ is called an optimal point.

The basic search procedure suggested by the tree structure requires that we start with some feasible solution and then search for an improvement from among "neighboring feasible solutions". Usually, "neighboring feasible solutions" are determined by their computational convenience, such as the first point offering an improvement in an ordered set $\mathcal{N}(p)$ as given in the illustrative cases (a) - (c). This process then continues until the algorithm (or successor function) can no longer find a better solution in the neighborhood of the best solution already found.

THEOREM 1

For any $(\mathcal{N}, s)$, there exists at least one locally minimal point in $\mathcal{P}$.

Proof:

Since there are only a finite number of points in $\mathcal{P}$, the minimum of F is attained on $\mathcal{P}$ at some point p^*. For this point we must have $s(p^*) = p^*$.

Q.E.D

THEOREM 2

An optimal point p^* is a locally minimal point with respect to every $(\mathcal{N}, s)$.

NOTE: The proof of this obvious result is omitted.

THEOREM 3

For any tree structure and each $p \in \mathcal{P}$, there exists an n, depending perhaps on p, such that $s^n(p) = s^{n+1}(p)$.

Proof:

Let $N = \text{Card}\,\mathcal{N}$, then we can have at most N distinct values in the set $F(p), F(s(p)), \ldots, F(s^k(p)), \ldots$. Since $F(s^j(p)) \leq F(s^{j-1}(p))$ for each j, where we define $s^0(p) = p$, there must be equality at some k, i.e., $F(s^k(p)) = F(s^{k-1}(p))$. From II - (c) of Definition 1, this implies that $s^k(p) = s^{k-1}(p)$, so choose $n = k-1$. Q.E.D

DEFINITION 3 For each $p \in \mathcal{P}$, $\hat{p}$ is called the ultimate successor of p, if for some n and all integer valued $t \geq 0$, $s^{n+t}(p) = \hat{p}$.

THEOREM 4

Every point in $\mathcal{P}$ has a unique ultimate successor.

Proof:

For given $p \in \mathcal{P}$, there exists, using Theorem 3, an n such that $s^n(p) = s^{n+1}(p)$. If we choose $\hat{p} = s^n(p)$, then it is easily verified that $s^{n+t}(p) = \hat{p}$ for all integers $t \geq 0$. The uniqueness follows simply from the observation that if $\hat{p}_1$ and $\hat{p}_2$ are both ultimate successors of p, then there exists integers n_1 and n_2 (assume $> n_1$) such that $\hat{p}_2 = s^{n_2}(p) = s^{n_1+(n_2-n_1)}(p) = \hat{p}_1$. Q.E.D.

THEOREM 5

If $(\mathcal{N}, s)$ and $(\mathcal{N}', s')$ are tree structures with Q and Q' the corresponding sets of locally minimal points, then $\mathcal{N} = \mathcal{N}' \Rightarrow Q = Q'$; that is, the set of locally minimal points is determined by $\mathcal{N}$ independently of s.

Proof:

Suppose $\mathcal{N} = \mathcal{N}'$, i.e., $\mathcal{N}(p) = \mathcal{N}'(p)$ for each $p \in \mathcal{P}$. Let $p \in Q$ then $s(p) = p$ and so $F(p) = F(s(p))$. Now, $s'(p)$ is chosen such that $F(s'(p)) \leq F(s(p))$; however, $s(p) = p$ implies, by II - (d) of Definition 1, that $F(p') \geq F(p)$ for all $p' \in \mathcal{N}(p)$. In particular, $F(s'(p)) \geq F(p)$ and so we must have $F(s'(p)) = F(p)$. From II - (c) of Definition 1 it follows that $s'(p) = p$ and so $p \in Q'$. Similarly, we can show that $Q' \subset Q$ and so we conclude that $Q = Q'$. Q.E.D.

We may interpret a tree structure in terms of a graph. The points of $\mathcal{P}$ are nodes, and an arc is directed from p to p' if p' is the successor of p. The resulting graph then consists of a number of disjoint (inverted) trees, each of which has a locally minimal point in $\mathcal{P}$ at its top or apex.

Choosing a point p to start with (either randomly or via some other selection process), the tree structure leads from p to a unique locally optimal point, namely the one at the top of the (unique) tree containing p. Unfortunately, there is no assurance that this search technique for a given neighborhood structure will yield a global minimum, however in the case of the neighborhood structure of case (c) the search technique will lead to the global minimum provided the Switch-and Check technique is applicable or, more specifically, Smith's Condition can be demonstrated.

To illustrate what can happen depending on the neighborhood structure chosen, let n=3 and suppose the objective function is defined as in equation (6). Let $a_1 = 2$, $a_2 = 1$, $a_3 = 3$, $b_1 = 3$, $b_2 = 6$ and $b_3 = 2$. Then we obtain

F(1,2,3) = 48	F(3,2,1) = 36
F(2,1,3) = 57	F(2,3,1) = 52
F(1,3,2) = 32	F(3,1,2) = 27

Clearly, the global minimum is given by the permutation (3,1,2) , which satisfies the condition $a_3/b_3 > a_1/b_1 > a_2/b_2$. Depending on the neighborhood structure used, the following ultimate successors are obtained:

Permutation		Neighborhood Structure (a)	(b)	(c)
(1,2,3)		(1,2,3)	(3,1,2)	(3,1,2)
(2,1,3)		(1,2,3)	(1,3,2)	(3,1,2)
(1,3,2)		(3,1,2)	(1,3,2)	(3,1,2)
(3,2,1)		(3,2,1)	(1,3,2)	(3,1,2)
(2,3,1)		(3,2,1)	(3,1,2)	(3,1,2)
(3,1,2)		(3,1,2)	(3,1,2)	(3,1,2)
Number of locally minimal points	=	3	2	1

VI. CONTIGUITY CONSTRAINTS

In the case of scheduling jobs on a machine, say $J_1, J_2, \ldots, J_n$, it sometimes happens that certain sets of jobs must be performed together, but without specification of the order in which the jobs within each set are to be processed. An example can be found in a machine shop in which all milling operations must be run together, all grinding operations together and all lathing operations together. Similarly, in the case of sequential testing of an item, it may happen that, depending upon the type of tests being performed, certain kinds of testing must be performed together such as all hardness tests or all electrical tests, etc. In other words, specified subsets of the integers 1, 2,..., n must appear together in the optimal permutation $(i_1, \ldots, i_n)$ for the objective function $F(i_1, \ldots, i_n)$. Requirements of this type are referred to as contiguity constraints in [15] and as proximity constraints in [18]. We shall choose to use the former term, namely, contiguity constraints.

Specifically, let the set of jobs $J_1, J_2, \ldots, J_n$ be partitioned into a collection $\mathcal{S} = \{S_t\}$ of disjoint subsets S_t. We require that all the jobs in a subset $S_t \in \mathcal{S}$ be run together, that is, if $J_k \notin S_t$ and $J_i, J_j \in S_t$, then the subsequences (J_i, J_k, J_j) and (J_j, J_k, J_i) are prohibited. Each subset $S_t \in \mathcal{S}$ is called a contiguity set and the collection constitutes the set of contiguity constraints which must be met.

When contiguity constraints are present, a technique to employ consists of replacing the elements of a given contiguity set by a hypothetical element which is in a sense "equivalent". Specifically, the concept desired here can be defined as follows:

DEFINITION 4 Suppose F is a function of permutations chosen from non-empty subsets of n elements. For a given such subset S, let $\mathcal{P}(S)$ be the set of all permutations of elements in S and let $p \in \mathcal{P}(S)$. Let m be an integer such that $2 \leq m \leq \text{Card } S$, then an element k' is said to be m-wise equivalent to an ordered m-tuple $(j_1, \ldots, j_m)$ in p if $F(p') - F(p) = C_m(j_1, \ldots, j_m)$ where p' is the permutation obtained from p by replacing the m-tuple $(j_1, \ldots, j_m)$ by k' in p and $C_m(j_1, \ldots, j_m)$ is independent of the order of the elements in p with the exception perhaps of the ordered m-tuple $(j_1, \ldots, j_m)$.

The benefit to be derived from replacing an ordered m-tuple by an m-wise equivalent element (thus reducing an n-dimensional problem to an (n-m+1) - dimensional one) can be demonstrated by the following theorem:

THEOREM 6

Suppose S is a non-empty subset of n elements, $\mathcal{P}(S)$ is the set of all permutations of elements in S and $p \in \mathcal{P}(S)$. Let k' be m-wise equivalent to an ordered m-tuple $(j_1, \ldots, j_m)$ in p. If a permutation p' minimizes F over all permutations in $\mathcal{P}(S')$ where $S' = (S - \{j_1, \ldots, j_m\}) \cup \{k'\}$, then p minimizes F over all permutations in $\mathcal{P}(S)$.

Proof:

By definition,

$$(42) \qquad F(p') - F(p) = C_m(j_1, \ldots, j_m)$$

and so the difference in the objective function values between p and p' is a constant which is independent of the order of elements in p, except for the implied ordering in $(j_1, \ldots, j_m)$. Consequently, if F(p') is a minimum over $\mathcal{P}(S')$, then F(p) must be a minimum over $\mathcal{P}(S)$.

Q.E.D.

To illustrate this concept of equivalence, let us consider first the case where the objective function F has the form given by equation (6) or, equivalently, when $F(i_1, \ldots, i_n)$ has the form

$$(43) \qquad F(i_1, \ldots, i_n) = \sum_{j=1}^{n} f(i_j) g_j(i_1, \ldots, i_j)$$

and each g_j is additively separable of the form

$$(44) \qquad g_j(i_1, \ldots, i_j) = \sum_{k=1}^{j} h(i_k) \; ;$$

or, equivalently, when

$$(45) \qquad F(i_1, \ldots, i_n) = \sum_{j=1}^{n} f(i_j) \sum_{k=1}^{j} h(i_k).$$

For the ordered m-tuple $(j_1, \ldots, j_m)$ we obtain the m-wise equivalent element k' by defining (see [15])

$$(46) \qquad f(k') = \sum_{k=1}^{m} f(j_k)$$

and

$$(47) \qquad h(k') = \sum_{k=1}^{m} h(j_k)$$

In particular, when m=2 it is easily shown that

$$(48) \qquad F(p') - F(p) = f(j_1)h(j_2) = C_2(j_1, j_2),$$

and, in general, for $m \geq 2$

$$(49) \qquad F(p') - F(p) = \sum_{k=1}^{m-1} f(j_k) \sum_{t=k+1}^{m} h(j_t) = C_m(j_1, \ldots, j_m)$$

When the objective function F has the form given by equation (7), or, equivalently, when $F(i_1, \ldots, i_n)$ has the form

$$(50) \qquad F(i_1, \ldots, i_n) = \sum_{j=1}^{n} f(i_j) g_j(i_1, \ldots, i_j)$$

and each g_j is multiplicatively separable of the form

$$(51) \qquad g_j(i_1, \ldots, i_j) = \prod_{k=1}^{j-1} h(i_k)$$

such that we can write

$$(52) \qquad F(i_1, \ldots, i_n) = \sum_{j=1}^{n} f(i_j) \prod_{k=1}^{j-1} h(i_k),$$

then for the ordered m-tuple $(j_1, \ldots, j_m)$ the m-wise equivalent element k' is defined by (see [18])

(53) $$f(k') = \sum_{k=1}^{m} f(j_k) \prod_{t=1}^{k-1} h(j_t)$$

and

(54) $$h(k') = \prod_{k=1}^{m} h(j_k),$$

where we require that $\prod_{t=1}^{0} h(j_t) \equiv 1$. In this case, one can show that

(55) $$F(p') - F(p) = 0;$$

that is,

(56) $$C_m(j_1,\ldots, j_m) \equiv 0.$$

For convenience, rather than say "k' is m-wise equivalent to a given ordered m-tuple $(j_1,\ldots, j_m)$" we shall say that "k' is equivalent to $(j_1,\ldots, j_m)$", provided, of course, that the requirements of Definition 4 are met.

An algorithm to find an optimal feasible permutation satisfying a set $\mathcal{S}$ of contiguity constraints when the objective function satisfies Smith's Condition or, in particular, has the form given by equation (22) and the associated conditions (A) and (B) are met, can be stated as follows:

(A) For each contiguity set $S_t \in \mathcal{S}$, apply (23) to find an optimal permutation for the elements of S_t.

(B) Replace the elements of each contiguity set $S_t \in \mathcal{S}$ by an element equivalent to the elements in the optimal permutation.

(C) Arrange the equivalent elements in an optimal permutation using (23) again.

(D) Replace each equivalent element in this optimal permutation by its equivalent optimal permutation. The resulting permutation is optimal for the original problem and also satisfies the contiguity constraints given by $\mathcal{S}$.

When F has the form given by equation (45), the special form of the algorithm is given in [15], and when F has the form given by equation (52), the special form of the algorithm is given in [18]. To illustrate the former case, suppose F is given by equation (52), then in the unconstrained situation we desire to order the elements by decreasing f/h ratios (see equation (30)). Suppose we have eight jobs and contiguity constraints of the form $\mathcal{S} = \{J_1, J_5\} \cup \{J_2, J_4\} \cup \{J_3, J_6, J_7\} \cup \{J_8\}$. Also, suppose that we have the following data:

Job J_i	Type	$f(J_i)$	$h(J_i)$	$f(J_i)/h(J_i)$
J_1	Lathe	20	100	.20
J_2	Drill	40	80	.50
J_3	Mill	25	150	.17
J_4	Drill	30	90	.33
J_5	Lathe	20	60	.33
J_6	Mill	30	120	.25
J_7	Mill	20	140	.14
J_8	Grind	11	110	.10

For $S_1 = \{J_1, J_5\}$ we obtain the optimal permutation (J_5, J_1) with equivalent job J_1' such that $f(J_1') = 40$, $h(J_1') = 160$ and $f(J_1')/h(J_1') = .25$. For $S_2 = \{J_2, J_4\}$, the optimal permutation is (J_2, J_4) with equivalent job J_2' such that $f(J_2') = 70$, $h(J_2') = 170$ and $f(J_2')/h(J_2') = .41$. For $S_3 = \{J_3, J_6, J_7\}$, the optimal permutation is (J_6, J_3, J_7) with equivalent job J_3' such that $f(J_3') = 75$, $h(J_3') = 410$ and $f(J_3')/h(J_3') = .18$. Finally, for $S_4 = \{J_8\}$, we set $J_4' = J_8$. From the set $\{J_1', J_2', J_3', J_4'\}$ we observe that the optimal permutation is given by (J_2', J_1', J_3', J_4') and so the optimal arrangement of all eight jobs is $(J_2, J_4, J_5, J_1, J_6, J_3, J_7, J_8)$. If the contiguity constraints had been ignored, we would have obtained $(J_2, J_4, J_5, J_6, J_1, J_3, J_7, J_8)$ or $(J_2, J_5, J_4, J_6, J_1, J_3, J_7, J_8)$.

VII. PRECEDENCE CONSTRAINTS

Frequently, technological or other types of considerations will dictate that, for example, in the ordering of jobs on a machine certain jobs must precede others or, in general, only those permutations $(i_1, \ldots, i_n)$ which observe certain prescribed precedence orderings among the components are allowable. If i must precede j, then this precedence relationship is denoted by $i \prec j$. A constraint of this type is commonly referred to as a precedence constraint (see References [3] and [15]). The optimization of functions of permutations in which precedence constraints are imposed is not, in general, an easy task, and special schemes in the unconstrained case, such as suggested by the use of equation (23), are no longer directly applicable, nor are they necessarily easily modified to be applicable. The use of search theory is a viable approach. In fact, the tree structure approach can be employed when precedence constraints exist merely by selecting the neighborhood points to be only those that meet both the rules for constructing the neighborhoods and the precedence constraints. Not only can the neighborhood points be chosen so as to be feasible relative to the precedence relationships, but can be chosen to meet any general type of constraint such as of the form given by equation (4).

To illustrate the use of the tree structure approach in the presence of precedence constraints, suppose n = 4. Assume $2 \prec 3$ and $1 \prec 4$. Thus the feasible neighborhood structures for each of the cases (a)-(c) in Section V are as follows:

(1) p = (1,2,3,4)

Case	$\mathcal{N}(p)$
(a)	$\{(2,1,3,4), (2,3,1,4), (1,2,3,4)\}$
(b)	$\{(1,2,3,4)\}$
(c)	$\{(2,1,3,4), (1,2,4,3), (1,2,3,4)\}$

(2) $p = (2,1,3,4)$

Case	$\mathcal{N}(p)$
(a)	$\{(1,2,3,4),\ (2,1,3,4)\}$
(b)	$\{(2,1,3,4)\}$
(c)	$\{(1,2,3,4),\ (2,3,1,4),\ (2,1,4,3),\ (2,1,3,4)\}$

(3) $p = (2,1,4,3)$

Case	$\mathcal{N}(p)$
(a)	$\{(1,2,4,3),\ (1,4,2,3),\ (2,1,4,3)\}$
(b)	$\{(2,1,4,3)\}$
(c)	$\{(1,2,4,3),\ (2,1,3,4),\ (2,1,4,3)\}$

(4) $p = (2,3,1,4)$

Case	$\mathcal{N}(p)$
(a)	$\{(2,3,1,4)\}$
(b)	$\{(1,2,3,4),\ (1,4,2,3),\ (2,3,1,4)\}$
(c)	$\{(2,1,3,4),\ (2,3,1,4)\}$

(5) $p = (1,2,4,3)$

Case	$\mathcal{N}(p)$
(a)	$\{(2,1,4,3),\ (1,2,4,3)\}$
(b)	$\{(1,2,4,3)\}$
(c)	$\{(2,1,4,3),\ (1,4,2,3),\ (1,2,3,4),\ (1,2,4,3)\}$

(6) $p = (1,4,2,3)$

Case	$\mathcal{N}(p)$
(a)	$\{(1,4,2,3)\}$
(b)	$\{(2,1,4,3),\ (2,3,1,4),\ (1,4,2,3)\}$
(c)	$\{(1,2,4,3),\ (1,4,2,3)\}$

If the objective function is defined as in equation (6) and we set $a_1 = 2$, $a_2 = 1$, $a_3 = 3$, $a_4 = 4$, $b_1 = 3$, $b_2 = 6$ and $b_3 = b_4 = 2$, then we obtain

F(1,2,3,4) = 100	F(2,3,1,4) = 104
F(2,1,3,4) = 109	F(1,2,4,3) = 98
F(2,1,4,3) = 107	F(1,4,2,3) = 76

Clearly, the global minimum is given by the permutation (1,4,2,3), however in terms of a choice of the neighborhood structure the resulting ultimate successors are as follows:

Permutation	Neighborhood Structure (a)	(b)	(c)
(1,2,3,4)	(1,2,3,4)	(1,2,3,4)	(1,4,2,3)
(2,1,3,4)	(1,2,3,4)	(2,1,3,4)	(1,4,2,3)
(2,1,4,3)	(1,2,4,3)	(2,1,4,3)	(1,4,2,3)
(2,3,1,4)	(2,3,1,4)	(1,2,3,4)	(2,3,1,4)
(1,2,4,3)	(1,2,4,3)	(1,2,4,3)	(1,4,2,3)
(1,4,2,3)	(1,4,2,3)	(1,4,2,3)	(1,4,2,3)
Number of locally minimal points =	4	5	2

Another possible approach to solving the problem in the presence of precedence constraints is to divide the set of possible permutations into subsets which we shall refer to as "form classes", that is, all elements of a particular subset will have (in a to be defined sense) the "same form". To illustrate this concept, suppose we are given a set of precedence constraints. Arbitrarily position the integers constrained by these precedence relationships in an n-tuple so that all the precedence constraints are observed. Leave all other entrics in the n-tuple unfilled. Denote this n-tuple form by T_1 and let S_1 be the set of all permutations of the integers 1,2,...,n which have the form prescribed by T_1. Now, repeat this process to generate distinct forms $T_2,\ldots, T_K$ and corresponding sets, called form classes, $S_2,\ldots, S_K$, where K represents the total number of distinct permutation forms that are possible due to the given precedence constraints. Define p_i^* for

$i=1,\ldots, K$ by the condition $F(p_i^*) = \min_{p \in S_i} F(p)$. Then, if $F(p^*) = \min_{p \in \bigcup_i S_i} F(p)$, we must have

(57) $$F(p^*) = \min_{i=1,\ldots,K} F(p_i^*);$$

that is, to find the optimal permutation p^* we need only to find the optimal permutation for each of the form classes $S_1,\ldots, S_K$.

If there is only one precedence constraint, say $i \prec j$, then the number of form classes is $K = \binom{n}{2}$. This means that for n=20 there are 190 subsets of permutations that need to be examined --- a potentially better starting point than examining each of the at most 20!/2 possibilities. If the precedence constraint is strict, denoted by $i \preceq j$, that is, i must immediately precede j, then one merely replaces the ordered pair (i,j) by a 2-wise equivalent element and solves the remaining problem as an unconstrained one involving n-1 integers. The potential difficulty here is in finding the equivalent element, because the objective function need not have one of the simple forms such as given by equations (45) or (52).

An important observation to be made that can be useful when considering form classes involves the ordering of elements between two fixed ones. Specifically, suppose we are given a permutation of the form $(i_1,\ldots, i_k,\ldots, i_m,\ldots, i_n)$ where the kth and m^{th} components are fixed at least one entry apart. If i_k and i_m are immediately adjacent, then we merely replace them by an (2-wise) equivalent element (or job) according to Definition 4. From equation (23) it is easily demonstrated that we can improve (in the sense of make smaller) upon the value of F merely by choosing the integers $i_1,\ldots, i_{k-1}, i_{k+1},\ldots, i_{m-1}, i_{m+1},\ldots, i_n$ to satisfy the following conditions:

(58) $$G(i_1)/f(i_1) \geq \ldots \geq G(i_{k-1})/f(i_{k-1})$$

(59) $$G(i_{k+1})/f(i_{k+1}) \geq \ldots \geq G(i_{m-1})/f(i_{m-1})$$

(60) $$G(i_{m+1})/f(i_{m+1}) \geq \ldots \geq G(i_n)/f(i_n)$$

where $G(i_j) = -h(i_j)$ when F is given by (45) and $G(i_j) = 1 - h(i_j)$ when F is given by (52). The preceding observation holds, of course, for the case when F is given by (22) and the associated conditions (A) and (B) are satisfied. In other words, when the Switch-and-Check technique is applicable, we order elements between two fixed elements according to the resulting rule based on the pair-wise interchange of adjacent elements. In particular, if there are only two fixed elements and in the form considered they appear at the endpoints of the permutation, we merely order the remaining n-2 elements according to equation (23), provided it is applicable, to obtain the optimal permutation p_i^* for that form class.

The next question to address, given that the terms are ordered according to (58)-(60), is whether or not we can make the value of F even smaller by switching terms on either side of i_k or on either side of i_m. This question is, in general, not easily answered, however, in a simple case, Smith's g-function concept finds application. Suppose first that F is given by equation (45) and observe that, if we interchange i_{k-1} and i_{k+1}, then we obtain the following difference:

$$(61)\qquad \begin{aligned} &F(i_1,\ldots, i_{k-2}, i_{k-1}, i_k, i_{k+1}, i_{k+2},\ldots, i_n) \\ &-F(i_1,\ldots, i_{k-2}, i_{k+1}, i_k, i_{k-1}, i_{k+2},\ldots, i_n) \\ &\doteq [h(i_k)f(i_{k+1}) + h(i_{k-1})(f(i_k) + f(i_{k+1}))] \\ &\quad - [h(i_k)f(i_{k-1}) + h(i_{k+1})(f(i_k) + f(i_{k-1}))] \end{aligned}$$

$$(62)\qquad = g(k+1, k-1) - g(k-1, k+1)$$

where we define the g-function by

$$(63)\qquad g(s,t) = h(i_k)f(i_s) + h(i_t)(f(i_k) + f(i_s))$$

The rule then is to switch i_{k-1} and i_{k+1} if $g(k+1, k-1) \geq g(k-1, k+1)$, and to leave i_{k-1} and i_{k+1} alone otherwise. In the case where F is given by equation (52), the corresponding difference is

$$(64)\qquad \begin{aligned} &F(i_1,\ldots, i_{k-2}, i_{k-1}, i_k, i_{k+1}, i_{k+2},\ldots, i_n) \\ &-F(i_1,\ldots, i_{k-2}, i_{k+1}, i_k, i_{k-1}, i_{k+2},\ldots, i_n) \end{aligned}$$

$$= \left(\prod_{j=1}^{k-2} h(i_j)\right)\left(f(i_{k-1}) + f(i_k)h(i_{k-1}) + f(i_{k+1})h(i_{k-1})h(i_k)\right.$$

$$\left. -f(i_{k+1}) - f(i_k)h(i_{k+1}) - f(i_{k-1})h(i_{k+1})h(i_k)\right)$$

(65) $$= \left(\prod_{j=1}^{k-2} h(i_j)\right)\left(g(k+1, k-1) - g(k-1, k+1)\right)$$

where, in this case, we define the g-function by

(66) $$g(s,t) = f(i_t) + h(i_t)(f(i_k) + f(i_s)h(i_k)).$$

Thus, the rule is as before, namely, switch i_{k-1} and i_{k+1} if $g(k+1, k-1) \geq g(k-1, k+1)$, and to leave i_{k-1} and i_{k+1} alone otherwise.

On the other hand, it may be more profitable to switch i_{k+1} with i_{k-2} or i_{k+1} with i_{k-3}, etc. One needs to explore whether or not i_{k+2} should be moved over i_k, as well as moving as many as possible of $i_{k+3}, i_{k+4}, \ldots, i_{m-1}$ over i_k. General rules for performing this jumping over fixed elements are not known, but the use of these partial results can be illustrated for a simple example.

Let the objective function F have the form given by equation (6) where n=4, the a_i's and b_i's are as given below and suppose we must have $1 \prec 4$.

i	a_i	b_i	a_i/b_i
1	2	3	2/3
2	1	6	1/6
3	3	2	3/2
4	4	2	2

In this case there are six possible forms, namely:

$$T_1 = (1, 4, -, -)$$
$$T_2 = (1, -, 4, -)$$
$$T_3 = (1, -, -, 4)$$

$$T_4 = (-, 1, -, 4)$$
$$T_5 = (-, -, 1, 4)$$
$$T_6 = (-, 1, 4, -)$$

Since $a_3/b_3 > a_2/b_2$, then by application of equation (30), it follows that $p_1^* = (1, 4, 3, 2)$ with $F(p_1^*) = 53$, $p_3^* = (1, 3, 2, 4)$ with $F(p_3^*) = 84$ and $p_5^* = (3, 2, 1, 4)$ with $F(p_5^*) = 88$. Jumping must be considered for the forms T_2, T_4 and T_6. In the case of T_2, we define

(67) $\qquad g(2,3) = b_4a_2 + b_3(a_4 + a_2)$

and

(68) $\qquad g(3,2) = b_4a_3 + b_2(a_4 + a_3)$

Since $g(3,2) - g(2,3) = 36$, we must have 3 precede 2; hence $p_2^* = (1, 3, 4, 2)$ with $F(p_2^*) = 62$. In the case of T_4, we define

(69) $\qquad g(2,3) = b_1a_2 + b_3(a_1 + a_2)$

and

(70) $\qquad g(3,2) = b_1a_3 + b_2(a_1 + a_3)$

Since $g(3,2) - g(2,3) = 30$, we must again have 3 precede 2; hence $p_4^* = (3, 1, 2, 4)$ with $F(p_4^*) = 79$. Similarly, replacing the ordered pair (1, 4) by an equivalent element and using (63), we obtain $p_6^* = (3, 1, 4, 2)$ and $F(p_6^*) = 57$. We conclude that $p^* = p_1^*$ since $F(p_1^*) = \min_{i=1,\ldots,6} F(p_i^*)$.

VIII. GENERAL CONSTRAINTS

The general optimization problem with constraints, as described by (3)-(5), is not readily solvable except in very special cases such as when there exist contiguity and/or precedence constraints. However, the special Lagrange multiplier method due to H. Everett (see [12]) has applicability. This method requires introducing non-negative real numbers $\lambda_1, \ldots, \lambda_m$ and minimizing the unconstrained objective function

$$F(i_1,\ldots, i_n) - \sum_{j=1}^{m} \lambda_j H_j(i_1,\ldots, i_n)$$

The optimal point, say $(i_1^*,\ldots, i_n^*)$, then provides the solution to the problem

(71) Minimize $F(i_1,\ldots, i_n)$

subject to

(72) $H_j(i_1,\ldots, i_n) \leq H_j(i_1^*,\ldots, i_n^*)$ for $j=1,\ldots, m$

(73) $(i_1,\ldots, i_n)$ is a permutation of the integers 1, 2,..., n.

The trick here is to choose (if possible) the non-negative λ_j's so that

(74) $H_j(i_1^*,\ldots, i_n^*) = C_j$ for $j=1,\ldots, m$.

Unfortunately, this unconstrained objective function may be even more difficult to handle than the original function $F(i_1,\ldots, i_n)$. Furthermore, the presently most useful tool represented by the Switch-and-Check technique may not be applicable. Of course, the search method given by the construction of a tree structure could be employed.

REFERENCES

[1] Bellman, R.E., Dynamic Programming, Princeton University Press, Princeton, N.J., 1957

[2] Boothroyd, H., "Least-Cost Testing Sequence", Oper. Res. Quart., Vol. 11 No. 3, Sept. 1960

[3] Bowdon, E.K., "Priority Scheduling in a Network of Computers", IEEE Trans. on Comp., Vol. C-18 No. 11, Nov. 1969, pp. 1021-1026

[4] Conway, R.W., W.L. Maxwell and L.W. Miller, Theory of Scheduling, Addison-Wesley, Reading, Mass., 1967

[5] Dean, B.V., "Low Cost-High Risk Sequencing of Research Tasks", Opsearch, Vol. 5 No. 3, Sept. 1968

[6] Demski, J.S., "Optimizing the Search for Cost Deviation Sources", Man. Sci., Vol. 16 No. 8, April 1970, pp. 486-494

[7] Denby, D.C., "Minimum Downtime as a Function of Reliability and Priority Assignments in a Component Repair", Jour. of Ind. Eng., Vol. 18 No. 7, July 1967, pp. 436-439

[8] Elmaghraby, S.E., "The Sequencing of "Related" Jobs", Nav. Res. Log. Quart., Vol. 15 No. 1, March 1968, pp.23-32

[9] Elmaghraby, S.E., "The Machine Sequencing Problem - Review and Extensions", Nav. Res. Log. Quart., Vol. 15 No. 2, June 1968, pp. 205-232

[10] Elmaghraby, S.E., "A Graph Theoretic Interpretation of the Sufficiency Conditions for the Contiguous-Binary-Switching (CBS)-Rule", Oper. Res. Report No. 68, Nov. 3, 1970, North Carolina State University

[11] Emmons, H., "One-Machine Sequencing to Minimize Certain Functions of Job Tardiness", Oper. Res., Vol. 17 No. 4, July-Aug. 1969, pp. 701-715

[12] Everett, H., "Generalized Lagrange Multiplier Method for Solving Problems of Optimum Allocation of Resources", Oper. Res., Vol. 2, 1963, pp. 399-417

[13] Firstman, S.I. and B. Gluss, "Optimum Search Routines for Automatic Fault Location", Oper. Res., Vol. 8 No. 4, July-Aug. 1960, pp. 512-523

[14] Flood, M.M., "The Traveling - Salesman Problem", Oper. Res., Vol. 4, 1956, pp. 61-75

- [15] Gapp, W., P.S. Mankekar and L.G. Mitten, "Sequencing Operations to Minimize In-Process Inventory Costs", Man. Sci., Vol. 11 No. 3, Jan. 1965, pp. 476-484

[16] Held, M. and R.M. Karp, "A Dynamic Programming Approach to Sequencing Problems", J. Soc. Indust. Appl. Math., Vol. 10 No. 1, March 1962, pp. 196-210

[17] Johnson, S.M., "Optimal Two-and Three-Stage Production Schedules with Setup Times Included", Nav. Res. Log. Quart., Vol. 1 No. 1, March 1954, pp. 61-68

[18] Mankekar, P.S. and L.G. Mitten, "The Constrained Least Cost Testing Sequence Problem", Jour. of Ind. Eng., Vol. 16 No. 2, Mar.-Apr. 1965, pp. 146-149

[19] Mitten, L.G., "An Analytic Solution to the Least Cost Testing Sequence Problem", Jour. of Ind. Eng., Vol. 11 No. 1, Jan.-Feb. 1960

[20] Rau, J.G., "Minimizing a Function of Permutations of n Integers" Oper. Res., Vol. 19 No. 2, Mar.-Apr. 1971, pp. 237-240

[21] Reiter, S. and G. Sherman, "Discrete Optimizing", J. Soc. Indust. Appl. Math., Vol. 13 No. 3, Sept. 1965, pp. 864-889

[22] Reynolds, G.H., "A Shuttle Car Assignment Problem in the Mining Industry", Man. Sci., Vol. 17 No. 9, May 1971, pp. 652-655

[23] Smith, W.E., "Various Optimizers for Single-Stage Production", Nav. Res. Log. Quart., Vol. 3, 1956, pp. 59-66

[24] Wagner, H.M., Principles of Operations Research, Prentice-Hall, Englewood Cliffs, N.J. 1969

DISCUSSION OF

"SELECTED COMMENTS CONCERNING OPTIMIZATION THEORY FOR FUNCTIONS OF PERMUTATIONS"

Kenneth R. Baker

North Carolina State University, Raleigh, North Carolina, USA

The overview presented in this survey paper is quite broad, and its major achievement is to draw together a variety of perspectives on the problem of determining an optimal permutation. By incorporating these perspectives in one review and by including an extensive array of examples, the paper depicts the richness and the scope of the problem. Nevertheless, the paper's coverage of the topic is uneven, and it avoids evaluating or interpreting the diverse material it has drawn together.

Seldom is it practical to solve combinatorial problems of even moderate size by complete enumeration, and this is the challenge facing optimization theory for function of permutations. The motivating problem is one of practicability and is posed quite succinctly in the opening paragraph by the interpretation of the number 20! But while the research in this area is aimed at developing more efficient methods, any review would be flawed if it did not begin to come to grips with the question of what is "efficient." A popular definition, for instance, is to designate a combinatorial optimization procedure as efficient if its computational requirement is bounded by a polynomial function of the problem size. Since no comparable definition is proposed in the paper, it is inappropriate to say (page 1) that "efficient algorithms essentially exist in all cases but the [traveling salesman problem]." For example, there are some job sequencing problems referenced in the paper for which only implicit enumeration algorithms have yet been demonstrated. On a broader level, the notion of efficiency may also involve a trade-off between the utility of rapid, suboptimal procedures and the marginal improvements which could be expected from optimizing techniques. This broader concept of efficiency certainly needs to be treated more explicitly before one can fully assess the search theory approach of Reiter and Sherman.

The collection of examples (pages 2-7) is an impressive set of combinatorial problems, though it could be expected that researchers with specialized interests could extend the reference list in their own area of expertise. For instance, the least-cost testing sequence model (Example 2) has also been adapted in information retrieval [1], manpower training [2], file organization [3], and sequencing computations [4].

The section on the "state-of-the-art" dramatizes how primitive that state currently is. Table 1 in the paper graphically indicates the lack of analytic results for the permutation problem, but the significance of the theorems due to Elmaghraby and to Rau (page 9) is much less obvious. The paper should have included some perspective on these two theoretical properties: Do they provide any useful insight beyond what is learned by using switch-and-check? Have they spawned new solutions to previously intractable problems? Just what is their contribution to the research on finding efficient solutions?

The section on solution approach describes the switch-and-check method and the dynamic programming procedure. Since the former is optimal only for certain special cases, while the latter is a very general solution procedure, the balanced treatment of the two may be misleading. There is more to be said about the computational demands of dynamic programming, and a more complete discussion would have been a good opportunity to help define "efficient" (or to define "inefficient").

The presentation of the search theory approach is useful, although it is really the properties (not their proofs) that are of primary importance. At the same time, the paper pays little attention to the fact that, for a given neighborhood structure, this technique will not guarantee a global optimum. This raises the interesting question of whether some neighborhood structures are "better" than others, and leaves open the development of effective computational strategies for the search theory approach in situations where many local optima exist.

When precedence constraints are added to the formulation, optimization is usually more difficult, but the importance of the problem cannot be understated. Obviously, the existence of precedence constraints reduces the feasible region, although it may render the systematic exploration of the feasible region more difficult. The hope would be to use the set of precedence relationships to reduce the enumerative search for an optimum. In fact, the work of Srinivasan [5] has exposed a remarkable link between precedence-related structures and the unconstrained problem. In his hybrid algorithm for the (unconstrained) single-machine tardiness problem, Srinivasan employed the theorems of Emmons in order to impose a partial ordering on the tasks to be sequenced. The resulting constrained problem was solved by a dynamic programming procedure that was especially adapted for the reduced feasible region. The success of Srinivasan's work suggests the following two-phase approach for certain unconstrained problems:

1. Develop a means for imposing a partial ordering on the elements of the problem in order to create a relatively small "sufficient region" in which an optimum is known to exist.
2. Solve the resulting constrained problem by taking advantage of the reduction in problem size obtained with the partial ordering.

This type of approach to the unconstrained problem will place a substantial premium on efficient solution mechanisms for the precedence-constrained problem.

Two significant topic areas have been entirely omitted from this discussion but are intimately related to the problem of finding "efficient" solutions. If one considers only optimizing procedures, a viable solution strategy is often the method of branch and bound, which has been widely used for solving permutation problems. Indeed, one of the landmark papers on the subject [6] dealt with the traveling salesman problem and reported a certain kind of efficiency. In addition, branch and bound techniques have been developed for a variety of problems involving optimal permutations (e.g., [7], [8], [9]).

Secondly, if the scope of the review is to include efficient solution techniques which do not guarantee optimality (such as the Reiter-Sherman approach) then another very important candidate is random sampling [10]. While the major accomplishments of sampling approaches to combinatorial problems have been reported in more complicated formulations (job shop scheduling [11], project scheduling [12], line balancing [13], etc.), the technique is quite suitable as a heuristic procedure for determining an optimal permutation. In fact, the Karg-Thompson heuristic for the traveling salesman problem [14] represents a well-known instance in which a sampling scheme was effectively used in solving a permutation problem. Such results suggest that research aimed at justifying the usefulness of any suboptimal procedure for this type of problem should include an experimental comparison with sampling as part of the justification.

In summary, for a review to do justice to the topic of finding an optimal permutation, it must deal directly with the concept of algorithm efficiency, and should certainly include branch and bound techniques. A thorough treatment of these two significant topics would have made the survey much more complete.

REFERENCES

1. Baker, N. R. "Optimal User Search Sequences and Implications for Information Systems Operation," American Documentation, Vol. 20, No. 3 (July 1969) pp. 203-212.

2. Grieshop, D. S. "An Analysis of the System Effectiveness of a Sequential Manpower Training Model," M.S. Thesis, School of Industrial and Systems Engineering, Georgia Institute of Technology (January 1972).

3. Merten, A. G. "Some Quantitative Techniques for File Organizations," Technical Report No. 15, University of Wisconsin Computing Center (June 1970).

4. Riesel, H. "In Which Order Are Different Conditions to be Examined?", BIT, Vol. 3, (1963), 255-256.

5. Srinivasan, V. "A Hybrid Algorithm for the One Machine Sequencing Problem to Minimize Total Tardiness," Naval Research Logistics Quarterly, Vol. 18, No. 3, (September 1971), pp. 317-327.

6. Little, J. D. C., Murty, K. G., Sweeny, D. W., and Karel, C. "An Algorithm for the Traveling-Salesman Problem," Operations Research Vol. 11, No. 6, (November 1963).

7. Ignall, E. and Schrage, L. "Application of the Branch-and Bound Technique to Some Flow-Shop Scheduling Problems," Operations Research, Vol. 13, No. 3, (May 1965).

8. Pritsker, A., Miller, L. and Zinkl, R. "Sequencing n Products Involving m Independent Jobs on m Machines, "AIIE Transactions, Vol. 3, No. 1, (March 1971).

9. Shwimer, J. "On the n-Job, One Machine, Sequence-Independent Scheduling Problem with Tardiness Penalties: A Branch-Bound Solution," Management Science, Vol. 18, No. 6, (February 1972), pp. 301-313.

10. Elmaghraby, S. E. "The Machine Sequencing Problem--Review and Extensions," Naval Research Logistics Quarterly, Vol. 15, No. 1, (March 1968).

11. Conway, R. W., Maxwell, W. L., and Miller, L. W. Theory of Scheduling, Addison-Wesley, Reading, Mass., 1970.

12. Wiest, J. D. "A Heuristic Model for Scheduling Large Projects with Limited Resources," Management Science, Vol. 13, No. 6, (February 1967), pp. B359-B377.

13. Mastor, A. A. "An Experimental Investigation and Comparative Evaluation of Production Line Balancing Techniques," Management Science, Vol. 16, No. 11, (July 1970).

14. Karg, R. L. and Thompson, G. L. "A Heuristic Approach to Solving Traveling Salesman Problems," Management Science, Vol. 10, No. 2, (January 1964).

AN OUT-OF-KILTER APPROACH FOR MACHINE SEQUENCING PROBLEMS

Said Ashour and R. G. Parker

Iowa State University, Ames, Iowa, USA
Georgia Institute of Technology, Atlanta, Georgia, USA

The primary objective of this paper is to develop an out-of-kilter algorithm to deal with the machine sequencing problem in which both job precedence and machine non-interference constraints are involved. This approach is inspired by the graph theoretical representation of the problem. In addition to the out-of-kilter concept, the algorithm employs a waiting feature and pertinent criteria for the selection of resolutions. The basic concepts of the approach are presented and the entire algorithm is then summarized and condensed to five basic steps. A sample problem is presented and manually solved to illustrate the technique to job-shop problems. Computational experiences as well as comparative evaluations of other existing techniques are included. Finally, a diagnostic view of the scheduling problem is reported with reference to specific characteristics which comprise a major share of the difficulty surrounding the problem.

INTRODUCTION

The job-shop scheduling problem arises whenever it is necessary to schedule a set of J jobs on M machines, so that the sequence of such jobs is optimal or near optimal with respect to a given measure of performance such as schedule time. The order in which the various operations concerning each job are to be performed is fixed by the technological process and the associated processing times are known.

In recent years there has been a great increase in the interest surrounding the shop scheduling problem from both its practical as well as theoretical view. Such a phenomena is due, largely to the massive computational effort associated with the problem arising from its inherent combinatorial property. Such an attribute and its corresponding ramifications have been diluted through the employment of the high speed computer. However, it still remains that the shop scheduling problem commands a computational degree of severity that renders many approaches to its solution, immobile.

The purpose of this paper is to report on the development of and computational experience with an algorithm for obtaining an optimal or near optimal solution to the shop scheduling problem. The algorithm, based upon certain graph-theoretic aspects, embodies an out-of-kilter concept, a waiting feature, and a specific criteria, based upon idle and delay times, for resolution of conflicts. Computational experience is reported such that various published problems have been solved and compared to other existing techniques employed in similar treatises. Finally, a diagnostic view of the shop scheduling problem is reported, with reference to specific characteristics which comprise a major share of the difficulty surrounding the problem. Moreover a critical appraisal of the proposed algorithm is suggested in order to illucidate its inherent capabilities and weaknesses.

AN OUT-OF-KILTER ALGORITHM

In order to derive an algorithm for solving the job-shop problem, it becomes necessary to view certain graph-theoretic concepts, the uses of which can be applied explicitly to the traditional shop scheduling formulation. Let $G(N, A)$ be a finite directed graph such that N is the set of nodes given by the operations comprising the specific problem and A, the set of arcs which define various relationships of precedence between nodes in N. Further, consider specific decompositions of the sets N and A. First, the set of nodes, N can be decomposed into particular subsets M_j and J_m such that M_j represents a set consisting of those nodes indicative of like-jobs and J_m, the set of nodes representing like-machines. Clearly, there will be J and M sets, M_j and J_m, respectively for a given problem of size J $\times$ M. It is obvious that all sets M_j as well as those, J_m are disjoint such that

$$M_1 \cap M_2 \cap \ldots \cap M_J = \phi,$$

$$J_1 \cap J_2 \cap \ldots \cap J_M = \phi,$$

The subsets arising from N are relatively uninteresting in that they have a nearly insignificant bearing upon the operative characteristics of the algorithm. However, the set A, as will become more explicit, embodies particular characteristics sharply related to the construction of and manipulation by the proposed algorithm. Explicitly, A can be decomposed into two sets, C and D. The set C consisting of conjunctive arcs, is constructed from the technological requirements given by the problem definition. Elements of C arise from specific relationships among those nodes in M_j, for all j. The set D consists of disjunctive arcs. Initially, D contains arcs connecting all pairs of nodes in J_m, for all m. It is clear that C and D are disjoint and that $C \cup D = A$ for a given problem. Moreover, it emerges that the set C will contain exactly $J(M + 1)$ arcs where N is supplemented with initial and terminal nodes. The total arcs in D is $2M\binom{J}{2}$.

Given the structure and composition of the sets N and A, it remains to be demonstrated that a particular problem can be constructed graphically from particular N and A. Consider a (3 x 3) problem illustrated in Figure 1. Let the sets M_j and J_m be given such that $M_1 = \{(1), (4), (7)\}$, $M_2 = \{(5), (2), (8)\}$, $M_3 = \{(3), (6), (9)\}$, $J_1 = \{(1), (2), (3)\}$, $J_2 = \{(4), (5), (6)\}$, and $J_3 = \{(7), (8), (9)\}$. Note that N has been updated to include (0) and (T). The graph given in Figure 1 is said to be a mixed graph since its structure contains arcs from C as well as from D. In general, any shop scheduling graph will be mixed, where alternative graphs indicative of other combinatorial models, may arise as either conjunctive or disjunctive, the implication of which is that they consist of arcs solely from C or from D, respectively.

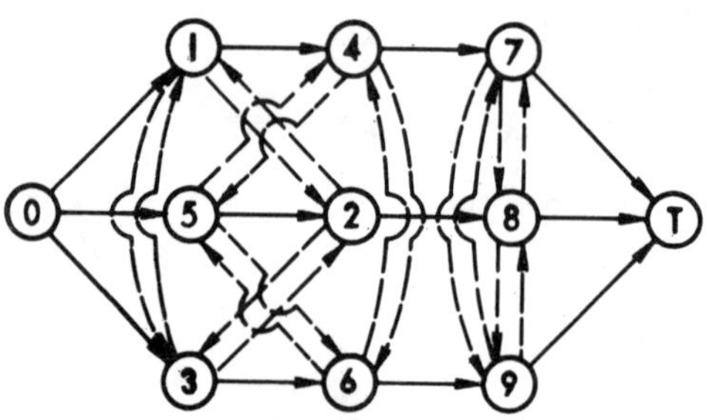

Fig. 1. Graph Formulation of (3 x 3) Sample Problem.

Given the initial structure of the shop scheduling problem as a mixed graph, $G(N, C, D)$ it remains to be conceptualized how a solution can be synthesized from the initial structure to one such that a particular conjunctive graph, $G(N, C)$ emerges and is indicative of the particular solution. Explicitly, an iterative scheme is introduced such that successive selections from D are performed based upon a particular selection criterion. The selections and corresponding deletions from a current set, A continues until a point is reached where a solution emerges from a particular current graph.

The synthesis suggested is accomplished by employing the concept of the kilter state. Specifically, a given kilter state for a node and hence, a graph is defined such that at any current state in the synthesis, a given node is either in-kilter or out-of-kilter. If any node is out-of-kilter, the current graph is said to be out-of-kilter. Only when all nodes are in-kilter does the iterative scheme terminate, such that a first point of feasibility arises. Further, a solution will have been attained and will be given by the final graph.

In order to define particular kilter states, it is necessary to consider graphically the form of a solution to the shop scheduling problem. Given that the nature of a solution is a feasible sequence of nodes in J_m for all m, it is clear that such a sequence can be given by a particular connected graph depicting transitenee.elac tionships arising from a given sequence of all nodes (k) in each J_m. Further, such graphs can be identified as a special form of a tree known as an arborescence. It is obvious that each arborescence will contain J-1 arcs and correspondingly, there will be M arborescences given by a particular solution. Nevertheless, given that certain arborescences are sought, it is clear that three particular node states arise which define a kilter state. Depicted in Figure 2, the states are such that they are determined by particular functions of incidence created by arcs in $\mathcal{D}$. For example, the first node state in the figure is out-of-kilter, since it is null of any incidence from arcs in $\mathcal{D}$. The second and third cases are also out-of-kilter, since they possess indices of incidence in excess of that allowed by definition of the particular arborescence structure indicative of a solution. Any node in a current graph possessing one of the states in Figure 2, is said to be out-of-kilter.

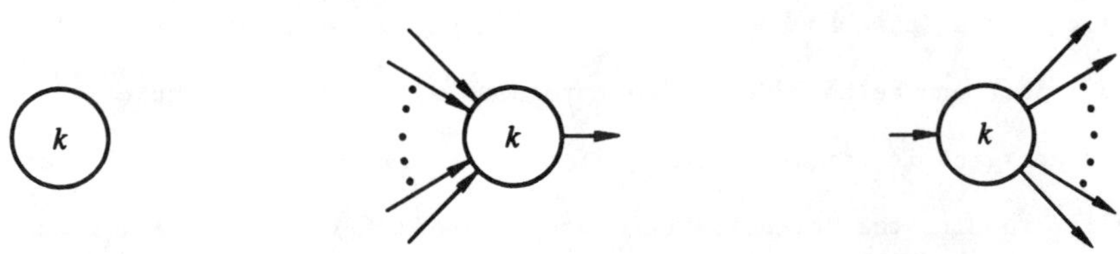

Fig. 2. Typical Out-of-Kilter States for Nodes in Shop Scheduling Formulation.

Otherwise, the node is in-kilter. Recall, that kilter states are derived from conditions created by current arcs in $\mathcal{D}$ only.

From an initial graph, consisting of only arcs in $\mathcal{C}$, an initial selection from $\mathcal{D}$ is performed. If exactly M(J - 1) arcs from $\mathcal{D}$ arise initially, the graph is said to be saturated. However, it is rare that at the initial stages of computation, a state of saturation arises. Hence, in most cases, a current graph will be unsaturated initially as well as at early stages of the iterative scheme. Ultimately, however, a point of saturation must arise such that exactly M(J - 1) arcs from $\mathcal{D}$ arise. It is clear that a solution can only be obtained from a saturated graph.

In essence, the computational algorithm, to be presented and demonstrated shortly, consists of two general phases. The first involves a consequetive updating scheme embodied by a series of successive deletions and/or additions of pertinent arcs from the disjunctive set. The second involves an evaluative phase such that after a current update to the graph, the kilter state of the graph is considered. If the kilter state of the graph is other than zero, the scheme proceeds through another iteration. Otherwise, the procedure terminates. It can be easily proven that the procedure will converge to a point of termination.

At any iteration n, an update to the current graph, $G_n(\mathcal{N}, \mathcal{A})$ is conceived by first identifying a particular candidate set of nodes, C_n of size J. It is evident that at any current state, C_n will be such that several nodes may be in conflict. A state of conflict in C_n exists if more than one node from given J_m is included in a current C_n. The resolution of such conflicts is accomplished by employing specific criteria, the more desirable evaluations of which with reference to each node in conflict, provides the basis for resolution. Clearly, the employment of a powerful conflict resolution measure is necessary.

An additional feature can be introduced which considers, rather than nodes in conflict, these that are not in conflict. Introduced as a "waiting feature", an added aspect can be embedded in the proposed algorithm such that a given node (k) not in conflict is removed from C_n of state n, if the resolution of certain nodes in conflict results the potential candidacy of node (ℓ) at state n + 1 such that (k), (ℓ) J_m. The employment of the waiting feature is considered in order to enhance the quality of solution rendered by the proposed scheme. However, it is clear that the computational demands may be increased somewhat by virtue of such inclusion.

The formal step-by-step computational algorithm can now be given as well as its demonstration with a sample problem.

Step 1: Construct the initial graph.

1.1 Construct J linear graphs, from the machine orderings such that

$(0) \ll (k) \ll (\ell) \ll .. \ll (T) \mid (k), (\ell) \in M_j; \forall j, (k, \ell) \in C,$

and (0) and (T) are initial and terminal nodes, respectively.

1.2 Label all arcs in A such that

$$f(k, \ell) = t(k), (k) = 1, 2, \ldots, N,$$

and

$$f(0, k) = 0, \forall (k).$$

1.3 Select the initial set of arcs (k, ℓ) such that

$$(k, \ell) \in D, (k), (\ell) \in J_m, \forall m,$$

where $(0, k)$ exists for single (k) in specific J_m.

1.4 Set $n = 0$, $C_n = \{\phi\}$, $N_1 = \{(0), (k)\}$, $N_2 = N - N_1$.

Step 2: Construct the candidate set C_n.

2.1 Set $n = n + 1$ and scan each linear graph for nodes (k) such that

$$(\ell) \ll (k), (\ell) \in N_1, (k) \in N_2.$$

Set (k) into C_n.

2.2 Update C_n such that nodes (x) are removed, where

$$(k) \in C_n; (k) \ll (h); (h), (x) \in J_m, N_2,$$

and (x) is the only node from specific J_m in C_n.

2.3 If more than one node (k) from a given J_m exists in C_n, resolve the conflict by a particular rule. Then update N_1 and N_2.

Step 3: Update the graph.

3.1 If $G_n(N, A)$ is unsaturated over given J_m, update the graph such that

$$A: = A + (k, \ell), (k) \in C_n, J_m, (\ell) \in J_m, N_2.$$

3.2 If $G_n(N, A)$ is saturated over given J_m, update the graph such that

$$A: = A + (k, \ell), (k) \in C_n, J_m, (\ell) \in J_m, N_2,$$

and

$$A: = A - (h, x), (h) \in J_m, (x) \in J_m, N_2, (x) \neq (k).$$

Step 4: Check the kilter state of the current graph.

4.1 Compute Q_n such that

$$Q_n = \sum_{(k) \in N} q(k), \ (k) \neq (0), (T),$$

where

$$q(k) = \begin{cases} d(k) - 1, & d(k) > 1, (k, \ell); (\ell) \in N_2, J_m, \\ 1 & , d(k) = 0, \\ 0 & , \text{otherwise.} \end{cases}$$

and $d(k)$ is the degree of incidence at node (k).

4.2 If $Q_n = 0$, go to Step 5. Otherwise, return to Step 2.

Step 5: Evaluate the final graph

5.1 Determine the final sequence for each set J_m, where

$$S(J_m) = \{(k) \ldots (\ell) \ldots (h)\} \ ; \ m = 1, 2, \ldots, M$$

and

$$s(k) < \ldots < s(\ell) < \ldots < s(h); \forall (k), (\ell), (h) \in J_m.$$

5.2 Compute the schedule time T(S) such that

$$T(S) = s(T),$$

where s(T) is the starting time of the terminal node (T). Further,

$$s(T) = \max_{(k)} [s(k) + t(k)]; \ (k, T).$$

SAMPLE PROGRAM

The above algorithm can now be demonstrated by considering a sample problem such that the following machine ordering and processing time matrices are given.

$$M = \begin{bmatrix} 11 & 12 & 13 \\ 22 & 21 & 23 \\ 31 & 32 & 33 \end{bmatrix}, \qquad J = \begin{bmatrix} 3 & 5 & 1 \\ 4 & 3 & 2 \\ 4 & 4 & 3 \end{bmatrix}$$

This sample problem is given in its graphic context in Figure 1. The particular sets M_j and J_m for all j and m arise as those previously stated.

<u>Step 1</u>. Construct the initial graph. The initial graph can be constructed such that each linear graph is connected to initial and terminal nodes as illustrated in Figure 3. Note the operation indices have been replaced by their node

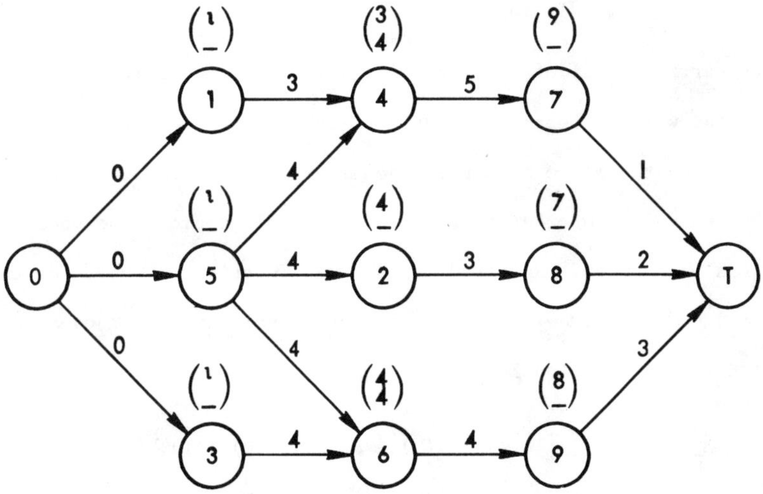

Fig. 3. Initial Graph for Sample Problem.

counterparts such that specific jm in M is replaced by specific (k). All arcs can be labeled such that

$$f(k, \ell) = t(k), \qquad (k) = 1, 2, \ldots, N,$$

and

$$f(0, k) = 0, \qquad (k) \in (0, k).$$

The initial set of arcs are chosen from D such that arcs (5, 4) and (5, 6) are selected. Note that a selection is made, since only node (5) exists alone in specific J_m, among those nodes considered by arcs (0, k). Hence, arcs incident to nodes (1) and (3) are not considered initially. The state or iteration index is set at zero, and the disjoint sets N_1 and N_2 are initialized at $\{(0), (5)\}$ and $N - N_1$, respectively. Note the graph is saturated over J_2.

<u>Step 2</u>. Construct the candidate set C_n. The iteration index is increased to 1 and each linear graph is inspected for those nodes in N_2 which are directly preceded

by nodes in N_1. From the first linear graph, given by the nodes in M_1 along with (0) and (T), it is clear that node (1) is a candidate. Likewise, node (3) becomes a candidate upon inspection of M_2, since both (1) and (3) are directly preceded by node (0). Also, node (2) becomes a candidate, since (5) is included in N_1. Hence, the candidate set can be given such that

$$C_1 = (1),\ (2),\ (3)\ .$$

The current candidate set contains only these nodes in a single set J_1. Consequently, the waiting feature is not applicable and no nodes are removed from C_1. However, since there are multiple nodes from a specific J_m, a conflict exists and must be resolved. In order to resolve the conflict among the nodes in C_1, a criterion Z_n is introduced. Referred to as a gap function, Z_n measures the total machine idle and job delay times created by potential selections of nodes in conflict in current C_n. The function Z_n can be computed such that

$$Z_n = \sum_{(k)\in N} |\ u(k) - v(k)\ |\ ,\quad (k) \neq (0),\ (T),$$

where u(k) and v(k) are the job and machine availability times of each node (k) given from a particular resolution. If the conflict is resolved in favor of node (1), two arcs (1, 2) and (1, 3) are affixed to $G_1(N, A)$ and the values of u(k) and v(k) are computed. The gap $Z_1(1)$ results as 35. By considering partial resolutions in favor of node (2) and (3) such that arcs $\{(2, 1), (2, 3)\}$ and $\{(3, 1), (3, 2)\}$ are affixed respectively, the values of the gap function can be computed such that $Z_1(2) = 68$ and $Z_1(3) = 34$. Hence, the conflict in C_1 is resolved in favor of node (3), since its inclusion into N_1 and its corresponding update with reference to pertinent arc selections, results in minimum Z_1 for all (k) in the conflict set. Hence, N_1 is updated to include (3) and nodes (1) and (2) remain in N_2.

Step 3. Update the graph. Since the current graph is unsaturated, it is clear that no arcs are considered for deletion; however, certain arcs are added with reference to the latest entry into N_1. Hence, the current set of arcs, A is updated such that arcs (3, 1) and (3, 2) are added. The resultant graph would appear as shown in Figure 3.

Step 4. Check the kilter state of the current graph. The kilter state of the graph given in Figure 3 can be computed as the sum of the kilter states of each node in the current graph. From the graph, it is clear that nodes (3), (5), (7), (8), and (9) are out-of-kilter. From the convention adopted previously, it is clear that each out-of-kilter node expresses a kilter number of 1. Hence, Q_1 can be computed such that

$$Q_1 = 0 + 0 + 1 + 0 + 1 + 0 + 1 + 1 + 1 = 5 .$$

Since $Q_1 > 0$, the procedure returns to step 2.

The procedure continues through four additional iterations such that the corresponding candidate sets are checked with reference to nodes in conflict as well as the applicability of the waiting feature. The computation is summarized in Table 1 and the graphs arising from each iteration are given by Figures 4-8.

Table 1. Summary Computation of Sample Problem: Gap Function Application.

n	C_n	Delete (k)	Updated C_n	$Z_n(k)$	N_1	Q_n	T(s)
0					(0),(5)	7	*
1	(1),(2),(3)		(1),(2),(3)	35,68,34	(0),(5),(3)	5	*
2	(1),(2),(6)	(6)	(1),(2)	40,43	(0),(5),(3),(1),(2)	4	*
3	(4),(6),(8)	(8)	(4),(6)	56,39	(0),(5),(3),(1),(2), (4),(6)	3	*
4	(7),(8),(9)		(7),(8),(9)	31,23,19	(0),(5),(3),(1),(2), (4),(6),(9)	1	*
5	(7),(8)		(7),(8)	22,17	(0),(5),(3),(1),(2), (4),(6),(9),(7),(8)	0	14

*Infeasible sequence evaluation; i.e., $Q_n > 0$.

Step 5. Evaluate the final graph. The resultant solution to the sample problem can be obtained by observing the final arborescences given for each set J_m. Hence, for the nodes in J_1, namely (1), (2), and (3), it can be seen from the final graph that an arborescence results such that

$$(3) \rightarrow (1) \rightarrow (2) .$$

Similarly, the nodes in J_2 and J_3 constitute the following arborescences, respectively:

$$(5) \rightarrow (6) \rightarrow (4) \ ,$$
$$(9) \rightarrow (8) \rightarrow (7) \ .$$

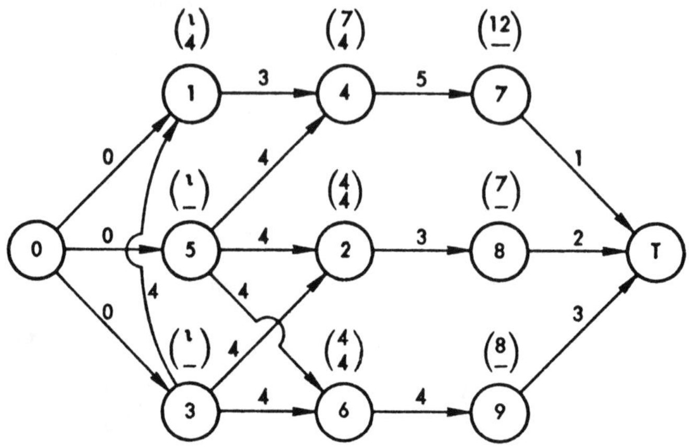

Fig. 4. Intermediate Graph of Sample Problemn n=1.

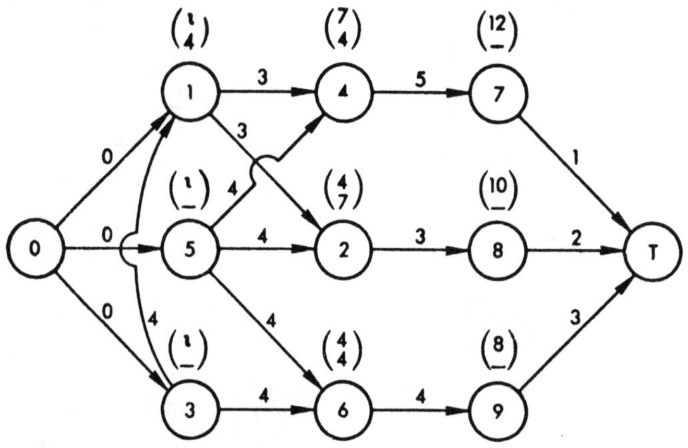

Fig. 5. Intermediate Graph of Sample Problem, n=2.

The evaluation of the structures rendered by the three resultant arborescences and depicted by $G_5(N, A)$, can be obtained by computing $u(k)$ and $v(k)$ for all $(k) \in N_1$, where $(k) \neq (0), (T)$, and observing the starting time of the terminal node $s(T)$ such that

$$s(T) = \max_{(k)} [\max [u(k), v(k)] + t(k)] \ ,$$

for all (k) such that (k, T) exist. Hence,

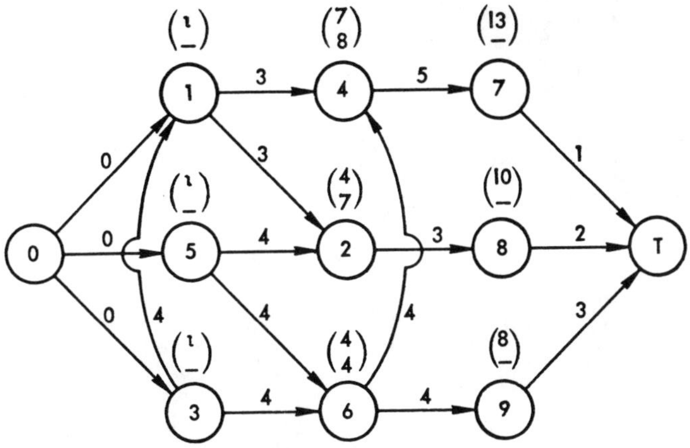

Fig. 6. Intermediate Graph of Sample Problem, n=3.

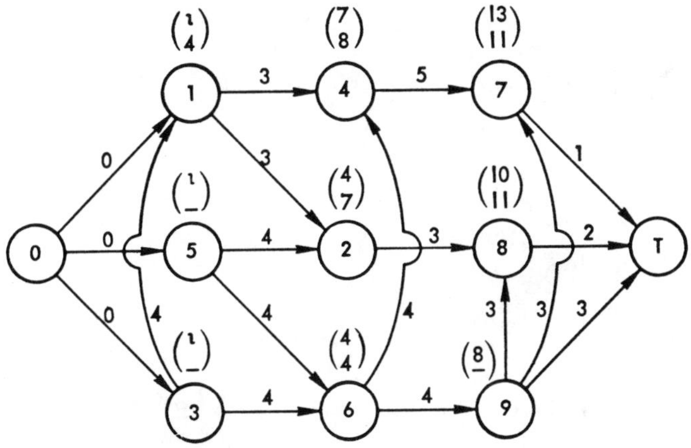

Fig. 7. Intermediate Graph of Sample Problem, n=4.

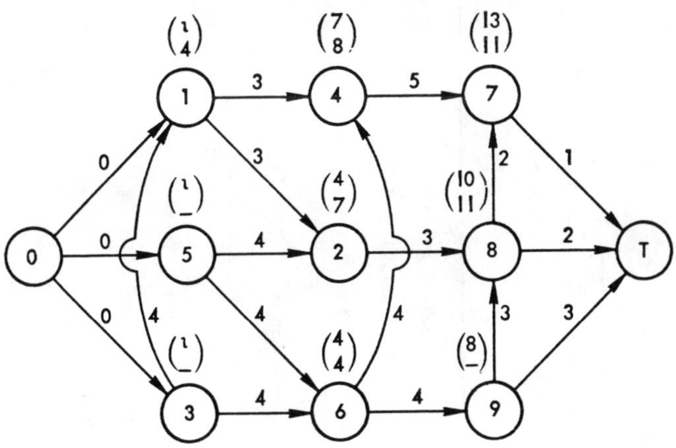

Fig. 8. Final Graph of Sample Problem, n=5.

Table 2. Comparative Evaluation of Various Techniques for Solving Published Sample Problems (Set I)

No.	Job Set	Reference	Number of Operations	Optimal Solution	Number of Nodes Explored						
					Brooks and White Method [9]	Bales Method [8]	Schrage Method [17]	Charlton and Death Method [10]	Ashour and Hiremath Method [5]	Ashour, *et. al.* Method [6]	Proposed Method **
1	(3x2) Flow-Shop	Balas [8]	6	31	7	12	6	7	7	5	4
2	(5x4) Job-Shop	Balas [7]	13	13	77	>149	>40	13	14	9	5*
3	(3x4) Flow-Shop	Brooks and White [9]	12	32	17	> 40	>23	20	17	13	7
4	(2x3) Job-Shop	Ashour [4]	6	13	4	3	6	4	4	2	2
5	(4x3) Job-Shop	Ashour and Hiremath [5]	12	27	52	–	20	>41	21	11	6
6	(2x2) Flow-Shop	Greenberg [14]	4	10	4	2	4	4	4	2	1
7	(2x2) Job-Shop	Greenberg [14]	4	9	2	2	4	2	2	2	0***
8	(2x3) Job-Shop	Greenberg [14]	6	14	2	3	6	2	2	2	1
9	(3x2) Job-Shop	Greenberg [14]	6	8	6	5	13	11	6	3	2

*Solution obtained is 14.

**Gap function.

***Initial graph is saturated such that an optimal solution occurs from initial covering.

Table 3. Comparative Evaluation of Various Techniques for Solving Published Sample Problems (Set II)

No.	Job Set	No. of Oper-ations	Balas [8]		Schrage [17]		Florian,et al[13]		Simulation[12,15]		Ashour, et al[6]		Proposed Methods*		
			Best Solution*	Iter-ations	First Solution	Iter-ations	First Solution	Iter-ations	Best Solution	Sample Size	First Solution	Iter-ations	Solution	Iter-ations	Computer Time(Sec.)
1	(6x6) Job-Shop	36	55	41	55	36	59	36	56	370	55	21	55	15	1.6
2	(10x10) Job-Shop	100	1177	156	1193	107	1041	100	1101	108	1018	55	1148	30	22.5
3	(20x5) Job-Shop	100	1231	89	1306	100	1252	100	1359	263	1188	94	1326	28	55.5
4	(20X10) Flow-Shop	200							150	3000	148	116	164	37	237.2
5	(20X10) Flow-Shcp	200							1598	5000	1590	103	1733	37	208.7
6	(30X5) Flow-Shop	150							16350	15250	16321	145	18046	37	169.0

*Best first solution considered from gap function.

$$s(T) = \max \begin{bmatrix} \max\,[u(7),\ v(7)] + t(7) \\ \max\,[u(8),\ v(8)] + t(8) \\ \max\,[u(9),\ v(9)] + t(9) \end{bmatrix}$$

$$= \max\,[14,\ 13,\ 11] = 14\ .$$

Hence, the schedule time for the sequence given by $G_5(N, A)$ is 14. It is of interest to note that this is an optimal solution to the given problem.

COMPUTATIONAL EXPERIENCE

The merit of the proposed out-of-kilter algorithm can be evaluated by considering its performance with reference to numerous published problems. Summarized in Tables 2 and 3, computational experience was gained such that the proposed algorithm was compared with reference to various other existing techniques. Further, the proposed algorithm is considered explicitly with reference to three measures: (1) number of iterations, (2) quality of solution, and (3) computational time. The algorithm was coded in FORTRAN IV and run on the UNIVAC 1108 computer.

Number of iterations. The proposed algorithm as can be seen from Tables 2 and 3, converges to a first pass solution in consistently fewer iterations than other techniques used for comparitive purposes. Such a resultant is especially worthy to note, in light of the embodiment of the waiting feature and its inherent ramification concerning increased computational demands. Moreover, it is clear that the number of iterations is influenced directly by the structure of subsequent conflict sets arising from the nature of particular problems. Certainly, a minimum of J-2 iterations must result for a given problem; however, most realistically sized problems will incur more than the minimum number of iterations as can be judged from Tables 2 and 3.

Quality of solution. The efficience of solutions rendered by the proposed algorithm is derived from the ratio of such solutions to the optimal or best known solution. As can be judged from Table 3, as the problem size increases, the quality of solution is reduced. Such a characteristic is not surprising, since the sizes of the conflict sets increase as the problem size, especially the number of job, increases. It is clear that the ultimate quality of solutions will depend heavily upon the particular criterion used to resolve conflicts that arise within current candidate sets. Although not reported in the present paper, additional criteria have been the subject of experimentation such that they have been used in the same capacity as the gap function reported herin.

Computational time. An in-depth analysis of the computational times required using the proposed algorithm, is relatively meaningless, comparitively. Clearly, such times are functions of not only specific computing hardware but perhaps more critically, of coding efficiency. Nevertheless, as can be seen from Table 3, the set of problems considered, commanded relatively low computational times. In fact, for problems 1 through 9, excluding the 5X4 problem of Balas, all of the problems were solved in 0.924 seconds. The remaining problems solved, required computational times indicated in the table. Recalling that the proposed algorithm renders a first pass solution, the relatively low computational times is especially significant should a backtracking component be considered to enhance solution quality.

From the diagnostic view, it is clear that particular problem structures illicit variable performances of the algorithm. Explicitly, flow shop structures command, in general, a greater number of iterations to arrive at a first pass solution. Characteristic of such an increase in iterative demands is the increased usage of the waiting feature. Another, critical aspect of the algorithm involves the specific use of the gap function. Although the algorithm is criterion independent, the gap function is considered relatively powerful, especially when its use is conceived in a saturated graph state. However, when in a state of unsaturation, the use of the gap function can sometimes dictate undesirable conflict resolutions. Such a phenomena is especially critical for large problems having processing times that vary greatly in magnitude.

SUMMARY AND CONCLUSIONS

The basic concepts of the proposed algorithm embody three phases: (1) construction of the shop scheduling problem in its graph theoretic context such that particular $G(N, A)$ results; (2) manipulation of the graph by an iterative scheme of arc additions and deletions based upon a specific criterion related to schedule time; and (3) use of an out-of-kilter concept to control the iterative process.

One of the primary advantages of the algorithm is its ability to yield first pass solutions of good quality with relatively inexpensive computational demands. Associated with such demands is the reduction in storage requirements rendered as well as relatively low computational times. It is worth noting that the algorithm is criterion independent in the sense that other, perhaps more powerful criteria could be employed which would enrich the solution quality capability of the algorithm, yet not alter the computational characteristics drastically.

The primary disadvantage of the proposed algorithm is its inability, in the current state, to yield other than first pass solutions. In order to enhance its ability to render better solutions, with present criteria concerned, a backtracking aspect could potentially be added. In lieu of present computational requirements, as well as the first pass solution quality, such an added embodiment would appear a feasible consideration.

REFERENCES

1. Ashour, S. and M. N., Quarishi, "Investigation of Various Bounding Procedures for Production Scheduling Problems," Int. Jnl. Prod. Research, Vol. 7 (1969), pp. 249-252.
2. Ashour, S., "A Branch-and-Bound Algorithm for Flow-shop Scheduling Problems," AIIE Transactions, Vol. 2 (1970), pp. 172-176.
3. Ashour, S., "An Experimental Investigation and Comparative Evaluation of Flow-Shop Scheduling Techniques," Operations Research, Vol. 18 (1970), pp. 541-549.
4. Ashour, S. and R. G., Parker, "A Precedence Graph Algorithm for the Shop Scheduling Problem," Operations Research Quarterly, Vol. 22 (1971), pp. 165-175.
5. Ashour, S. and S. R., Hiremath, "A Branch-and-Bound Approach to the Job-Shop Scheduling Problem," Int. Jnl. Prod. Research, Vol. 9 (1972).
6. Ashour, S., K. Y. Chiu, and T. E. Moore, "An Optimal Schedule Time of a Job Shop-Like Disjunctive Graph," submitted to Networks, January (1972).
7. Balas, E., "Discrete Programming by the Filter Method," Operations Research,

Vol. 15 (1967), p. 915.

8. Balas, E., "Machine Sequencing via Disjunctive Graphs: An Implicit Enumeration Algorithm," Operations Research, Vol. 17 (1969), pp. 941-957.

9. Brooks, G. H. and C., White, "An Algorithm for Finding Optimal or Near Optimal Solutions to the Production Scheduling Problem," Jnl. Ind. Engg., Vol. 16 (1965), pp. 34-40.

10. Charlton, J. M. and C. C., Death, "A Generalized Machine-scheduling Algorithm," Operations Research Quarterly, Vol. 21 (1970), pp. 127-134.

11. Charlton, J. M. and C. C., Death, "A Method of Solving for General Machine-Scheduling Problems," Operations Research, Vol. 18 (1970), pp. 689-707.

12. Fisher, H. and G. L., Thompson, "Probabilistic Learning Combinations of Local Job-shop Scheduling Rules," Chapter 15 in Industrial Scheduling, Prentice-Hall Inc., Englewood Cliffs, N. J. (1963).

13. Florian, N., P. Frepant, and G., Mcmahon, "An Implicit Enumeration Algorithm for the Machine Sequencing Problem," Management Science, Vol. 17 (1971), pp. B-782 – B-792.

14. Greenberg, H., "Solution of the General Scheduling Problem," Operations Research, Vol. 16 (1968), p. 353.

15. Heller, J., "Some Numerical Experiments for an MXJ Flow-Shop and its Decision-Theoretical Aspects," Operations Research, Vol. 8 (1960), pp. 178-184.

16. Moore, T. E., K. Y. Chiu, and S. Ashour, "An Implicit Enumeration Algorithm for the Non-Preemptive Shop Scheduling Problem," submitted to Operational Research Quarterly, February (1972).

17. Schrage, L., "Solvlng Resource-constrained Network Problems by Implicit Enumeration-Nonpreemptive Case," Operations Research, Vol. 18 (1970), pp. 263-278.

TRADING RULES FOR A DECENTRALIZED EXCHANGE ECONOMY*

Gordon H. Bradley

Yale University, New Haven, Connecticut, USA

Constructive proofs of the existence of trading rules for a decentralized monetary economy and for a non-decentralized barter economy are developed. It is proven that there is no trading rule for a decentralized barter economy.

Introduction

Recently, Ostroy and Starr (5) have investigated constructions that show how a decentralized economy might accomplish the exchange of goods in a Walrasian model of pure exchange after equilibrium prices have been announced. Assuming that every economic unit meets pairwise with every other economic unit exactly once, they described a "trading rule" for a decentralized monetary economy and for a non-decentralized barter economy. Ostroy and Starr conjectured that there is no decentralized trading rule for the barter economy; below, with an additional condition on decentralized trading, we establish the conjecture for economies with at least five economic units and at least four goods. This shows the critical role of money in a decentralized economy.

A linear programming model to construct a trading rule for the non-decentralized barter economy that minimizes exchange costs is described.

The existence of the various trading rules is shown constructively using some basic ideas from graph theory and combinatorics. Unlike the usual machine scheduling problem, the basic sequence of activities is given. The problem is to give a simple method to specify a complex set of actions that will work for all possible sequences.

Related problems on moving freight and on sorting computer records are described.

---

*Supported by National Science Foundation Grant GP-32158X.

Equilibrium Models of Exchange

In a simple exchange model of an economy, each individual economic unit (an individual or a firm) is given an initial vector of goods. Given prices for the goods, each economic unit makes plans to buy and sell goods in a way that will maximize its satisfaction (utility or profit). However, it is not possible, in general, for the members of the economy to fulfill their plans because the aggregate supply does not equal the aggregate demand for each good. The economic variables, namely the prices and the amount of each good that each economic unit wishes to buy or sell, are interrelated; therefore, it is necessary to coordinate the choice of the variables. A vector of prices are called equilibrium prices if each economic unit can maximize its own satisfaction and aggregate supply equals aggregate demand for each good.

Leon Walrus, a French economist, developed the notion of equilibrium economy in the 1870's. He stated the equations that must be satisfied in order to have equilibrium prices. His contribution to economic theory is often compared to Newton's contribution to the physical sciences. Although Walrus formulated the idea of equilibrium economy, the existence of equilibrium prices for a general economy was not settled until the 1950's. The theory was given a solid mathematical foundation when Arrow and Debrew, McKinzie, Gale and Nikiado used fixed point theorems to establish the existence of equilibrium prices under very general assumptions. More recently, Scarf developed a constructive proof of the existence of equilibrium prices that allows finite computation of approximate equilibrium prices. See Nikiado (4) and Scarf (6,7) for a discussion of these results.

The exchange economy is a decentralized model of an economy. That is, given prices, the individual economic units make their own decisions on the amounts they wish to buy and sell. Decentralization in this sense is an important characteristic of a capitalistic economy and of many planned economies. The Walrasian model uses the notion of equilibrium prices to show how a decentralized economy might reconcile the difference between aggregate supply and aggregate demand.

Ostroy and Starr point out that although the determination of how much of each good to buy and sell is decentralized in a Walrasian economy, the usual story of the actual exchange of goods is centralized. In the usual story, the economic units simultaneously give all the goods they plan to sell to a central clearing house which gives them the goods they plan to buy. Ostroy and Starr propose an exchange economy where the exchange of goods is to take place in pairwise meetings of the economic units. Their results, together with results developed here, show that [1] there is a decentralized exchange procedure for a monetary economy, [2] there is a non-decentralized exchange procedure for the barter economy, and [3] with an additional condition it is impossible to find a decentralized procedure for barter economies with at least five economic units and at least four goods.

Notation and Definitions

For an equilibrium model of pure exchange where no goods are created or destroyed, let n be the number of traders (i.e., economic units) and m be the number of goods. The ith trader has as initial endowment an m-vector of goods a_i. The aggregate initial endowment is $a = \sum_{i=1}^{n} a_i$, a is assumed positive in all components. Each trader has a utility function that is strictly increasing on R^m. Given prices $p > 0$, the value of the initial endowment of trader i is pa_i. Each trader decides what goods he would like to have by maximizing his utility function subject to his budget constraint of pa_i. Let $x_i(p)$ be the m-vector of goods that trader i would like to hold with prices p. We assume that a vector of equilibrium prices, $\bar{p}$, is known (see Nikaido (4) for an example of sufficient conditions on the utility functions in order that equilibrium prices will always exist). By definition of equilibrium, the aggregate demand, $x(\bar{p}) = \sum_{i=1}^{n} x_i(\bar{p})$, is equal to a. Supressing $\bar{p}$, the excess demand vector for trader i is $z_i = x_i - a_i$; z_i is positive for goods where additional quantities are wanted and negative for goods that are to be traded away. Under the assumptions, $\bar{p}z_i = 0$, that is, each trader satisfies his budget constraint with equality. For the whole economy $\Sigma_{i=1}^{n} z_i = 0$.

In the trading part of the model, the traders meet two at a time for possible exchange. They meet as in a round-robin tournament. In a play of the tournament, every pair of traders meet exactly once. Each meeting is called an encounter. When n is even, the $n(n-1)/2$ encounters are split into $n-1$ rounds of $n/2$ encounters where each trader appears in one encounter each round. When n is odd, an extra round is required and one trader "sits out" each round. The sequence of encounters is arbitrary but it is known in advance.

A trading rule is a 2m valued function that for each encounter in a tournament specifies what vectors of goods (if any) are to be exchanged. A trading rule is said to achieve full execution if after a play of the tournament, each trader has the goods he wants, i.e., $x_i(\bar{p})$.

Non-Decentralized Trading Rule

Given equilibrium prices and given the corresponding excess demand functions z_i , Ostroy and Starr (5) are concerned with the existence of trading rules that achieve full execution and that satisfy the following conditions.

C1: No trader may transfer goods that he does not have. Since it is possible that a trader initially have only excess supplies, this condition will be imposed by allowing the traders to transfer only the excess supplies that they have at the time of the encounter.

C2 (*quid pro quo*): Exchanges may only involve vectors of goods of equal value. That is, if traders i and j exchange y_i for w_j then $\bar{p}\, y_i = \bar{p}\, w_j$.

C3 (Decentralization): The vectors of goods exchanged in an encounter are a function of only the excess demand vectors of the two traders.

A possible trading rule is: exchange when both traders receive goods that help satisfy their excess demands (i.e., positive components on their z_i vectors). The following example shows that this rule may not achieve full execution.

Example 1: 3 traders, 3 goods, $\bar{p} = (1,1,1)$, $z_1 = (1,0,-1)$, $z_2 = (-1,1,0)$, $z_3 = (0,-1,1)$. For each encounter, one trader has excess demand for a good and the other has no excess supply for that good. Thus under the rule, no exchange is possible.

Another possible trading rule is for a central controller to announce that good k is a special good and that the traders should always make trades that would satisfy excess demands for this good with C2 being satisfied by transferring other goods. After the tournament is completed, all traders will have their desired quantity of the special good but will not necessarily have the desired quantities of the other goods. It would take m-1 plays of the tournament with a different special good each play to guarantee that every trader obtains his x_i .

If condition C2 is relaxed, then full execution can be achieved by transferring goods whenever one trader has an excess supply and the other has an excess demand of a good. Although C2 is violated for individual encounters, at the completion of the tournament, each trader has his x_i which is the same value as his initial endowment a_i . In the framework of our model, this trading rule is merely a time extended version of the central clearing house model we are trying to improve.

If the decentralization condition C3 is relaxed, it is possible to construct a trading rule that will achieve full execution. The proof is constructive. The proof is divided into three parts; the first two parts are due to Ostroy and Starr, the third part is due to the author. We give a graph theoretic proof of the second part to replace a matrix argument of Ostroy and Starr.

Theorem 1 (Ostroy and Starr with help from the author): In the barter economy, there is a trading rule satisfying conditions C1 and C2 that achieves full execution.

Definition: An equal value cycle is a list of traders (w.o.l.g. traders 1,2,...,k) each with a bundle of goods. Trader i has a bundle of goods that is needed by trader i+1 , i=1,2,...,k-1 and trader k has a bundle needed by trader 1. By "needed" we mean that it will help satisfy excess demand, i.e., the bundle held by trader i will help trader i+1 obtain x_{i+1} . Also, the value of all bundles using the prices $\bar{p}$ are equal. The equal value cycle may be viewed as:

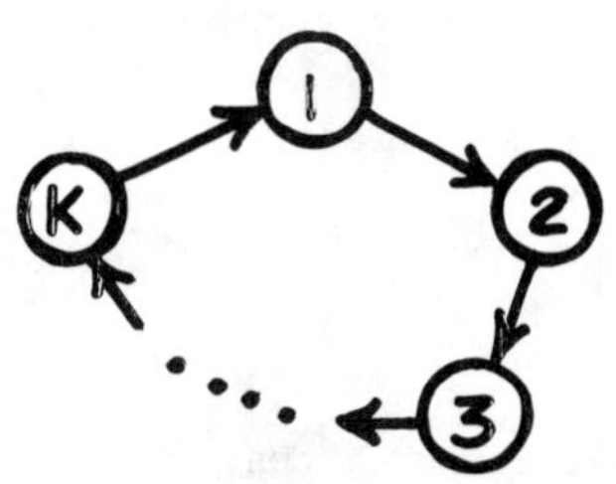

where the nodes are traders and the arc from node i to node i+1 is the bundle of goods held by i and needed by i+1 .

The trading rule of the theorem is constructed by decomposing the problem into a finite number of equal value cycles and then specifying a trading rule for an equal value cycle. It is convenient for the exposition to assume the units for the goods have been adjusted so that the equilibrium price of each good is 1; we can then use the number of units and the value of goods interchangeably.

Construction 1: Construct a directed labeled graph as follows: Make a node for each trader. For good 1, there is a subset of the traders that have excess supply $(z_i)_1 < 0$ and a disjoint subset of traders with excess demand $(z_j)_1 > 0$; the aggregate excess supply equals the aggregate excess demand. Arbitrarily assign the excess supply to traders with excess demand by constructing a directed arc from a trader with supply to a trader with demand with the label being the value (or units) of good 1. Do this for all goods on the same graph. If there is more than one arc from node i to node j, the arcs may be combined by adding the labels. Since the value of the excess demand vector is zero, the sum of the labels on incoming arcs will equal the sum of the labels on outgoing arcs for each node.

Construction 2: Decompose the above graph into a finite number of equal value cycles. The procedure is: Pick any node with arcs incident to it and walk from the node on any outgoing arc to another node. Continue the walk by walking out on any outgoing arc. It is always possible to find an outgoing arc since any node with an incoming arc must have an outgoing arc. Terminate the walk when for the first time, a node is revisited. The part of the walk from the first time the terminal node was visited to the terminal node again is a directed cycle in the graph. Let α be the minimum label among the arcs of the cycle, remove an equal value cycle of value α from the graph by subtracting α from all labels on arcs of the cycle and removing arcs with label 0. The new graph still has the property that the sum of the labels on incoming arcs equals the sum of the labels on outgoing arcs for each node. The new graph has at least one less arc. The procedure is applied again and again until after a finite number of steps the graph has no arcs.

This procedure is an obvious generalization of Euler's theorem for finding a walk through a graph that traverses each arc exactly once (see for example Harary (2)).

For the third construction it will be assumed that the encounters take place in a sequence. In the model, all encounters in a round take place simultaneously, however, for the purpose of specifying a trading rule, we may assume the encounters in a round are given an ordering.

<u>Construction 3</u>: Trading rule for an equal value cycle: When traders i and j meet they exchange their appropriate bundles of goods if they are both in the same equal value cycle.

This trading rule will be called rule N (for non-decentralized). If i and j exchange goods, the equal value cycle is broken into two equal value cycles. Assume without loss of generality that $i < j$, the result is:

If $j = i+1$ one cycle is the trivial cycle of j having the bundle of goods j needs. Notice that if i and j are not adjacent in the cycle, then neither obtains the bundle of goods that it needs.

It now must be shown that repeated application of the rule will result in full execution in a single tournament. The proof is by induction on k the number of traders in the original equal value cycle. For $k=2$, the result follows. Now assume that if you have an equal value cycle with $k-1$ or fewer traders and the sequence of encounters remaining includes all possible encounters between the members of the cycle, then by following the rule full execution follows. For an equal value cycle with k traders, let the encounter between i and j be the first encounter in the sequence between members of the cycle. As above, the cycle is decomposed into two cycles with $k-1$ or fewer traders. By construction the i-j encounter was the first between members of the original cycle so the sequence of remaining encounters includes all possible encounters in each of the two new equal value cycles. The induction hypotheses then gives the result.

The trading rule of the theorem is obtained by decomposing the problem into a finite number of equal value cycles, as done in construction 1 and 2. Viewing the encounters as a sequence, the trading rule when i and j meet, is that they should exchange the bundles of goods indicated by <u>all</u> equal value cycles they are both in. The actual exchange is the sum of the bundles of goods of all the equal value cycles they both appear in. Condition C1 follows from construction 1 and condition C2 follows from the definition of equal value cycles. Q.E.D.

Example 1 redone: Construction 1 gives

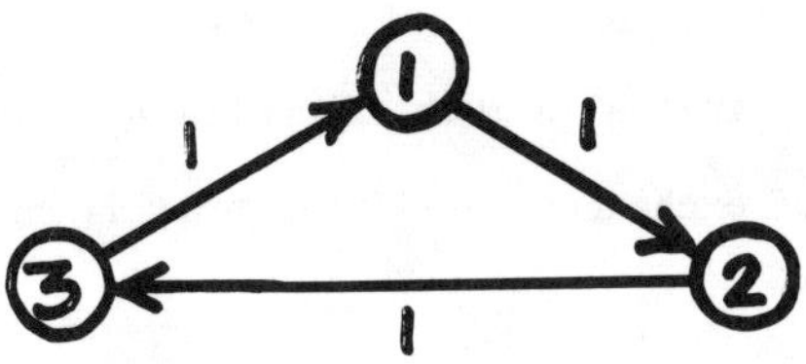

which also gives construction 2. For sequence of encounters 1-2, 2-3, 1-3, the exchanges are: 1 gives 2 (0,0,1) in exchange for (1,0,0); 2 gives 3 (0,0,1) in exchange for (0,1,0); 1 and 3 do not trade.

In construction 3, if when i and j trade they are not adjacent in the cycle, then neither obtains a bundle of goods that it needs. They exchange so they can later exchange for a bundle of goods they need.

Constructions 1 and 2 may, in general, be done in many different ways. However, given constructions 1 and 2, it is now shown that rule N yields a minimum number of exchanges and rule N may be ignored only at the risk of failing to get full execution.

<u>Proposition 1</u>: For a single equal value cycle with k traders, trading rule N achieves full execution with k-1 exchanges.

Proof: Each exchange increases the number of equal value cycles by one. There is full execution when there are k equal value cycles (i.e., each trader has the bundle he needs). Q.E.D.

<u>Proposition 2</u>: For a single equal value cycle with k traders, any trading rule that keeps the bundles of goods intact and that achieves full execution will require k-1 + 2j exchanges where j is a non-negative integer.

Proof: Every exchange that satisfies trading rule N (i.e., both traders in the same equal value cycle) increases the number of equal value cycles by one. Every exchange that violates this rule (i.e., the traders are in different equal value cycles) decreases the number of cycles by one. To see this, let i and j exchange in violation of rule N.

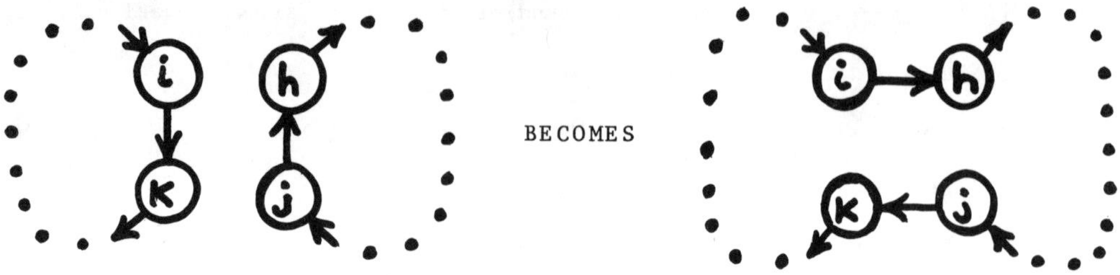

From the hypothesis, the initial number of cycles is 1 and the final number is k . Let i be the number of exchanges satisfying trading rule N and j the number violating it.

Then i-j = k-1 and i+j = k-1 + 2j as was to be shown. Q.E.D.

Thus, given constructions 1 and 2, construction 3 is minimal and every exchange violating rule N must be undone by an extra exchange that satisfies the rule. It is easy to construct different examples to show that [1] if rule N is violated for some encounter, it is still possible to achieve full execution; [2] if rule N is violated for some encounter, it is impossible to achieve full execution; [3] if rule N is not followed (that is, an indicated exchange is not carried out) for some encounter, then it is still possible to achieve full execution; and [4] if rule N is not followed for some encounter, then it is impossible to achieve full execution.

Since it is essential to increase the number of equal value cycles and this happens only if rule N is used, it might appear that if rule N were violated for some encounter, then the best policy to achieve full execution would be to follow rule N from that point on. The following example shows this is not the case.

Example 2:

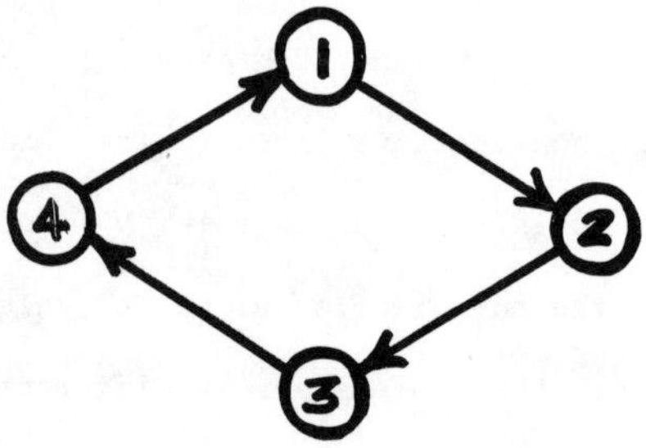

Encounters 1. 1-4 No Exchange (Violates Rule N)

2. 2-3 No Exchange (Violates Rule N)

3. 1-3 Exchange

4. 2-4 No Exchange

5. 1-2 No Exchange

6. 3-4 No Exchange

Encounters 3 to 6 followed rule N but did not achieve full execution. If rule N is not followed on encounter 3 (i.e., don't exchange) and rule N is followed for encounters 4 to 6, then full execution is achieved.

The examples show that given construction 1 and 2, rule N must be used to guarantee full execution.

Monetary Economy

A monetary economy will be the same model as above except that one good, say good m , will be used as money. It will be assumed that everyone knows which good is money. In addition, it is assumed that each trader has a sufficient initial supply of money to finance his purchases, where sufficient means

$$\bar{p}_m (a_i)_m \geq \sum_{j=1}^{m-1} \bar{p}_j \max (0,(z_i)_j)$$

Theorem 2 (Ostroy and Starr) In a monetary economy where each trader has a sufficient supply of money, there is a trading rule satisfying conditions C1, C2, C3 that achieves full execution.

Proof: The trading rule: For each good where one trader has excess demand and the other has excess supply, transfer that good until one trader has excess supply zero or the other trader has excess demand zero. Condition C2 is fulfilled by using money. Q.E.D.

Decentralized Trading Rule

In the monetary economy we saw that a decentralized trading rule was possible if each trader has enough money. Before it can be claimed that money makes it possible to decentralize the trading activity, it must be shown that in a barter economy it is impossible to find a decentralized trading rule. Ostroy and Starr conjectured that no such rule exists.

Conjecture (Ostroy and Starr (5)): In a barter economy, there is no trading rule that will satisfy conditions C1, C2 and C3 and that will achieve full execution.

Before we can resolve the conjecture, it is necessary to specify more completely what is meant by decentralization. Replace condition C3 by:

C3a. In encounter i-j only the present excess demand vectors of i and j are known.

C3b. The trading rule may not be a function of a particular labeling of the goods. Specifically, given the two excess demand vectors, the trading rule must make the same decision if the goods are renamed or relabeled. The rule may not be of the form: if one trader has an excess supply of shoes and an excess demand for hats, ...

With the replacement of condition C3 by C3a and C3b, the following theorem resolves the conjecture for all but one case.

Theorem 3: In the barter economy with n traders and m goods, there is a trading rule satisfying conditions C1, C2, C3a and C3b that achieves full execution if

A1. m is 2 and n is arbitrary.

A2. n is 2 or 3 and m is arbitrary,

and there is no such rule if

B1. $n \geq 5$ and $m \geq 4$.

The theorem is proven for A1 and A2 by giving trading rules and for B1 by giving a counterexample for $n=5$ and $m=4$.

Proposition 3: For case A1, the trading rule is: if there is a good for which one trader has excess supply and the other excess demand, transfer that good until one trader has excess supply zero or the other has excess demand zero (the exchange is financed with the other good).

Proof: Obvious. At least one trader must end up with his z vector equal zero since $\bar{p}\, z_i = 0$. The economy acts like a monetary economy; since there are only two goods it is not necessary to specify which is money.

For case A2 with $n=2$ the result is obvious.

Proposition 4: For case A2 with n=3 , the trading rule is: For each good where one trader has excess supply and the other excess demand, the good is transferred until one trader has excess supply zero or the other has excess demand zero. After this is done, goods that are in excess supply are transferred to satisfy condition C2.

Proof: When the two traders meet in Round 1, using their excess demand vectors they can compute the excess demand vector of the other trader. They can then do the construction of Theorem 1 to determine how they should exchange. The above trading rule is a specific example of construction 1 and 2 where the traders form as many arcs as possible between them and as large an equal value chain as possible between them.

Case B1 is established by constructing two examples with n=m=4 . For an encounter in Round 2 of the first example, it is shown that the traders must exchange unique bundles of goods in order for the economy to achieve full execution. In the second example, there is a Round 1 encounter where the traders have the exact same excess demand vectors as in the first example, however, if they exchange the bundle of goods required to achieve full execution in the first example it is impossible to achieve full execution. The two examples are then extended to examples with five traders by the addition of a dummy trader. In the extended examples, the contradictory trading behavior occurs in Round 2 of both examples.

In all the following examples the trading order is: round one, encounters 1-2 and 3-4; round two, encounters 1-4 and 2-3; round three, encounters 1-3 and 2-4. In all the examples $\bar{p} = (1,1,1,1)$.

For the examples it will be critical to specify exactly what some traders will exchange in a given situation.

Proposition 5: For an economy with n = m = 4 , if traders i and j meet in Round 1 with excess demand vectors $x_i = (8,-8,0,0)$ and $x_j = (0,0,8,-8)$, then in order to achieve full execution it is necessary for i and j to exchange goods so that after trade their excess demand vectors are $x_i = (8,-4,0,-4)$ $x_j = (0,-4,8,-4)$.

Proof: The proof follows from considering two examples.

Example 3: $x_1 = (8,-8,0,0)$, $x_2 = (0,0,8,-8)$, $x_3 = (-8,8,0,0)$ and $x_4 = (0,0,-8,8)$. In the Round 1 exchanges let $a \geq 0$ be the amount of good 2 that is transferred from trader 1 to trader 2 and let $b \geq 0$ be the amount of good 1 that is transferred from trader 3 to trader 4. After the Round 1 exchanges, the excess demand vectors are

$x_1 = (8,-8+a,0,-a)$, $x_2 = (0,-a,8,-8+a)$, $x_3 = (-8+b,8,-b,0)$ and $x_4 = (-b,0,-8+b,8)$. In the Round 2 encounter 1-4, trader 1 needs the b units of good 1 held by trader 4; trader 1 must obtain them in this encounter because the one remaining encounter offers him no possibility to obtain these goods. Similarly, trader 4 must obtain the a units of good 4 held by trader 1. Also, no units of good 3 held by trader 4 may be transferred to trader 1; good 3 is needed only by trader 2 and it can only be obtained by trader 2 in Round 3 from trader 4. Similarly, no units of good 2 held by trader 1 may be transferred to trader 4. Under these restrictions, after 1 and 4 trade the excess demand vectors are $x_1 = (8-b,-8+a,0,0)$ and $x_4 = (0,0,-8+b,8-a)$. From the condition $\bar{p}x_i = 0$, it follows that $a - b = 0$ is necessary to achieve full execution.

We now look at a new example that differs from the above example only in the demands for goods 2 and 4 by traders 3 and 4: $x_3 = (-8,0,0,8)$ and $x_4 = (0,8,-8,0)$. Define b as above. Condition C3b requires that b take on the same value as above since the new excess demand vectors for traders 3 and 4 could have resulted from a relabeling of goods 2 and 4. As above, it can be shown that after 1 and 4 trade the excess demand vectors must be $x_1 = (8-b,0,0,-a)$ and $x_4 = (0,a,-8+b,0)$ in order to achieve full execution. From the condition $\bar{p}x_i = 0$, it follows that $a+b-8=0$. Together with $a-b=0$, this implies that $a = b = 4$ as was to be shown Q.E.D.

Since traders 1 and 2 don't know the distribution of demand among traders 3 and 4, they must trade as the proposition indicates in order to achieve full execution. Condition C3b was very critical in the proof of Proposition 5. It is possible to specify a decentralized trading rule for example 3 that violates C3b.

We now consider the first example to be used in the counterexample.

Example 4: $x_1 = (8,-8,0,0)$, $x_2 = (0,0,8,-8)$, $x_3 = (-1,1,0,0)$ and $x_4 = (-7,7,-8,8)$. Proposition 5 specifies the 1-2 exchange. Again let b be the amount of good 1 transferred from trader 3 to trader 4. After traders 3 and 4 exchange, the excess demand vectors are $x_3 = (-1+b,1,-b,0)$ and $x_4 = (-7-b,7,-8+b,8)$. It turns out that any value of b, $0 \leq b \leq 1$ has a chance of achieving full execution. We can assume $b < 1$ since if $b = 1$ we can switch the demands of traders 1 and 2 for goods 1 and 3 so that the net effect is that $b = 0$ for the original problem. In Round 2, encounter 1-4, the beginning excess demand vectors are $x_1 = (8,-4,0,-4)$ and $x_4 = (-7-b,7,-8+b,8)$. To achieve full execution it is necessary that [1] trader 1 transfer all 4 units of good 4 to trader 4, [2] trader 4 transfer all 7+b units of good 1 to trader 1 and [3] trader 4 transfer no units of good 3. The amount of good 2 transferred is determined by $\bar{p}x_i = 0$. The unique exchange necessary to achieve full execution gives $x_1 = (1-b,-1+b,0,0)$ and $x_4 = (0,4-b,-8+b,4)$.

We now consider another example where the same 1-4 encounter as above occurs in Round 1.

Example 5: $x_1 = (8,-4,0,-4)$, $x_2 = (-7-b,7,-8+b,8)$, $x_3 = (-1+b,-3,8-b,-4)$ and $x_4 = (0,0,0,0)$. Recall $0 \leq b < 1$. Since traders 1 and 2 have exactly the same excess demand vectors as did the traders in encounter 1-4 above, we assume they exchange as above. That is, after trade, $x_1 = (1-b,-1+b,0,0)$ and $x_2(0,4-b,-8+b,4)$. Trader 2 needs the $1-b > 0$ units of good 2 held by trader 1. The only possible way for trader 2 to get these goods would be for trader 1 to transfer them to trader 4 in Round 2 who would then transfer them to trader 2 in Round 3. Since we may not assume that trader 4 has sufficient assets to finance this exchange, we conclude that full execution is impossible.

To get the counterexample with $n=4$ we had to assume that the traders in Round 1 of example 5 would act exactly as the traders in Round 2 of example 4. We can remove this assumption by extending the examples to 5 traders where the extra trader 5 is a dummy with a zero excess demand vector. In example 4, Round 1 is encounters 1-2, 3-4 with 5 "sitting out"; Round 2 is 1-4, 3-5 with 4 sitting out. In example 5, Round 1 is 1-5, 3-4 with 2 sitting out; Round 2 is 1-2,3-5 with 4 sitting out. The critical encounter is in Round 2 of both extended examples.

Thus, we have shown that under the conditions of the theorem even if the traders know how many traders are in the economy, know which round they are in, and know the full sequence of encounters, they cannot make decentralized exchange decisions that will guarantee full execution.

It is still an open question if the conjecture is true with condition C3b dropped. Although C3b can be given a meaningful economic interpretation, it is of interest to economists to settle this question. The case of $m=3$, n arbitrary for theorem 3 is still open, this question is of much less interest.

Minimizing Exchange Costs

One of the disadvantages of a barter economy is that it may be necessary to pass goods through many hands before full execution is achieved. The monetary economy is often said to be more "efficient" because the cost of passing money is assumed to be small relative to the cost of passing other goods. Given costs for passing goods from one trader to another, the problem of constructing a non-decentralized trading rule to yield minimum total exchange costs can be formulated as a linear programming problem. This problem may also be of interest for a monetary economy where the cost of passing money is not insignificant (e.g., cost of currency exchange) or where it may be cheaper to pass goods through a third trader rather than passing goods directly (e.g., it may be cheaper to sell stock through a broker than to search for a buyer).

If y_{ijk} is the amount of good k passed from trader i to trader j in encounter i-j and c_{ijk} is the unit cost of passing the good, then the total exchange cost is defined to be $\sum_{i=1}^{n} \sum_{\substack{j=1 \\ j \neq i}}^{n} \sum_{k=1}^{m} c_{ijk}\, y_{ijk}$. Let the vector of goods passed from i to j be $y_{ij} = (y_{ij1}, y_{ij2}, \ldots, y_{ijm})$. The conditions C1 and C2 are expressed as linear constraints; if trader i meets trader j in Round 1 and trader k in Round 2 the constraints include

$$a_i - y_{ij} \geq 0$$

$$a_j - y_{ji} \geq 0$$

$$a_i - y_{ij} + y_{ji} - y_{ik} \geq 0$$

$$\bar{p}\, y_{ij} - \bar{p}\, y_{ji} = 0$$

$$y_{ij};\ y_{ji},\ y_{ik} \geq 0$$

For the final round, the constraint is

$$a_i - y_{ij} + y_{ji} + \ldots - y_{ip} + y_{pi} = x_i$$

There are $n(n-1)m$ variables and $n(n-1) / 2 + n(n-1)m$ constraints. Theorem 1 is a constructive proof that this linear program has a feasible solution; the solution set is bound, thus an optimal solution exists.

Related Problems

The above constructions can be used to solve some different problems. A possible freight problem involves the shipment of cargo from several West Coast cities to several East Coast cities. Each truck (trader) begins its trip with cargo for several East Coast cities (goods); the trucks would like to exchange cargo so that each truck would have to visit one or at the most two East Coast cities (i.e., final goods vector with one or two positive elements). Possible exchanges take place at various locations across the country where the trucks meet. The linear programming model described above could be used to construct a minimum cost solution.

For a different class of problems, construction 3 may be viewed as a method to rearrange a list of k elements where the elements may only be exchanged two at a time and there is a given sequence of all possible exchanges. In general, given a permutation $p_1, p_2, \ldots, p_n$ of the integers $1, 2, \ldots, n$ we want to rearrange the permutation into numerical order with the minimum number of pairwise exchanges. Construct a directed graph with nodes associated with the numbers $1, 2, \ldots, n$ and a directed arc from node i to node p_i for $i = 1, 2, \ldots, n$. This graph has j directed cycles (loops, i.e., $i=p_i$, count as cycles) where j may be any integer from 1 to n. The cycles are called the cycles of the permutation (see for example Hall (1)). Each cycle corresponds to one of our equal value cycles. For example, if the list (13,19,10,2) is to be rearranged by pairwise exchanges to increasing order, the permutation is (3,4,2,1) and the associated graph is:

If the order of possible exchanges is 1-3, 2-4, 1-4, 2-3, 1-2, 3-4, then the list is ordered by E,E,N,N,E,N, where E means exchange and N means no exchange. Construction 3 gives an algorithm that will accomplish the rearrangement with a minimum number of exchanges (the number of exchanges is n-j). The discussion following construction 3 shows that this is the only construction that will guarantee the desired result.

A possible scenario for this problem is that you are processing a computer job that has n large records in auxillary storage. The job involves bring two records into fast memory, working on them, and then returning them to the same locations or to the reserve locations. If in addition, you wish to reorder the n records, construction 3 specifies when to exchange (i.e., to return the records to the reverse locations).

The problems discussed here allow only pairwise exchanges; if it is also possible to move goods temporarily to a location (a warehouse in the trucking example or keep in fast memory in the computer example), then other algorithms can be used (see for example Knuth (3), p. 161).

Acknowledgements

I would like to thank Ross Starr for introducing me to this problem and for contributing many insightful comments and suggestions. The conference referees, Pierre Robillard, Universite de Montreal and Donald Ratiff, University of Florida, provided helpful comments and suggestions.

References

(1) Hall, M., *Combinatorial Theory*, Waltham, Mass: Blaisdell, 1967.

(2) Harary, F., *Graph Theory*, Reading, Mass: Addison-Wesley, 1969.

(3) Knuth, D., *The Art of Computer Programming, Volume I*, Reading, Mass: Addison-Wesley, 1969.

(4) Nikaido, H., *Introduction to Sets and Mappings in Modern Economics*, New York: American Elsevier, 1970.

(5) Ostroy, J., and Starr, R., Money and the decentralization of exchange, manuscript, February 18, 1972.

(6) Scarf, H., *The Computation of Economic Equilibrium* (to appear, Yale Press).

(7) Scarf, H., The approximation of fixed points of a continuous mapping, *SIAM Journal of Applied Mathematics*, 1967, 15, pp. 1328-1343.

(8) Starr, R., The structure of exchange in barter and monetary economies, *Quarterly Journal of Economics*, May 1972.

SCHEDULING WITH EARLIEST START AND DUE DATE CONSTRAINTS *

Paul Bratley, Michael Florian, and Pierre Robillard

Département d'Informatique

Université de Montréal, Montréal, Canada

ABSTRACT

We consider the scheduling of n tasks on a single resource. Each task becomes available for processing at time a_i, must be completed by time b_i, and requires d_i time units for processing. The aim is to find a schedule that minimizes the elapsed time to complete all jobs. We present solution algorithms for this problem when job splitting is permitted and when job splitting is not permitted. Then we consider several scheduling situations which arise in practice where these models may apply.

DISCUSSION OF THE PAPER "SCHEDULING WITH EARLIEST START AND DUE DATE CONSTRAINTS"

The purpose of this note is to report on some remarks made on the contents of the above paper. The problem considered is that of sequencing n items on one or more processors with ready times (earliest starts) a_i, due dates b_i and durations d_i in order to minimize the total completion time. In the paper we suggested a labeling type algorithm that consists of applying several max-flow type calculations in order to solve the problem when preemption is allowed. Professors David Rubin and William Maxwell pointed out that, on a single processor, it is much simpler to use an earliest due-date rule:

at each a_i, consider all jobs available and not yet completed; work on these in earliest due-date order. It can be shown that this rule minimizes the maximum lateness, and that it produces the minimum total elapsed

* To appear in NRLQ, Vol. 18, vo. 4.

time schedule if a feasible schedule exists. David Rubin suggests the following proof.

PROOF: The machine operates continuously, except when no jobs are present. Thus idle time is minimized, so the schedule is minimal if it is feasible. Suppose some feasible schedule exists. In that schedule, suppose jobs i and j are available at time t, that $b_i > b_j$, and that job i is processed first. Let v_{it} and v_{jt} be the remaining processing times on jobs i and j. Since the schedule is feasible $t + v_{it} \leqslant b_i$, $t + v_{it} + v_{jt} \leqslant b_j < b_i$. Hence $t + v_{jt} \leqslant b_j$, $t + v_{it} + v_{jt} \leqslant b_i$. Thus switching jobs i and j does not affect the feasibility of the schedule, and it follows that the earliest due date schedule is feasible if there is any feasible schedule whatsoever.QED. While the above algorithm is clearly superior to that given in the paper, it has the drawback that it does not extend to multiple processors, as demonstrated by the following counter example on 2 processors.

item	ready time (a_i)	duration (d_i)	due date (b_i)
1	0	1	2
2	0	1	2
3	0	2	3

The earliest due-date rule results in a schedule of duration 3, while the best schedule has a duration of 2.

We have assumed that the natural extention to 2 processors would be to apply the earliest due-date rule and load all available processors with the selected items. Thus, the max-flow type algorithm must be applied when several processors are considered.

THE SCHEDULING OF A MULTI-PRODUCT FACILITY

Salah E. Elmaghraby and Arup Mallik

North Carolina State University and the University of West Virginia

Raleigh, North Carolina, USA

We are concerned with obtaining the optimal schedule of N products over a single facility in the finite and infinite horizons. This is a generalization of the classical cyclical "Economic Manufacturing Quantity" formulations. Through number-theoretic arguments we discuss questions of feasibility, the existence of communicating classes, and sensitivity to "saturation". In the finite horizon case, we present an algorithm that generates the production plan, and another algorithm that generates the production schedule through a shortest-path interpretation. In the infinite horizon case, we do likewise: the optimal production plan is a Turnpike cycle. Due to the immensity of the state space, approximate procedures are also discussed.

Historical Perspective

This paper is concerned with obtaining the optimal schedule of N products over a single facility in the finite and infinite horizons. In a real sense, it is a generalization of the classical cyclical EMQ, for which we substitute a cycle of a different nature, but a cycle nevertheless.

The scheduling of a single multi-product facility for the purpose of controlling the production and inventory of N products is a problem that boasts a long history, dating back to the famous "square root" formula of Harris (1915) and Raymond (1931)[23]. The "EMQ" formula, as it has come to be known, received wide dissemination, and was elaborated upon in a variety of ways to incorporate price-quantity discounts, backlogging of demand, finite horizon, etc. For almost forty years it reigned supreme as <u>the</u> model for economic production and stock management.

This research was partially supported by ONR Contract No. N00014-70-A-0120-0002 and NSF Grant No. P1K1470-000.

This aura of omnipotence of this analytical model was shattered by Salveson[25] in 1953, who also attempted to rectify its deficiences. Several other people attempted the same thing in the subsequent years, notably Rogers[24], Hanssmann[16], Eilon[9], Maxwell[22], Bomberger[2], Brown[3], Doll[7] and Madigan[19]. A brief review of some of these approaches, and a critical evaluation of their solutions may be found in Elmaghraby[10], who also offered an exact solution to a limited version of the problem. His approach was expanded in the paper by Elmaghraby, Mallik and Nuttle[12], which may be viewed as the precurser of the present research.

§1. COMMUNICATING CLASSES AND THE PARTITIONING ALGORITHM

The problem of concern to us here is that of scheduling N products on a single facility when the demand for item j is a fixed rate, $r_j \geq 0$, integer, per period $j=1,2,\ldots,N$. The limited capacity of the facility is a major restriction since in any period t, the facility is devoted to the production of one, and only one, product. But while the production of that product is in progress for one or several periods, demand for the other products is continuing, hence their inventory is depleted at their respective demand rates.

Let p_j be the production rate per period, assumed integer, when the facility is devoted to the production of item j. To have a meaningful problem we assume that $p_j > r_j$ for all j. Furthermore, let I_j^o be the starting inventory of j: $I_j^o > 0$ implies a stock condition and $I_j^o < 0$ implies a backorder condition.

The restriction of limited capacity of a single facility serving several products provides a point of departure of this problem from other previously-treated inventory control models. For instance, in the case of one product, any inventory state I^h (at time h) can be reached from any initial inventory state I^o at time 0 if the gcd(p,r) = 1[†]. Indeed, we have

$$I^h = I^o + p\ x - r\ h$$

where x is the number of periods of production, $0 \leq x \leq h$; or

$$\Delta = I^h - I^o = p\ x - r\ h\ ;$$

a Diophantine equation which possesses nonnegative solutions (actually, an infinity of them) if $(p,r) = 1$.

Such is not true in the case of two or more products. Furthermore, the case of several products raises the spectre of infeasibility, which was completely absent in the single product case (as long as $p > r$).

These, and other questions, indicate that there exist some novel considerations peculiar to the problem under investigation. In particular, we must consider the following:

1. Under what conditions is the idea of "satisfying all demand at some future time" be possible? This is the "feasibility" problem.
2. Can any state $I^h = \langle I_1^h, I_2^h, \ldots, I_N^h \rangle$ be "reachable" from any state $I^o = \langle I_1^o, I_2^o, \ldots, I_N^o \rangle$ at time 0? This is the "communicating states" problem.
3. How many divisions of the state space are there, and do they exhaust the whole space or are there some states which are not reachable from <u>any</u> state? This is the "partitioning" problem.

Naturally, once these questions are settled there still remains the question of obtaining the minimal cost policy. This problem will be dealt with in the following section.

[†]The greatest common divisor (gcd) of a set of integers a,b,...,w will be written as (a,b,...,w).

Notation

We recapitulate the definitions and notation introduced thus far, and add a few more:

r_j: constant rate of demand for j; j=1,2,...,N, integer.

p_j: rate of production of j when facility is devoted to it; $p_j > r_j$, all j, integer.

x_j (or y_j): number of periods (in a given planning horizon) devoted to the production of item j; $x_j \geq 0$, all j. We shall call the set $<x_1, x_2, \ldots, x_N>$ a production plan.

I^t: vector of inventories at time t, the state vector, $I^t = <I_1^t, I_2^t, \ldots, I_N^t>$; with I^o the starting state at time 0. In the absence of explicit designation of time, we denote the inventory by I, I', etc.

$\Omega = Z^N \times Z^+$, the state space.

$(a_1, a_2, \ldots, a_M)$: the gcd of the integers $a_1, a_2, \ldots, a_M$

$a|b$: notation for "a divides b" ; a,b, integers

$a \nmid b$ a does not divide b.

$[a_1, a_2, \ldots, a_M]$: the ℓcm (least common multiplier) of the integers $a_1, a_2, \ldots, a_M$.

For the sake of economy of notation, let

$d_j = (r_j, p_j)$; i.e., d_j is the gcd of r_j & p_j , j=1,...,N

$\mu_j = r_j/d_j$; i.e., $r_j = \mu_j d_j$; μ_j integer

$\eta_j = p_j/d_j$; i.e., $p_j = \eta_j d_j$; η_j integer

$\rho = \sum_j (r_j/p_j)$.

Clearly, the space of the state vector I at any time t is the (denumerable) space of integers in N dimensions, Z^N. The complete characterization of the state must take time into consideration, hence $I^t \in Z^N \times Z^+$, where I^t is the end-of-the-t^{th} - period inventory.

Decision may be viewed to occur at the beginning of a period and its execution is instantaneous. Clearly, the decision space at any time t has at most N+1 points: the N decisions corresponding to: "produce item j; j=1,...,N", and the decision not to produce anything, the so-called "do-nothing" decision. We shall indicate the decision space by D^t .

Structure of the State Space

We now proceed to resolve the three issues raised above.

Theorem 1 Suppose I and I' belong to the same communicating class under ρ, and let $\{j_1, j_2, \ldots, j_u\}$ be a subset of the indices $\{1,2,\ldots,N\}$; then the vectors

$$< I_{j_1} , I_{j_2} ,\ldots, I_{j_u} > \text{ and } < I'_{j_1} , I'_{j_2} ,\ldots, I'_{j_u} >$$

belong to the same communicating class under $\rho\langle j_1, j_2, \ldots, j_u\rangle = \sum_{r=1}^{u} (\mu_{j_r} / \eta_{j_r})$.

Proof Obvious.

Lemma 1 (The Feasibility Lemma) From any (finite) starting inventory I^o , a nonnegative inventory $I^h \geq 0$ is realizable in some finite planning horizon $h \geq 0$ iff $\rho = \sum_j \mu_j/\eta_j < 1$.

Proof If $I^o \geq 0$ then the assertion is trivially true at $h = 0$. Hence, our concern is with $I^o < 0$.

The necessity of the condition is obvious for otherwise the available capacity in any finite h is less than the demand, hence it is impossible to satisfy all the demand in that horizon, let alone compensate for any initial backorders.[†]

The proof of sufficiency is by construction: we demonstrate that the condition is sufficient to generate at least one schedule with the desired properties. Set

$$h = [\eta_1, \eta_2, \ldots, \eta_N] \; ; \tag{1.1}$$

and

$$x_j = (\mu_j / \eta_j) \; h.$$

The condition $\rho < 1$ immediately yields that $\sum x_j < h$. Denote the slack $(= (h - \sum x_j) > 0)$ by s.

By construction, $p_j \; x_j = r_j \; h$, all j. Hence producing over $\sum x_j$ is sufficient for all the demand in h. Consequently, producing any item in the slack periods $s > 0$ must go towards either reducing any initial backorders, or accumulating stock. Hence, any finite initial backorders $I^o < 0$ will be eliminated over a finite number of horizons of duration h. Thereafter positive accumulation of stock must take place and a vector $I^t > 0$ is achievable for large enough t. Q.E.D.

[†]A more formal argument proceeds as follows: it is desired that $I_j^o + p_j \; x_j - r_j \; h \geq 0$, all j, $I_j^o \leq 0$ with at least one $I_j^o < 0$, where x_j is the number of periods (in h) devoted to item j. Thus $p_j x_j - r_j h \geq |I_j^o|$, or $x_j - (r_j/p_j) \; h \geq |I_j^o|/p_j$, all j. Summing over all j, $\sum x_j - h \; \sum(r_j/p_j) \geq \sum(|I_j^o|/p_j) > 0$. Thus $\sum x_j > h$, contradicting the condition that $\sum x_j \leq h$.

This completely settles the question of feasibility. Next, we treat the problem of "reachable states" from any given initial state.

Theorem 2 (Communicating States Theorem) Assume that $\rho = \sum_j (\mu_j/\eta_j) < 1$, and suppose that there exists a feasible schedule from initial state I to another state I'. Then there also exists a feasible schedule from I' to I in a finite number of periods. (The two states are said to "communicate" with each other.)

Proof Existence of a feasible schedule from I to I' implies the existence of a feasible solution in nonnegative integers to the set of equations:

$$I' = I - \begin{bmatrix} r_1 & -p_1 & & & \\ r_2 & & -p_2 & & \\ \cdot & & & \cdot & \\ \cdot & & & & \cdot \\ \cdot & & & & \cdot \\ r_N & & & & -p_N \end{bmatrix} \begin{bmatrix} x_o \\ x_1 \\ \cdot \\ \cdot \\ \cdot \\ x_N \end{bmatrix} \tag{1.2}$$

satisfying the constraints.

$$x_o \geq \sum_j x_j \; ; \quad x_j \geq 0, \text{ all } j \; ; \tag{1.3}$$

in which the length of the horizon is denoted by x_o (instead of h). We shall refer to Eqs. (1.2) as the state transformation equations. For brevity, denote the $N\times(N+1)$ matrix in (1.2) by A and the column vector of x's by $<x_o,X>$.

Clearly, $- <x_o,X>$ is a solution to the system of equations

$$I = I' - A. <x_o,X> \; ;$$

the jth term of which is

$$- r_j x_o + p_j x_j = I'_j - I_j \; ; \quad j=1,\ldots,N \; . \tag{1.4}$$

Hence, by an elementary theorem of number theory, it must be true that

$$d_j \mid (I'_j - I_j) \ ;$$

(recall that $d_j = (r_j, p_j)$). Furthermore, $-x_o + t\,p_j$ and $x_j + tr_j$ also satisfy (1.4) for all integer t. In particular, let

$$P = [p_1, p_2, \ldots, p_N] \ ;$$

then $-x_o + t\,p_j(P/p_j) = -x_o + tP$ and $x_j + t\,r_j(P/p_j)$ satisfy (1.4) for all j. Since P is > 0, and the term $-x_o + t\,P$ is independent of j, then for sufficiently large t we can make the horizon $h = x_o + tP$, as well as all the y_j's ; $y_j = x_j + tr_j(P/p_j)$, strictly positive.

Now it remains to show that this solution also satisfies the "planning horizon constraint":

$$\sum_j y_j = \sum_{j=1}^{N} (x_j + tr_j(P/p_j)) \leq x_o + tP = y_o \ . \tag{1.5}$$

From (1.4), it is easy to deduce that (by dividing throughout by p_j and then summing over all j),

$$\xi = \sum_j x_j = x_o \sum_j (r_j/p_j) + \sum_j (I'_j - I_j)/p_j \ . \tag{1.6}$$

Equation (1.6) is that of a straight line with "intercept" equal to

$$\sum_j (I'_j - I_j)/p_j = c \ , \text{ say,}$$

a constant, and a "slope" equal to

$$0 < \sum_j (r_j/p_j) = \rho < 1 \ , \quad \text{by hypothesis.}$$

I.e., (1.6) is $\xi = \rho x_o + c$. (1.6a)

Clearly, c is either positive, negative, or = 0. In any case, there is a value of $\underline{x}_o \geq 0$ at which $0<\xi \leq x_o$ for the <u>first time</u>, and s.t. ξ remains $< x_o$ <u>for all</u> $x_o > \underline{x}_o$. (This is illustrated in Fig. 1.1). Hence for large enough t in (1.5), $\xi \leq x_o$; i.e., the vector $Y = \langle y_o, y_1, \ldots, y_N\rangle = \langle x_o + tP,\ x_1 + tr_1(P/p_1), \ldots, x_N + tr_N(P/p_N)\rangle$ is s.t.,

$$\nu = \sum_j y_j = \sum_j (x_j + tr_j(P/p_j) \leq y_o = x_o + tP;$$

which was to be shown. This completes the proof. Q.E.D.

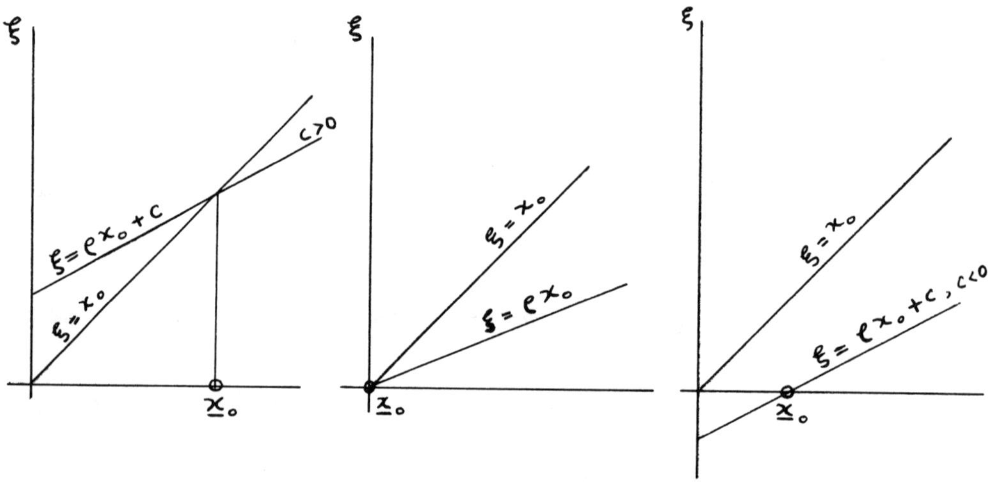

<u>Figure 1.1</u>

Theorem 2 leads immediately to the

Definition: Communicating Class: A subset of states $S \subseteq \Omega$ is said to be a communicating class (for a given set of parameters (r_j) and (p_j)) if there exists a feasible schedule between any pair of states in it. I.e., for any two states I, $I' \in S$, there exists a positive vector (x_o, X) s.t., (1.2) and (1.3) are satisfied, and there exists a (possibly different) vector (y_o, Y) s.t. again (1.2) (with the positions of I and I' interchanged) and (1.3) are satisfied.

Corollary 2.1 Communicating classes are mutually disjoint.

Proof If any two communicating classes share but one state, then one can use the common state to "jump" from one class to another. But this implies that the states in the union of the two classes communicate with each other, which is the definition of a communicating class. Hence, they could not have been designated as two "different" classes to start with: a contradiction. Hence, the two classes must be disjoint. Q.E.D.

Corollary 2.2 The set of communicating classes is a partition of the state space Ω.

Proof Obvious.

Having established the structure of the state space, it is essential for theoretical as well as computing purposes to determine the number of communicating classes for any given set of parameters (p_j), (r_j). To this end, define M_N to be the number of disjoint communicating classes in Ω.

Theorem 3 (The Partition Theorem) Under the assumption of finite N and $\rho < 1$, M_N is finite and is given recursively by

$$M_n = M_{n-1} \times \max_{j=1,\dots,n-1} \{(r_n \eta_j \, , \, p_n \mu_j)\} \; ; \quad n=2,3,\dots,N \tag{1.7}$$

with $$M_1 = d_1 = (r_1, p_1) \; . \tag{1.8}$$

<u>Proof</u> The theorem is easily established by induction on n and recognizing that each subspace for n-1 products must itself be subdivided further for the $n^{\underline{th}}$ product. Due to its length, the proof will be omitted. The interested reader may consult ref. (12) or ref. (22).

<u>The Partitioning Algorithm</u>

We now present the algorithm to generate the generic M_N vectors, one from each of the M_N communicating classes. The algorithm generates a tree having M_N terminal nodes and each branch represents a vector from a communicating class.

<u>Step 0</u> Put $k = 1$, $M_k = q_k = d_k = (r_k, p_k)$. Label the initial node as Ω.

<u>Step 1</u> From each unbranched node, branch to q_k nodes and label them as $I_k = 0, 1, \ldots, q_k - 1$.

<u>Step 2</u> Put $k = k + 1$; if $k = N+1$, stop; otherwise go to Step 3.

<u>Step 3</u> Compute $q_k = \max\limits_{j=1,\ldots,k-1} \left((r_k\eta_j \ , \ p_k\mu_m)\right)$ and $M_k = M_{k-1}\, q_k$. Go to Step 1.

<u>Sensitivity Analysis</u>

It is intuitively obvious that if we wish to build up stock to a level "higher" than its level at any time t then there should be enough "slack" between the available horizon x_o and the number of productive periods $\xi = \sum x_j$, both of which satisfy the homogeneous set of Diophantine equations

$$\begin{gathered} - r_j\, x_o + p_j x_j = 0 \quad ; \qquad j=1,2,\ldots,N \\ \sum_j x_j \leq x_o \quad ; \qquad x_j \geq 0 \, . \end{gathered} \tag{1.9}$$

Equally obvious, such a task increases in difficulty as $\rho \to 1$, since then the production facility approaches full utilization just satisfying the demand, let alone exceeding it. We term the study of the impact of ρ approaching 1 on x_o "sensitivity analysis".

Interest in the behavior of the system as $\rho \to 1$ is really meaningful only when the desired state, I' is "equal to, on, or above" the initial state, I. The precise characterization of what is meant by "on, or above" proceeds as follows. Certainly, if $I'_j \geq I_j$ for all j, with at least one i s.t. $I'_i > I_i$, then I' is "above" I. However, this is a stronger requirement than is needed.

Recalling the discussion leading to Eq. (1.6a), the intercept c was defined by

$$c = \sum_j (I'_j - I_j)/p_j \quad . \tag{1.10}$$

We now <u>define</u> I' <u>to be "above"</u> I if c > 0. This is a weaker requirement than before, for now some of the coordinates of the state vector may in fact be <u>smaller</u> for I' than I. However, these are more than compensated for by the other coordinates. The end result is an <u>increase</u> in the total demand on the system.

We now define I' to be "on" I if c = 0 in (1.10).

The case c < 0 is not interesting in sensitivity analysis because it implies a <u>relaxation</u> of the requirements on the system. Hence, the fact that $\rho \to 1$ is irrelevant. For instance, if the system is not penalized for backorders, there is no <u>real</u> demand irrespective of the value of ρ!

Returning to (1.6a) and Fig. 1.1, the straight line $\xi = \rho x_o + c$ must intersect the lines $\xi = 0$ and $\xi = x_o$ sooner or later. Let the point of intersection with $\xi = x_o$ be denoted by $x_o^{(1)}$; i.e., let

$$x_o^{(1)} = c/(1-\rho) \quad . \tag{1.11}$$

Furthermore, let $\xi = 0 = \rho\, x_o + c$ at $x_o^{(2)}$; i.e., let

$$x_o^{(2)} = -\, c/\rho \quad . \tag{1.12}$$

Since $0 < \rho < 1$, then $x_o^{(1)}$ and $x_o^{(2)}$ are either of opposite signs, or $x_o^{(1)} = 0 = x_o^{(2)} = c$; see Fig. 1.2. Let

$$\bar{x}_o = \max \langle x_o^{(1)}, x_o^{(2)} \rangle \; ; \quad \bar{x}_o \geq 0 \tag{1.13}$$

and let h be the smallest integer $\geq \bar{x}_o$ which generates a feasible production schedule X from initial state I to final state I'. (Throughout this section we assume that states I and I' belong to the same communicating class.) Clearly, h is the shortest horizon in which I' can be reached from I.

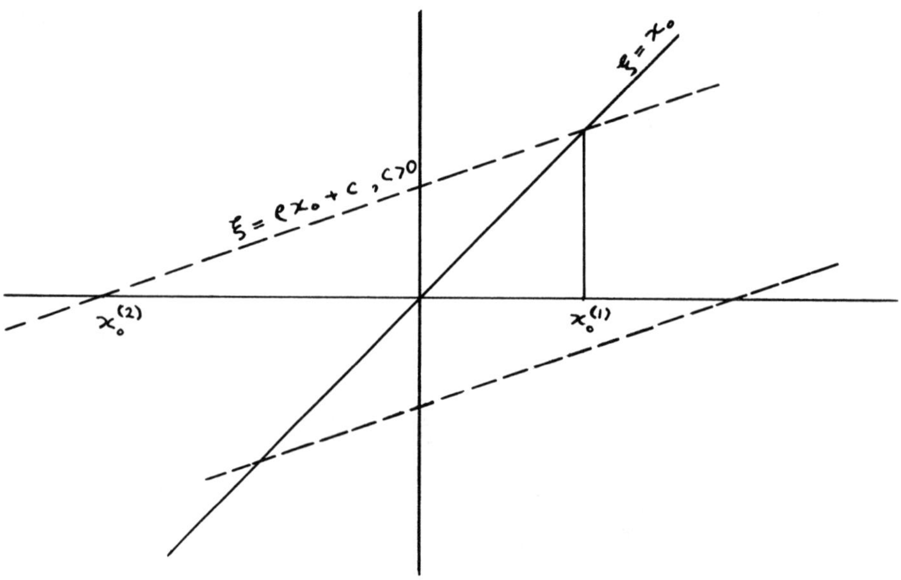

Figure 1.2

Consider first the case $c > 0$. Equation (1.11) indicates that $x_o^{(1)} \to \infty$ as $\rho \to 1$. Hence $h(\geq \bar{x}_o \geq x_o^{(1)})$ also $\to \infty$ as $\rho \to 1$. This result coincides with the intuitive notions presented at the beginning of this discussion.

Consider next the case $c = 0$. There are two circumstances under which this can occur:

(i) when $I' = I$; the so-called "first return problem" ;

(ii) when $I' \neq I$, but due to compensating terms, the sum $\sum_j (I_j' - I_j)/p_j = 0$, the so-called "I' on I" case.

We start by analyzing case (i). Here, $\bar{x}_o = 0$, and we shall refer to h under these conditions as the cycle length. Recalling Lemma 1, and in particular the proof of sufficiency, we may now assert the following:

Theorem 4 In the first return problem, if $\rho < 1$, then the horizon h given by (1.1) is the smallest cycle length. Moreover, any other cycle must be an integer multiple of h.

Proof That h is the smallest cycle length is evident from the definition of η_j and ℓ.c.m. Furthermore, putting $x_j = (\mu_j/\eta_j)h$ guarantees that the state transformation equations (1.9) are satisfied. These equations are also satisfied by any vector (y_o, Y) in which $y_o = \alpha h$ and $y_j = \alpha x_j$; $\alpha = 1,2,\ldots$, and this is the only set of vectors $\{(y_o, Y)\}$ that satisfy (1.9)[†], which proves the theorem. Q.E.D.

Corollary 4.1 In the "first return problem", the necessary and sufficient condition for feasibility is that $\rho \leq 1$. Furthermore, h is always given by (1.1) irrespective of the value of ρ.

[†]This is established by contradiction: if one subtracts integer multiples of the vector (h,X) from (y_o,Y), one would eventually be left with a horizon smaller than h in which equations (1.9) are satisfied - an impossibility, by the minimality of h.

<u>Proof</u> Follows from the fact that h depends <u>only</u> on the set (η_j) and not on the ratios (μ_j/η_j). Thus ρ can achieve its supremum with no impact on feasibility or on the value of h. Q.E.D.

The case "I' on I" is slightly more complicated. Substitution of $c = 0$ in (1.6a) yields

$$\xi = \rho x_o \ . \tag{1.14}$$

Since $c = 0$, it must be true that some of the components of the difference $e = I' - I$ are < 0, and some others are > 0; with at least one component in each category. For economy of notation, let

$e_j = I_j' - I_j \quad ; \quad j=1,2,\ldots,N;$

$j = 1,2,\ldots,k \ ; \ 1 \le k \le N-1,$ be the index of the components $e_j < 0$; and denote the sum $\sum_{j=1}^{k} (\mu_j/\eta_j)$ by $\rho^- < 1$;

$j = k+1, k+2,\ldots,\ell \ ; \ k+1 \le \ell \le N,$ be the index of the components $e_j > 0$; and denote the sum $\sum_{j=k+1}^{\ell} (\mu_j/\eta_j)$ by $\rho^+ < 1$;

$j = \ell+1, \ell+2,\ldots,N$ be the index of the components $e_j = 0$; and denote the sum $\sum_{j=\ell+1}^{N} (\mu_j/\eta_j)$ by $\rho^o < 1$.

$P = [p_1,p_2,\ldots,p_N]$

$P^{(1)} = [p_1,p_2,\ldots,p_k] \ ; \quad P_j^{(1)} = P^{(1)}/p_j \ ; \quad j=1,\ldots,k$

$P^{(2)} = [p_{k+1}, p_{k+2},\ldots,p_\ell] \ ; \ P_j^{(2)} = P^{(2)}/p_j \ ; \quad j=k+1,\ldots,\ell$

$P' = \min \{P^{(1)}, P^{(2)}\}$.

Write $P^{(1)} = a_1P'$; $a_1 \ge 1$, integer, and $P^{(2)} = a_2P'$; $a_2 \ge 1$, integer, with either a_1 or a_2 exactly equal to 1. Then the condition $c = 0$, combined with the conditions: $e_j < 0$ for $j=1,\ldots,k$; $e_j > 0$ for $j=k+1,\ldots,\ell$, imply that

$$a_2 \sum_{1}^{k} P_j^{(1)} e_j = a_1 \sum_{k+1}^{\ell} P_j^{(2)} e_j \tag{1.15}$$

As long as $\rho < 1$, I' will be reached from I in a finite horizon since the state transformation equations are assumed feasible.

When $\rho = 1$, h still remains finite between any two states in the same communicating class. In this case, (1.14) reduces to

$$\xi = x_o ;$$

which, unfortunately, yields no clue to either x_o or the manner of its division among the products. The following analysis should assist in providing such clues.

Let $w_j = e_j/p_j$, all j. Consider the components in the first subset j=1,...,k: their state transformation equations are of the form

$$x_j - (\mu_j/\eta_j)x_o = w_j < 0 \quad ; \qquad j=1,2,\ldots,k$$
$$\sum x_j \leq x_o ; \tag{1.16}$$

and possess a finite (first) integer solution since $\rho^- < 1$, by construction, and the vectors $\langle I_1, I_2, \ldots, I_k \rangle$, $\langle I_1', I_2', \ldots, I_k' \rangle$ belong to the same communicating class under ρ^- by Theorem 1. Furthermore, for all $t \geq 0$, integer, the vector (y_o, Y) defined by

$$y_o = x_o + t\, P^{(1)} ,$$
$$y_j = x_j + t\, r_j(P^{(1)}/p_j) \; ; \quad j=1,2,\ldots,k, \tag{1.17}$$

also satisfies the state transformation equations (1.16) and the admissibility constraint $\sum y_j \leq y_o$.

Next, consider the components in the second set of indices $j=k+1,k+2,\ldots,\ell$: their state transformation equations are of the same form as (1.16) with the sole difference that $w_j > 0$ for $j=k+1,\ldots,\ell$. Again, this system of linear Diophantine equations possesses a finite (first) integer solution, since $\rho^+ < 1$ (by construction) and the vectors $\langle I_{k+1}, I_{k+2}, \ldots, I_{\ell} \rangle$, $\langle I'_{k+1}, I'_{k+2}, \ldots, I'_{\ell} \rangle$ are in the same communicating class, by Theorem 1. Furthermore, for all $t \geq 0$, integer, the vector (z_o, Z) defined by

$$z_o = x'_o + t\, P^{(2)}$$

$$z_j = x'_j + t\, r_j\, (P^{(2)}/p_j) \; ; \qquad j=k+1,\ldots,\ell; \tag{1.18}$$

also satisfies the state transformation equations and the admissibility constraint $\sum z_j \leq z_o$; where x'_o , $(x'_j$, $j=k+1,\ldots,\ell)$ are the (first) solution to the state transformation equations.

Finally, for the components in the third set of indices $j=\ell+1, \ell+2,\ldots,N$ we know, by Theorem 4, that the smallest horizon, h, and the corresponding production schedule (v_j) are given by

$$h = [p_{\ell+1}, p_{\ell+2}, \ldots, p_N]$$

$$v_j = (\mu_j/\eta_j)\, h \; ; \qquad j' = \ell+1,\ldots,N . \tag{1.19}$$

The vector (h,v) satisfies the set of state transformation equations defined on the state vector $\langle I_{\ell+1}, \ldots, I_N \rangle$ and $\langle I'_{\ell+1}, \ldots, I'_N \rangle$, which belong to the same communicating class under ρ^o .

If all three sets of state transformation equations are to be satisfied simultaneously, then it must be true that there exists a $t \geq 0$, integer, s.t.

(i) $y_o = z_o$, or, equivalently,

$$\alpha_o + t\, P^{(1)} = \beta_o + t\, P^{(2)} \tag{1.20}$$

$$\alpha_o = x_o + u\, P^{(1)} \;,\; \beta_o = x'_o + v\, P^{(2)} \; ; \qquad u,v \text{ integer} \geq 0.$$

(ii) y_o (or z_o) is an integer multiple of h of (1.19). (1.21)

The above argument indicates that the horizon will be infinite if: (a) $P^{(1)} = P^{(2)}$ but $\alpha_o \neq \beta_o$; (b) there exists no t satisfying (1.20) and yielding a y_o which satisfies (1.21).

We have just proven the

Theorem 5 For the case I' is "on" I , there exists a finite horizon u_o and a corresponding production schedule U which satisfy the state transformation equations

$$p_j u_j - r_j u_o = e_j \; ; \qquad j=1,\ldots,N \; (e_j \text{ is defined on page 15})$$

$$\sum u_j \leq u_o$$

iff there exists a finite integer $t \geq 0$ which satisfies conditions (1.20), (1.21), independently of the value of ρ.

§2. THE CASE OF FINITE HORIZON

The previous section addressed itself to questions of existence and feasibility of a production schedule. This, and the following section, address themselves to the generation of optimal schedules. Of course, this necessitates the introduction of the value system.

Under the assumption of continuing deterministic demand, the gross income from sales is fixed, and hence the maximization of profit is equivalent to the minimization of cost. In this respect we shall be concerned with three costs: the holding cost, the stocking cost and the changeover cost. Let

σ_j: cost of holding stock j, per average unit, per period

β_j: cost of backorder of j, per average unit, per period

δ_{ij}: cost of changeover when j follows i

As before, let x_o, unknown, denote the length of the planning horizon, and x_j, also unknown, the number of periods in x_o devoted to the production of j; with $\sum_j x_j \leq x_o$. Furthermore, I_j is the starting inventory of j and I_i' is its terminal inventory at the end of x_o.

It is desired to determine the smallest planning horizon x_o and the set of production periods (x_j), as well as their temporal sequence, which move the system from state I to state I' and minimize the total cost of holding, backorder and changeover.

It is easy to see that the state space is <u>finite</u> and is bound by $I_j - r_j \times (x_o - x_j)$ from below and by $I_j + (p_j - r_j)\, x_j$ from above.

<u>The Planning Algorithm</u>

Our first order of business is to discuss the algorithmic method by which the values $x_o, x_1, x_2, \ldots, x_N$, i.e., the production plan can be determined. In the sequel we distinguish between two procedures: the one that determines the values of $x_o, x_1, \ldots, x_N$, assuming feasibility and subject to the given system parameters, which will be referred to as the <u>planning algorithm</u>, and the one that determines the precise temporal arrangement of the production periods $x_1, x_2, \ldots, x_N$ within the horizon x_o. This latter will be referred to as the <u>scheduling algorithm</u>. This two-step procedure is necessitated by the fact that the planning algorithm is independent of the cost parameters (and hence does not involve any cost minimization) while the scheduling algorithm optimizes the given production plan w.r.t. the specified cost parameters.

In this section we develop the planning algorithm.

The state transformation equations

$$p_j\, x_j - r_j\, x_o = \alpha_j \quad ; \qquad j=1,\ldots,N \text{ (recall } \alpha_j = I_j' - I_j) \tag{2.1}$$

can be written as a set of linear congruence relations

$$r_j\, x_o \equiv -\alpha_j \pmod{p_j} \quad ; \qquad j=1,\ldots,N \tag{2.2}$$

This is a system of linear congruences which may, or may not have a solution. It is well known[†] that they possess a solution iff $(p_i,p_j)|(\alpha_i-\alpha_j)$ for each pair of subscripts i, j. If this is the case and if h is a solution, then the general solution is given by

$$x_o \equiv h \pmod{[p_1,p_2,\ldots,p_N]} \; . \tag{2.3}$$

The method of obtaining h may be found in any introductory book to Number Theory. However, we wish to emphasize that h of (2.3) must satisfy the condition

$$\sum_{j=1}^{N} x_j \leq x_o = h \quad ; \tag{2.4}$$

in which the x_j's are obtained from (2.1). We refer to the resultant production plan as the "shortest production plan".

The Scheduling Algorithm

As a preliminary to the general formulation, consider first the case of <u>no change-over cost</u>. This simpler case gives rise to Transportation-type constraints with a nonlinear objective function. We do not exhibit this formulation.[*] But the exact solution of the problem is achieved through a network interpretation of a dynamic programming (DP) model. As it stands, the solution is achieved through a network representation and a shortest path calculation. We shall refer to this network as the <u>schedule network</u>. Furthermore, the structure of the network model suggests a heuristic approach which: (i) is extremely fast; (ii) yields an objective value at, or very close to, the optimal on the first trial; (iii) provides a good starting point for searching the space of feasible points for the optimum.

---

† See, e.g., James E. Shockley, "Introduction to Number Theory", Holt, Reinhart Winston, 1967, p. 47.

* The interested reader may consult ref. [12].

Let the states be represented by nodes $s_1, s_2, \ldots, s_M$; where state s_1 is the given initial inventory I at time 0 and state s_M is the terminal desired inventory I' at time x_o. In general, $s_m(t)$ occurs at time t (i.e., at end-of-period t). There exists a directed arc from $s_m(t-1)$ to $s_n(t)$ if the $i^{\underline{th}}$ coordinate of $s_n(t)$ is exactly r_j units less than the $i^{\underline{th}}$ coordinate of $s_m(t-1)$ for all i (the case of 'do nothing' in period t), except possibly for the $j^{\underline{th}}$ coordinate which is larger by $p_j - r_j$ (the case of producing item j in period t).

<u>Theorem 6</u> Given the shortest production plan $x_o, x_1, \ldots, x_N$, the resulting network is directed, acyclic, and with one "source", $s_1(0) \triangleq I$ and one "terminal", $s_M(x_o) \triangleq I'$, in which the length of every path from s_1 to s_M is exactly x_o .

The value of M, the number of nodes in the network, is easily obtained from the production assignment. Let x_{N+1} denote the slack in the horizon, $(= x_o - \sum_{j=1}^{N} x_j)$; then it is easy to verify that $M = \prod_{j=1}^{N+1} (x_j + 1)$.

It remains to determine the "cost" of each arc of the network where, in this context, the "cost" of an arc joining nodes s_m and s_n and falling in period t is the cost incurred in the period, given by

$$c_{mn}(t) = \sum_{j=1}^{N} [\sigma_j(S_{j,t-1} + S_{jt})/2 + \beta_j(B_{j,t-1} + B_{jt})/2] \tag{2.5}$$

where S_{jt} is the "stock" at the end of period t and B_{jt} is the "backorder", with $S_{jt}\, B_{jt} = 0$ for all t. We remark that the only difference between the analysis of the case in which the cost of change-over is taken into account and the case in which it is not lies in the evaluation of $c_{mn}(t)$. In particular we must then add a term representing the change-over cost to (2.5), see Eq. (2.6) below.

It is now evident that the minimal cost schedule is given by the shortest path from s_1 to s_M. Since the network is acyclic, that determination is accomplished in exactly one pass and x_o iterations. Assuming s_1 is at time 0 and s_M at time x_o, the minimum cost is determined recursively from:

$$v(s_1(0)) = 0$$

$$v(s_n(t)) = \min_{s_m \varepsilon P(s_n)} [v(s_m(t-1)) + c_{mn}(t)] \ ; \ t=1,2,\ldots,x_o \ ; \ n=2,3,\ldots,M$$

where $v(s_n(t))$ is the minimal cost of node $s_n(t)$, and the set $P(s_n)$ is the set of all nodes that "immediately precede" node s_n.†

As an illustration of the scheduling algorithm, consider the problem of producing two items with the following parameters:

Product	1	2
p_j	4	3
r_j	2	1
I_j	0	1
I'_j	6	2
σ_j	1	1.5
β_j	4	3
x_j	7	4; $x_o = 11$

The last row of the table gives the production plan, determined by the approach described in the previous section. Since the slack x_{N+1} is = 0, the network contains $\prod_{j=1}^{2} (x_j+1) = 40$ nodes. Figure 2.1 enumerates the nodes of the network; the right hand scale indicates the time periods $t=1,2,\ldots,x_o = 11$. To each node there are attached three numbers: the node number (running from 1 to 40), the state of inventory represented by that node (the vector(I_1^t, I_2^t)), and the minimal cost of the node, $v_n(t)$, (determined following (2.11)). The shortest

---

†These are nodes which are connected to s_n with arrows leading into s_n.

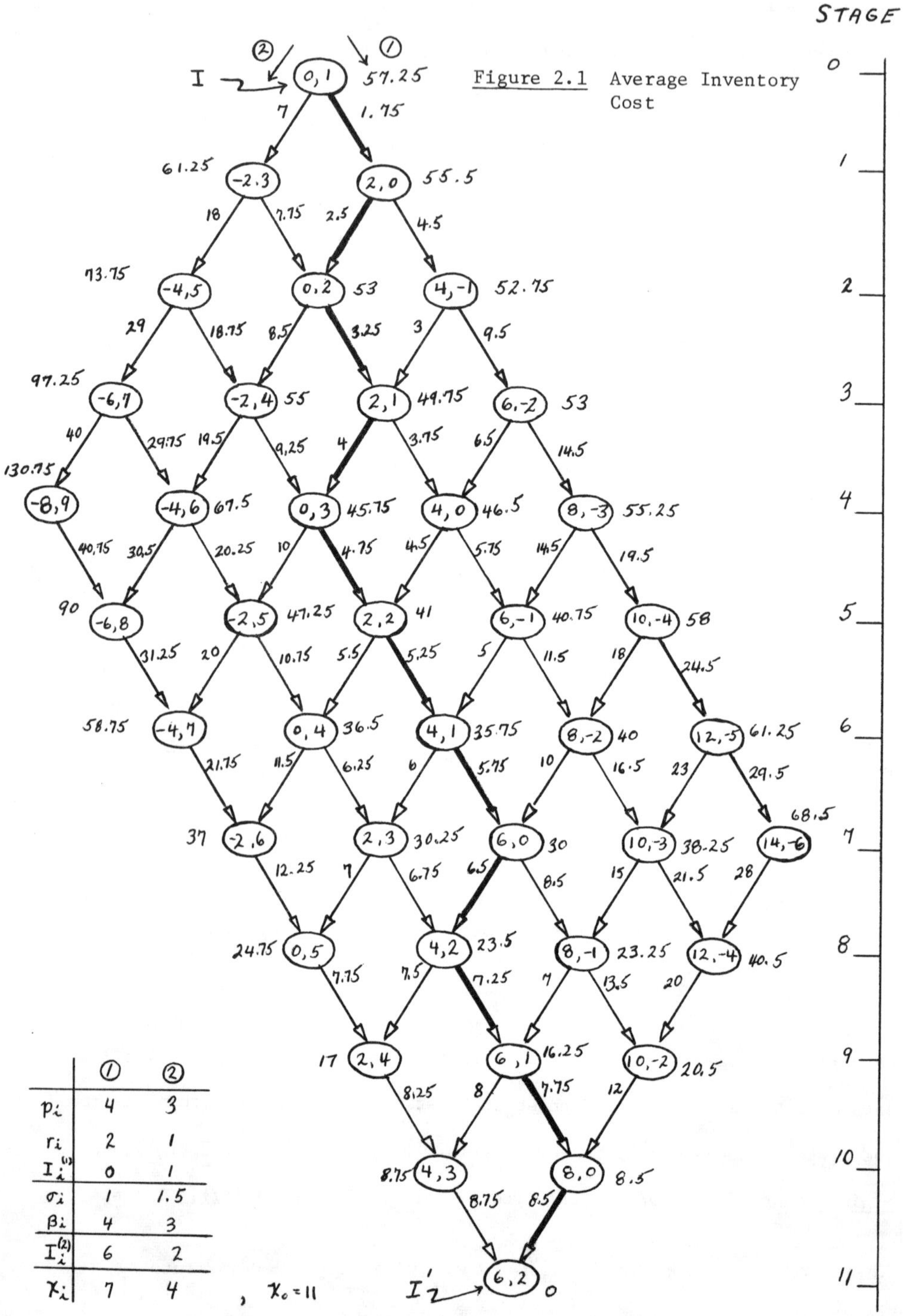

Figure 2.1 Average Inventory Cost

	①	②
p_i	4	3
r_i	2	1
$I_i^{(1)}$	0	1
σ_i	1	1.5
β_i	4	3
$I_i^{(2)}$	6	2
χ_i	7	4

, $\chi_0 = 11$

path is indicated by the heavy lines. The minimum is seen to be 57.25 units, and the optimal schedule is given by: 1,2,1,2,1,1,1,2,1,1,2.

Characteristics of the Schedule Network and an Approximate Procedure

It is easy to see that the size of the schedule network may be inordinately large while the optimal path utilizes only a miniscule fraction of the total number of paths (for a plan with $x_o = 100$, $x_j \simeq 10$ for $j=1,\ldots,10$; $M = (11)^{10} \simeq 225(10)^8$ nodes, while the optimal path utilizes only 101 nodes, or approximately $0.2\ (10)^{-9}$ of the nodes!). Thus it is easily seen that the construction of the complete network, even if it were feasible, provides a great deal of redundant information. Furthermore, it is evident that the relative magnitude of the cost parameters would give an indication that "outlying" nodes will be excessively expensive and consequently will be dominated by nodes within a "narrow band" about the optimal. It is only the states in this narrow band, which we shall term the core states, that should be investigated - the overwhelming majority of the nodes of the network are irrelevant. This would reduce the computing burden considerably.

We now make these notions precise.

By a path of the network we mean a directed path from node s_1 to node s_M, of exactly x_o arcs. Evidently, a path represents a production schedule, which, in turn, can be represented by a permutation of the numbers 1 to N, in which j appears x_j times. Two paths are said to be adjacent if they can be obtained from each other by a contiguous binary switch (CBS) in their permutation representation, i.e., if the order of two different products which are adjacent is switched. In this respect we note that a CBS is graphically represented by a diamond shape, and that successive CBS's involve "diamond shapes" of various sizes; see Fig. 2.2.

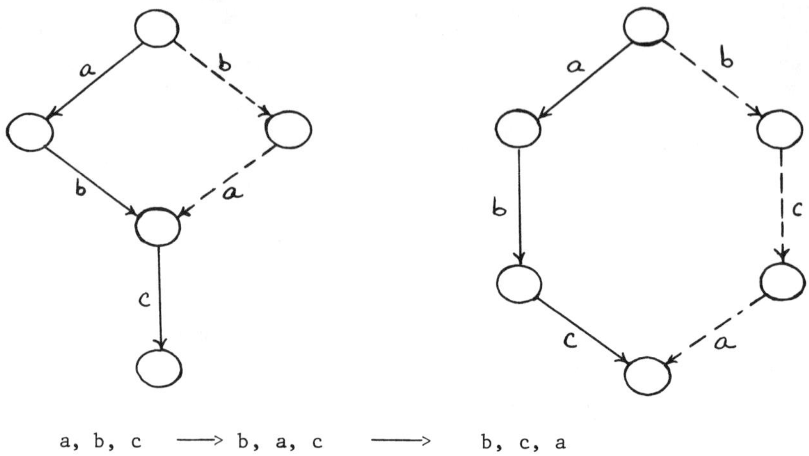

Figure 2.2

It is equally evident that there is a one-to-one correspondence between the paths of the schedule network and the permutations. Consequently, there are exactly $x_o!/\prod_{j=1}^{N+1} x_j$ paths and each path has at most x_o-1 adjacent paths; since there are so many CBS's in any permutation in which the same product does not appear in two contiguous positions. Furthermore, any path is accessible from any other path by a series of CBS's.

These properties are summarized as follows:

(a) Any path in the schedule network can be obtained from any other path by a sequence of CBS's;

(b) There exists at least one optimal path in the network.

Unfortunately, it is not evident to us at the moment that every non-optimal path has at least one adjacent path which is cheaper, although our (limited) numerical experience to date indicates that this conjecture was realized in every case we treated.† Still, we recognize that this does not constitute a proof of anything, and the property stands as an unverified conjecture.

Nevertheless, it indicates that perhaps a reasonable procedure to follow in approximating (and possibly achieving) the optimum is to start with a "reasonable" schedule then proceed to improve on it by a series of CBS's until no further improvement is possible.

A reasonable heuristic to follow in constructing the starting trial schedule is to sequence next the product of least immediate cost (including 'do-nothing') provided such production is permissible (i.e., the production of j is less than x_j periods thus far). Improvement on this starting schedule by the application of the CBS operation may proceed along any of the following rules (among several possible):

1. The Myopic Rule: investigate only CBS's; perform the switch at the first contiguous pair of products which shows a reduction in cost. Repeat with the new schedule, until no CBS causes any reduction in cost.
2. The Extended Rule: investigate only CBS's, but search the whole path for the most economical switch, then perform it. Repeat until no CBS is profitable.
3. The Look-Ahead Rule: instead of considering only two consecutive periods, consider four consecutive periods, and let the items produced be a,b,c,d, respectively. Compare the two binary switches

†If this conjecture is validated, this would establish the graph constructed on the the paths (i.e., the graph each node of which represents a path) as a convex graph, see ref. [13]. One would then immediately conclude that starting from any arbitrary feasible schedule one can arrive at the optimal schedule in a finite number of steps characterized by paths of non-increasing values. This idea is the basis for the approximate procedure presented.

ba + dc versus the single switch cb. If the former yield the greater reduction, perform the switch ba only, then proceed with the remainder of the schedule a,c,d,... . If the latter single binary switch yields the greater reduction, leave product a in its first position and proceed with the investigation of the remainder of the schedule b,c,d,...

This approximate procedure was applied to a few numerical examples with excellent results. It still remains to demonstrate its behavior in large size problems.

Considering Change-Over Cost

When change-over cost, δ_{ij}, is taken into account, the above discussion is modified in two major respects:

(i) There is need for identification of the last item produced on the machine, and identification of the occurrence of a change-over on the schedule-network.

(ii) The new cost must be reflected in an appropriately modified criterion function.

To accommodate both changes, write the state as $s_m(t,j)$ to indicate that it is "node number m", m=1,2,...,M; occurring in period t, and that product j was the last item produced on the machine. Note that due to the possible presence of idle periods, the j in the state designation may actually refer to the item produced several periods before.

Let $\delta[s_m(t-1,j),\ s_n(t,k)]$ denote the cost of change-over of the arc joining the two nodes $s_m(t-1,j)$ and $s_n(t,k)$; then

$$\delta[s_m(t-1,j)\ ,\ s_n(t,k)] = \delta_{jk} \quad ;$$

as given by the change-over cost matrix.

Let π denote a path in the schedule network. The cost of π is then given by

$$C'(\pi) = \sum_{\substack{t=1 \\ t,j \text{ on } \pi}}^{x_o} \sum_{j=1}^{N} [\sigma_j(S_{j,t-1} + S_{jt})/2 + \beta_j(B_{j,t-1} + B_{jt})/2]$$

$$+ \sum_{t=1}^{x_o} \sum_{m,n\in\pi} \delta[s_m(t-1,j),\ s_n(t,k)] \ ; \tag{2.6}$$

where the "source node" of inventory I is denoted by $s_1(0,0)$ and the "terminal node" of inventory I' is denoted by $s_M(x_o,j_M)$, for some $j \in N$; and in which the stocks S_{jt} and backorders B_{jt} are taken along the path π.

Let $f[s_m(t,j)]$ denote the minimal cost from node $s_m(t,j)$ to the terminal node $s_M(x_o,j_M)$. Then we have the DP functional equation:

$$f[s_m(t,j)] = \min_{k\in D_{t+1}} [(\text{cost of producing } k \text{ in period } t+1)$$

$$+ (\text{opt. cost of proceeding thereafter})]$$

$$= \min_{k\in D_{t+1}} \{c'_{mn}(t+1) + f[s_n(t+1,k)]\} \ ; \tag{2.7}$$

where $s_n(t+1,k)$ is a legitimate "descendent" of state $s_m(t,j)$; the space D_{t+1} is the space of possible products at stage t+1; and $c'_{mn}(t+1)$ is one term of (2.6).

The straightforward application of the DP formulation of (2.7) is onerous. Fortunately, it bears direct interpretation within the framework of schedule networks. Consequently, by a simple modification of the previous treatment one may still use the same basic principles.

In particular, we continue representing the state $s_m(t,j)$ by the vector of inventories I_t, _in conjunction with the last product produced_, j. This necessitates repeating each previous state (see Fig. 2.1, for example) at most N times, corresponding to the N possible products by which the same state could be generated. The schedule-network with no change-over cost can be used to generate the schedule-network with change-over cost. For the arrows in the former network are still valid in the latter network, except that they must now be repeated for each node in the family of nodes (I_t, j) for $j=1,2,\ldots,N$. In fact the previous costs (obtained by disregarding change-over costs) may form the basis for calculating the new costs when change-over cost is taken into account.

The shortest path calculations in the new schedule-network remain exactly the same as before.

For illustrative purposes, the two-product example of Fig. 2.1 was treated again assuming a non-symmetric change-over cost: $\delta_{12} = 5$, $\delta_{21} = 2$. The resulting optimal schedule is 1,1,2,2,1,1,1,1,2,2,1,1 ; and the corresponding cost is 73 units. (Notice the difference between the two optimal schedules, and the impact of the high change-over cost from product 1 to product 2.)

As a final remark, two comments are pertinent. First, the approximating procedure that seems to be so successful when change-over cost is ignored does not seem to carry over to this case with near equal validity. This is not surprising in view of the nature of the problem. Thus, there seems to be no escape from constructing the complete network, except through branch-and-bound approaches, which are discussed in the next section.

Second, the size of the network has increased approximately N-fold. This, coupled with the previous remark emphasizes the need for implicit enumeration approaches for the solution of real life problems.

§3. THE CASE OF INFINITE HORIZON

When the planning horizon grows without bound, it is meaningless to speak of reaching some state I' from some initial state I - such a task may be accomplished in a _finite_ horizon (assuming, of course, that I' communicates with I) and the unboundedness of the horizon is then irrelevant. A more pertinent question, in the case of unbounded horizon, is the following: what is the most advantageous strategy to follow first to get the system in a "favourable" inventory position and thereafter maintain its "good standing" forever? (The terms "favourable" and "good standing" will be made more precise below.)

Our approach to the resolution of this problem may be summarized in three basic steps. _First_, we demonstrate that although the state space is denumerable, we may limit ourselves to a _finite_ subset of it with no loss of optimality whatsoever. This seems intuitively appealing once it is realized that we insist, as indeed it is reasonable to do so, that both stockage and shortage incur a positive cost. _Second_, having thus delimited the state space we appeal to the rich theory of finite-state Markov chains to derive the properties of the optimal strategy. In particular, we specialize these results to degenerate Markov processes (since we are dealing with a _deterministic_ system) and establish that the optimal strategy must be a stationary policy, in particular, a policy that defines a _Turnpike cycle_. _Third_, we achieve an important link between the DP approach to such problems on the one hand and the LP solution (and others) of "least expensive cycle" problems on the other.

Since our research is continuing on the computational aspects of the proposed approach, and the results of such research will be reported on in a future report, we content ourselves here with a brief resume of its salient features.

It is easy to see that even if we restrict our attention to the states in a single communicating class, the number of states in this class is still denumerable. However, we assert that for optimality considerations, we are only interested in a finite subset of states, assuming that the stockage cost, σ_j, and the backorder cost, β_j, are positive. (The absence of this condition renders the problem vacuous.) The proof of this assertion anticipates the conclusion that the optimal solution is a finite cycle over the states; viz., the Turnpike cycle. Each state of that cycle must have finite coordinates since an infinite coordinate will incur an infinite cost, which is obviously nonoptimal. Hence, if one were to imbed such optimal cycle in a "rich enough" subset of states - which is still finite - one may restrict his attention to this subset without any loss of optimality. We refer to this subset of states as the "core states". (Of course, the sufficiency of the subset is conditional upon its containing the optimal cycle.)

One can estimate this set of core states relatively easily based on the cost parameters and the fact that the optimal production plan (the vector $(x_o^*, x_1^*, \ldots, x_N^*)$) is known.

Assuming that we restrict ourselves to a finite space of core states S, which is "rich enough" to contain the optimal strategy, we proceed to characterize the structure of such optimal strategy. To this end, we appeal to some well-known results in the theory of finite-state, finite-decision Markov Dynamic Programming. We take as our objective the minimization of the average total cost over the infinite horizon.

Under the hypotheses of completely ergodic Markov chain with finite state and decision spaces, and the criterion of minimum average cost, we conclude[6] that there exists an optimal strategy that is stationary.

Specializing this conclusion to the case of degenerate Markov Processes, we obtain the following results:

(i) The class of feasible policies F is s.t. in any state $s \in S$, the set $\{f(s);\ f \in F\}$ is nonempty; i.e., each state must have a successor state under each f.

(ii) It follows from the finiteness of the state space and the stationarity of f over the infinite horizon that the optimal stationary policy must be a cycle, the so-called Turnpike Cycle. Once the system reaches any state on this Cycle, it stays "locked" onto the Cycle forever thereafter.

(iii) The "value" of any state in the Turnpike Cycle is defined to an additive constant - hence it is the difference between the values at two states on the Cycle that is important, not the absolute measure at either state. This difference measures the differential worth of "starting" in one state versus "starting" in the other.

Policy iteration within the formalism of the theory of Dynamic Programming leads to a set of inequalities and a test for improvement that may be neatly interpreted within the framework of the theory of Linear Programming. Our resultant LP bears curious resemblance to the LP model for determining the cycle in a graph which minimizes the ratio of cost to time, due to Dantzig, Blattner & Rao[4]. (It is also a specialization to degenerate Markov processes of the LP's of Manne[21], Wagner[28] and de Ghellinck[5].) Hence their results are applicable to our case except for the following important difference: their LP model did not stem from a Markovian decision process (degenerate or otherwise) as ours did, hence an "outgoing" arrow from each state in the optimal solution was not necessary. But, for our case, to have a bona fide stationary policy f^{∞} we insist that the set of successors to each state $s \in S$ cannot be empty. Hence, we may use the approach proposed by Dantzig, Blattner & Rao to generate the Turnpike Cycle, which is indeed a prized result, and follow that up with solving for the optimal stationary policy at all states not on the Cycle.

Any algorithm (LP, shortest path, or policy iteration) will yield a Turnpike cycle of a fixed length given by multiples of h of Eq. (1.1).

At the time of this writing, it is not clear to us how to determine the correct multiple of h, other than by exhaustive enumeration of lengths h, 2h, 3h, etc., and stopping the search at the first increase in cost. Such an exhaustive procedure is clearly not computationally acceptable, since each iteration involves the determination of the optimal Turnpike cycle of that particular length; a no minor feat. This, as well as other questions listed below, are currently under investigation.

REFERENCES

1. Blackwell, D., "Discrete Dynamic Programming", Ann Math. Stat., 33, (1962), 719-726.

2. Bomberger, E. E., "A Dynamic Programming Approach to a Lot Size Scheduling Problem", Mgt. Sc., Vol. 12, No. 11, July 1966.

3. Brown, Robert G., Management Decisions for Production Operations, Dryden Press, Inc., Hinsdale, Ill. (1971).

4. Dantzig, G. B., W. O. Blattner, and M. R. Rao., "Finding a Cycle in a Graph with Minimum Cost to Time Ratio with Application to a Ship Routing Problem", Operations Research House, Stanford Univ., Technical Report No. 66-1.

5. de Ghellink, G. T., "Les Problemes de Decisions Sequentielles", Cahiers Centre d'Etudes Rech. Operationnelle, 2, (1960), 161-179.

6. Derman, G., "On Sequential Decisions and Markov Chains", Mgt. Sc., 9, 1 (Oct. 1962), 16-24.

7. Doll, C. Loren, "An Integer-Constrained EOQ-Based Single Machine Scheduling Heuristic", Paper No. 337, Oct. 1971, The Inst. for Res. in the Behavioral, Economic and Mgt. Sciences, Purdue Univ., Lafayette, Ind.

8. Dooley, J. E. and Ben Bernholtz, "A Many Product Single-Machine Recursive Scheduling Procedure", Can. Oper. Res. Soc. Jour., 7,3, (1969), 165-176.

9. Eilon, S., "Economic Batch Size Determination of Multi-Product Scheduling", Operational Research Quarterly, (England), Vol. 9, No. 4, Dec. 1958 .

10. Elmaghraby, S. E., "The Machine Sequencing Problem: Review and Extensions", Nav. Res. Log. Quart., Vol. 15, No. 2, June 1968.

11. ________ and A. Mallik, "The Scheduling of a Multi-Product Facility", O.R. Report No. 76, N. C.State Univ., April 10, 1972.

12. ________, A. Mallik and H.L.W. Nuttle, "The Sequencing of Lots on a Single Facility", Ind. Eng. Transactions, Vol. 2, No. 3, Sept., 1970.

13. ________, "A Graph Theoretic Interpretation of the Sufficiency Conditions for the Contiguous-Binary-Switching (CBS) Rule, Nav. Res. Log. Quart., 18, 3, 339-364 (Sept. 1971).

14. Glassey, C. R., "An Algorithm for a Machine Loading Problem", Jour. of I.E., Oct. 1967, Vol. XVIII, No. 10.

15. Haessler, R. W., "A Note on Scheduling a Multi-Product Single Machine System for an Infinite Planning Period", Letter to Ed., Mgt. Sc., 18, 4, (Dec. 1971), B240-B241).

16. Hanssmann, Fred, "Operations Research in Production and Inventory", John Wiley, 1962, pp. 158-160.

17. Harris, Ford, Operations and Cost (Factory Management Services), A. W. Shaw Co., Chicago, (1915), pp. 48-52.

18. Krone, L. W., "A Note on Economic Lot Sizes for Multi-Purpose Equipment", Mgt. Sc., 10, 3, 461-464 (Apr. 1964).

19. Madigan, J. G., "Scheduling a Multi-Product Single Machine System for an Infinite Planning Period", Mgt. Sc., 14, 11, (July 1968), 713-719.

20. Mallik, A., "The Scheduling of a Single Processor Under Deterministic Demand for Several Products", Unpublished Ph.D. Dissertation, N. C. State Univ., Raleigh, N. C., August, 1972.

21. Manne, A. S., "Linear Programming and Sequential Decisions", Mgt. Sc., 6, 3 (Apr. 1960), 259-267.

22. Maxwell, William L., "The Scheduling of Economic Lot Sizes", Nav. Res. Log. Quart., Vol. 11, Nos. 2 & 3, June-Sept. 1964.

23. Raymond, F. E., Quantity and Economy in Manufacture, McGraw-Hill, 1931.

24. Rogers, J., "A Computational Approach to the Lot Scheduling Problem", Mgt. Sc., Vol. 4, No. 3, April, 1958.

25. Salveson, M. E., "Mathematical Theory of Production", Jour. of Ind. Eng., Vol. 4, No. 1, Feb. 1953.

26. Shapiro, J. F., "Shortest Route Methods for Finite State Space Deterministic C.P. Problems", SIAM,Vol. 16, No. 6, Nov. 1968.

27. Vazsonyi, A., "Economic Lot Size Formulas in Manufacturing",Oper. Res., 5, 1 (Feb. 1957, 28-44.

28. Wagner, Harvey, Principles of Operations Research, Prentice-Hall, N. J., Chapter 12, (1969), pp. 425-430.

THE TWO-MACHINE JOB-SHOP WITH EXPONENTIAL PROCESSING TIMES

Hamilton Emmons

Case Western Reserve University, Cleveland, Ohio , USA*

Using a generalized random walk model, we analyze the problem of processing a fixed number of jobs on each of two machines. The order of processing is unconstrained, and the processing times are exponential random variables. Rather simple formulas are obtained for the machine idle time distribution and the mean makespan.

1. Introduction

The mathematical theories of queueing and scheduling presumably were once quite distinct. Like many other pairs of disciplines, their expanding frontiers have long since met, and the interface continues to be a rich research area.

The problem to be discussed here lies somewhere along that border, but we choose to use job-shop terminology. We consider a fixed number, n, of jobs to be processed on two machines. The jobs are considered to be indistinguishable, either because they are indeed all of the same type, or because our advance information is insufficient to distinguish among them. Also, there are no technological constraints: any job can be processed on either machine first. The times for each operation are random variables, each independent of all others, and exponentially distributed with parameters λ and μ on machines 1 and 2, respectively. There is no question of sequence here, the problem being to calculate the

---

*At the time of the Symposium, the author was at Cornell University, Ithaca, New York.

expected makespan and other measures of interest.

2. The Random Walk Approach

It will be convenient to visualize the process to be considered as a random walk over the integer lattice points of the positive quadrant of the plane (see Feller (1957), Milch and Wagonner (1970)). The state, (x,y), of the system at any time represents the number of jobs that have been processed on each machine up to that time. Thus, we start at the origin, and move one place to the right or one place up each time a job is completed on machine 1 or 2, respectively. Since all processing times are assumed to be exponential, the transition probabilities and times are obtained from the following well-known results.

If A and B are independent exponential random variables with parameters λ and μ, then:

(1) $\min(A,B)$ is exponentially distributed with parameter $\lambda + \mu$.

(2) The conditional distribution of $A - B$ given $B < A$ equals the distribution of A.

(3) $P(A < B) = \lambda/(\mu + \lambda)$.

Hence, if jobs with processing times A and B are either continuing or starting on machines 1 and 2, respectively, at the time we enter state (x,y), then we remain in that state an expected time $1/(\mu + \lambda)$ and then go to state $(x + 1, y)$ or $(x, y + 1)$ with probabilities $\lambda/(\mu + \lambda)$ and $\mu/(\mu + \lambda)$, respectively. If either server is idle, the corresponding parameter is zero. In this way, the random walk moves in $2n$ steps from $(0,0)$ to (n,n), where all jobs are completed on both machines.

3. Problem Formulation

Milch and Waggoner (1970) have considered the problem described above (they refer to it as a shutdown queue, the jobs being customers, etc.), but with an additional constraint: a predetermined number, n_1, of the jobs are initially assigned to machine 1, while the other $n_2 = n - n_1$ jobs are assigned to machine 2. Then as jobs are completed on one machine, they join the end of the queue at the other. This constraint is, of course, reasonable if there are indeed reasons why some jobs must first be processed on one machine and some on the other. If not, one would rather hold the jobs in a single queue, feeding them to the machines as vacancies arose and taking advantage of random fluctuations to put more or less work on a machine.

The computational advantage of predetermining two initial queues is that thereafter the state of the system at any time can be unambiguously specified by the numbers (x,y) of jobs that each machine has processed in that time. Thus, the random walk approach can be used straightforwardly. Nevertheless, considerable complications are encountered and the final formulas are quite messy (and of course they depend on n_1 and n_2).

We shall consider all n jobs to be initially held in a common queue, and issued to the two machines as needed. When all have been processed once, the second operations are performed as required. (For convenience, we shall refer to jobs that have not completed processing on either machine as type-0 jobs, and jobs that have been processed on one machine only as type-1 jobs.) By carefully defining the state space, it will be possible to use a modified random walk. Beyond this, we shall follow the procedure of Milch and Waggoner, obtaining finally a comparatively simple formula for the expected makespan.

First, observe where the new difficulty arises. As long as type-0 jobs remain unassigned, it can easily be seen that (x,y) is a complete state description. When the last such job is assigned to its first machine, we will have $x + y = n - 2$ and the last two type-0 jobs will be on the machines (it will be useful here and in what follows to consult Figure 1, which illustrates the modified random walk for the case $n = 5$). The next completion will leave only one type-0 job. As long as it remains uncompleted, a potential bottleneck exists: if the other machine completes all its type-1 work, it will have to wait for that one job, and this delay will affect the expected makespan. As soon as the last type-0 job completes its first operation, we can again completely specify the system state by (x,y), and each machine proceeds independently with its remaining.

It thus becomes necessary to specify a pair of additional states for the lattice points (x,y) where $x + y > n - 2$. Let $(x,y)'$ and $(x,y)''$ be the states where there is still a first-run job on one machine (the number of primes denotes which machine) which could potentially delay, or is currently delaying, the other machine (in Figure 1, these are the offcenter states at lower right and upper left of each lattice point, respectively). We denote by $[x,y]$ the states above $x + y = n - 2$ (centered at the lattice points in Figure 1) which have all first runs completed on both machines. When referring to the three states at a lattice point collectively, we shall retain the notation (x,y).

We now conclude that the difficulty comes at the points $(x, n - 1)$ and $(n - 1, y)$. If we reach $(x, n - 1)$ from the point (x,y) on the diagonal $x + y = n - 2$, that is, from the point $(x, n - 2 - x)$, then we are in state $(x, n - 1)'$ and the next transition can only be to $(x + 1, n - 1)$, in expected time $1/\lambda$. If we come to $(x, n - 1)$ from any other point $\{(x - k, n - 2 - x + k), k = 1(1)x\}$ on the diagonal, we reach state $[x, n - 1]$ so that we can also go to (x,n) and the expected time is $1/(\lambda + \mu)$.

4. The Number of Paths of Each Type

We now classify the paths from $(0,0)$ to (n,n) by the number, i, of upper boundary (UB) points: the circled points in Figure 1, where there is a forced transition to the right; and the number, j, of right boundary (RB) points, boxed in Figure 2, where an upwards transition is forced. Let R_{ij} be the set, and r_{ij} the number, of paths with i UB points and j RB points. Since the expected times and the probabilities of traversing any of the paths in R_{ij} are the same, it will be a simple matter to calculate the expected makespan once we evaluate each r_{ij}.

First define $r(x,y)$, $r(x,y)'$, and $r(x,y)''$ to be the number of paths from $(0,0)$ to (x,y), $(x,y)'$, and $(x,y)''$, respectively. First, it is a well-known fact that

$$r(x,y) = \binom{x+y}{x} .$$

We can then see with reference to Figure 1 that

$$r(x,\ n-1)' = r(n-1,\ x)'' = \binom{n-2}{x} \qquad (x = 0(1)n-2)$$

since the only way to reach $(x,\ n-1)'$ is to go first to $(x,\ n-2-x)$ and then follow the unique path straight up.

Now define $r[x,\ n-1]$ to be the number of paths from $(0,0)$ to $[x,\ n-1]$ which avoid all boundary states $(k,\ n-1)'$ and $(n-1,\ k)''$. Then

$$r[x,\ n-1] = r(x,\ n-1) - \textstyle\sum_{k=0}^{x} r(k,\ n-1)' \qquad (x = 1(1)n-2)$$

with

$$r[n-1,\ n-1] = r(n-1,\ n-1) - \textstyle\sum_{k=0}^{n-2} r(k,\ n-1)' - \sum_{k=0}^{n-2} r(n-1,\ k)''$$

since, for $x < n - 1$, any path to a boundary point $(k, n-1)'$, $k \leq x$, contributes one path to $(x, n-1)$ and no path goes through more than one such boundary point.

Thus for $x < n - 1$

$$r[x, n-1] = \binom{n-1+x}{x} - \sum_{k=0}^{x} \binom{n-2}{k}$$

while

$$r[n-1, n-1] = \binom{2n-2}{n-1} - 2\sum_{k=0}^{n-2} \binom{n-2}{k}$$

$$= \binom{2n-2}{n-1} - 2^{n-1}$$

By symmetry, similar results hold for the points $[n-1, y]$. Now, with the aid of Figure 1, we get

$$r_{01} = r_{10} = r[n-1, n-1] = \binom{2n-2}{n-1} - 2^{n-1}$$

and, for $i = 2(1)n - 1$:

$$r_{0i} = r_{i0} = \sum_{k=0}^{n-i} r(k, n-1)' + r[n-i, n-1]$$

$$= \sum_{k-0}^{n-i} \binom{n-2}{k} + \binom{n-1+n-i}{n-1} - \sum_{k=0}^{n-i} \binom{n-2}{k} = \binom{2n-i-1}{n-1}$$

This formula also works for $i = n$, giving $r_{0n} = r_{n0} = 1$. It is easy to see that no other types are possible.

5. Calculation of Expected Makespan

Given all the r_{ij} values, we can now calculate all quantities of interest following Milch and Waggoner. It will be convenient to introduce the notation:

$$p = \frac{\lambda}{\lambda + \mu}, \quad q = 1 - p = \frac{\mu}{\lambda + \mu}.$$

First, the probability that the path, π, followed by an arbitrary realization will lie in the set R_{ij} is

$$p(\pi \varepsilon R_{ij}) = r_{ij}p^{n-i}q^{n-j}$$

since transitions from the boundary points have probability one, and every path must have n transitions to the right and n transitions up.

Then, the time to complete processing, given $\pi \varepsilon R_{ij}$, is the sum of $2n$ independent exponential random variables, i with parameter λ, j with parameter μ, and $2n - i - j$ with parameter $\lambda + \mu$. Its mean value is therefore $m_{ij} = i/\lambda + j/\mu + (2n - i - j)/(\lambda + \mu)$. Finally, the expected makespan is:

$$m = \sum_i \sum_j p(\pi \varepsilon R_{ij})m_{ij}$$

$$= \sum_{i=1}^{n} \binom{2n - i - 1}{n - 1} p^{n-i}q^{n} \left(\frac{i}{\lambda} + \frac{2n - i}{\lambda + \mu}\right)$$

$$+ \sum_{j=1}^{n} \binom{2n - j - 1}{n - 1} p^{n}q^{n-j} \left(\frac{j}{\mu} + \frac{2n - j}{\lambda + \mu}\right)$$

$$+ 2^{n-1}\left[p^{n-1}q^{n-1}\left(\frac{1}{\lambda} + \frac{1}{\mu} + \frac{2n - 2}{\lambda + \mu}\right) - p^{n}q^{n-1}\left(\frac{1}{\mu} + \frac{2n - 1}{\lambda + \mu}\right) - p^{n-1}q^{n}\left(\frac{1}{\lambda} + \frac{2n - 1}{\lambda + \mu}\right)\right]$$

Letting $k = n - i = n - j$ in the two summations, and simplifying the final term:

$$m = \sum_{k=0}^{n-1} \binom{n-1+k}{k} \left[p^k q^n \left(\frac{n-k}{\lambda} + \frac{n+k}{\lambda+\mu} \right) + p^n q^k \left(\frac{n-k}{\mu} + \frac{n+k}{\lambda+\mu} \right) \right] + \frac{(2pq)^{n-1}}{\lambda+\mu}$$

If we write (to illustrate one of two cases):

$$\frac{n-k}{\lambda} + \frac{n+k}{\lambda+\mu} = \frac{n}{\lambda(\lambda+\mu)} \left(2\lambda + \mu - \mu \frac{k}{n} \right)$$

$$= \frac{n}{\lambda+\mu} \left(\frac{p+1}{p} - \frac{\mu}{\lambda} \frac{k}{n} \right)$$

and introduce the negative binomial distribution

$$B(x;\, n,\theta) = \sum_{k=0}^{x} \binom{n-1+k}{k} (1-\theta)^k \theta^n \qquad (0 < \theta < 1)$$

where x and n are positive integers, then:

$$m = \frac{n}{\lambda+\mu} \left[\frac{p+1}{p} B(n-1;\, n,q) - \frac{\mu}{\lambda} \sum_{k=1}^{n-1} \frac{k}{n} \binom{n-1+k}{k} p^k q^n \right.$$

$$\left. + \frac{q+1}{q} B(n-1;\, n,p) - \frac{\lambda}{\mu} \sum_{k=1}^{n-1} \frac{k}{n} \binom{n-1+k}{k} p^n q^k \right] + \frac{(2pq)^{n-1}}{\lambda+\mu}$$

For the two remaining summations, use the identities

$$\frac{k}{n} \binom{n-1+k}{k} = \binom{n-1+k}{k-1}, \quad \frac{\mu}{\lambda} p^k q^n = p^{k-1} q^{n+1}$$

and substitute $j = k - 1$, to obtain finally:

$$m = \frac{n}{\lambda+\mu} \left[\frac{p+1}{p} B(n-1;\, n,q) - B(n-2;\, n+1,\, q) \right.$$

$$+ \frac{q+1}{q} B(n-1; n,p) - B(n-2; n+1, p)\Bigg] + \frac{(2pq)^{n-1}}{\lambda + \mu}$$

Since extensive tables (Pearson, 1956) exist for the incomplete Beta function:

$$I_x(a,b) = \frac{(a+b-1)!}{(a-1)!(b-1)!} \int_0^x t^{a-1}(1-t)^{b-1} dt \qquad (a,b = 0,1,\ldots)$$

we can use the identity

$$B(x; n,\theta) = 1 - I_{1-\theta}(x+1, n)$$

to express m:

$$(\lambda + \mu)m = n\left[\frac{1}{p} + \frac{1}{q} - \left(1 + \frac{1}{p}\right) I_p(n,n) + I_p(n-1, n+1)\right.$$

$$\left. - \left(1 + \frac{1}{q}\right) I_q(n,n) + I_q(n-1, n+1)\right] + (2pq)^{n-1}$$

Using the identity $I_x(a,b) + I_{1-x}(b,a) = 1$, which implies $I_p(n,n) + I_q(n,n)=1$, we get the further simplification:

$$(\lambda + \mu)m = n[1 + I_p(n,n)/q + I_q(n,n)/p - I_p(n+1, n-1) - I_q(n+1, n-1)]$$
$$+ (2pq)^{n-1}.$$

We note that the final term, which cannot exceed $(.5)^{n-1}$, can be neglected for large n. A more important observation is that, although m depends on three parameters (n,λ,μ), $(\lambda + \mu)m$ depends on only two (n,p). Thus, a family of curves can be computed, plotting $(\lambda + \mu)m$ versus p for various values of n. Using them the expected makespan can quickly be obtained for any problem.

In Figure 2, several such curves are plotted. Since $(\lambda + \mu)m$ is symmetric in p about $p = 0.5$, we need only consider $0 < p \leq 0.5$. For small p (say $p < 0.3$), especially for large values of n, we have to a very good approximation $(\lambda + \mu)m = n/p$, or $m = n/\lambda$. This is because, when one server is much slower than the other, the fast server has negligible effect on the makespan. At the other extreme, when $\lambda = \mu$, $p = q = 0.5$, and $I_p(n,n) = I_q(n,n) = 0.5$, then the formula simplifies to

$$m = [n(1.5 - I_{.5}(n + 1, n - 1)) + (.5)^n]/\lambda \quad .$$

For example, if we have $n = 20$, $1/\lambda = 3.0$, and $1/\mu = 3.5$ (an example considered by Milch and Waggoner), then $p = \lambda/(\lambda + \mu) = .538$. Looking above the equivalent $p = .462$ in Figure 2, we find for $n = 20$ that $(\lambda + \mu)m = 45.9$, so that $m = 74$. This is essentially equal to the minimum achieved over various initial queue sizes, n_1 and n_2, by the above authors, which occurs at $n_1 = n_2$. It is not surprising that little time is saved by the additional freedom to assign work, since Milch and Waggoner noted the insensitivity of the makespan to a considerable range of variation of n_1 and n_2.

It is interesting to observe how the approximation n/p (the dashed lines in Figure 2) behaves for increasing n. We can see from Figure 2 that, as n ranges from 3 to 20 jobs, the error incurred by using this approximation (that is, the difference between the dashed and solid lines) decreases at $p = 0.4$, remains roughly stationary at $p = 0.45$, and increases at $p = 0.5$. This phenomenon is clarified by Figure 3, where the difference $(\lambda + \mu)m - n/p$ is plotted against n for $p = 0.4$, 0.45, and 0.5. It appears (though it remains unproven) that, for any $p < 0.5$, the difference approaches zero for large enough n, though it may increase for a while. For $p = 0.5$,

$$(\lambda + \mu)m - n/p = n[1 - 2I_{.5}(n+1, n-1)] + (0.5)^{n-1}.$$

It can be shown that, neglecting the last term, this equals

$$\frac{2n\Gamma(2n)}{\Gamma(n)\Gamma(n+1)}(0.5)^{2n-1} = \frac{3\cdot 5\cdot 7\cdot \ldots \cdot(2n-1)}{2\cdot 4\cdot 6\cdot \ldots \cdot(2n-2)}\,.$$

This is clearly increasing in n, and indeed can be shown to diverge. Thus, as $n \to \infty$, the error involved in using n/λ to approximate the makespan increases without bound when $\lambda = \mu$, but goes to zero for all $\lambda \neq \mu$.

6. The Idle Time of a Machine

We close with a brief discussion of machine idle time: the total time that a server is idle, prior to the completion of all its work. For simplicity, we speak hereafter of machine 1 idle time, without loss of generality. We quickly see that there are only two possibilities: zero idle time, or a single period of idle time exponentially distributed with parameter μ. The second case corresponds to traversing one of the states $(n - 1, y)''$, $y = 0(1)n-2$; i.e., that half of the RB points (boxed in Figure 1) which have $x = n - 1$. Let S be the set of paths that include one of these points (and note that no path includes more than 1), and for an arbitrary path π let

$$\delta = P\{\pi \in S\} = \Pr\{\text{Positive machine 1 idle time}\}.$$

We enumerate the paths, using the results already obtained. With the aid of Figure 1 we see that S includes half of the r_{11} paths in R_{11}, plus $\sum_{k=0}^{n-j} r(n-1, k)$" paths in R_{oj}, $j = 2(1)n$. Since the probability of realizing any path in R_{ij} is $p^{n-i}q^{n-j}$,

$$\delta = 2^{n-2}p^{n-1}q^{n-1} + \sum_{j=2}^{n} p^{n}q^{n-j} \sum_{k=0}^{n-j} \binom{n-2}{k}$$

interchanging the order of summation in the last term, and simplifying, we get:

$$\delta = p(1-q^2)^{n-2} \qquad (n = 2,3,\ldots) \quad .$$

Thus, with probability $1 - \delta$ there is no idle time, and with probability δ there is exponentially distributed idle time, with mean $1/\mu$. Not surprisingly, δ goes to zero very rapidly with n, so that idle time is usually negligibly small.

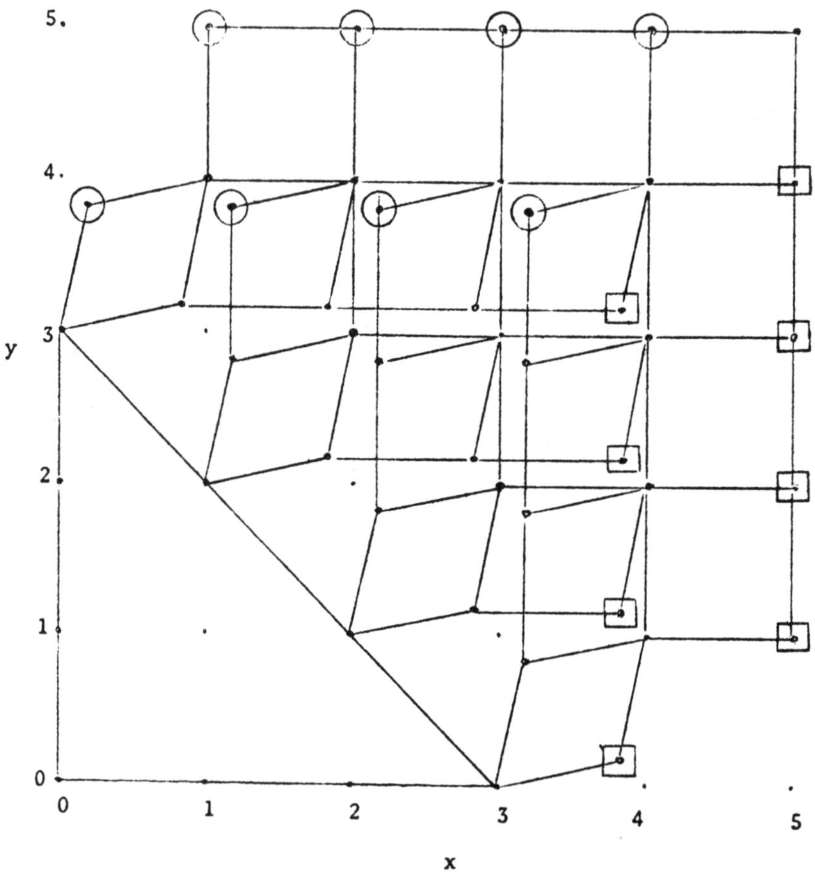

Figure 1

The Random Walk for the Five-Job Two-Machine Job-Shop

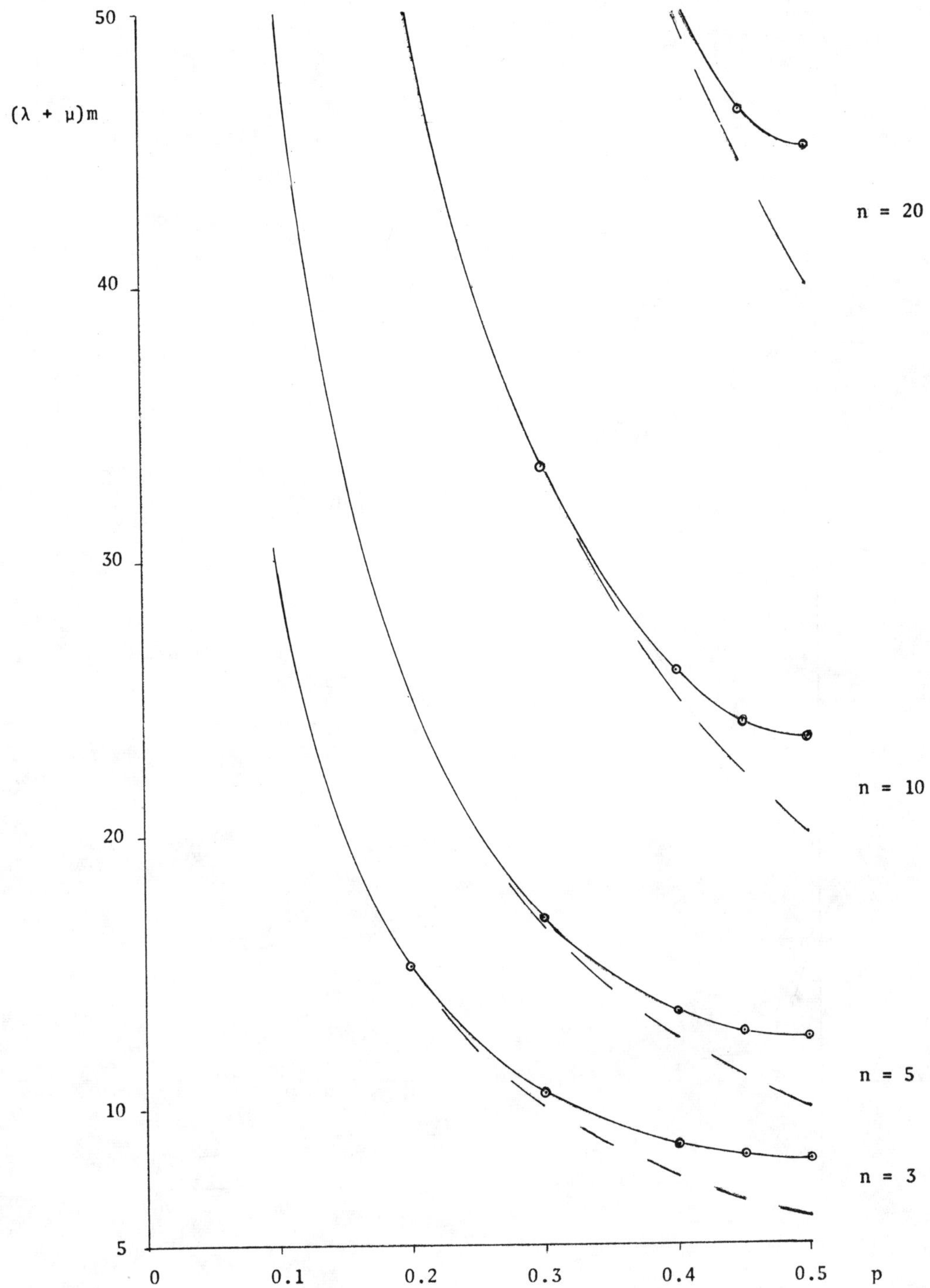

Figure 2

$(\lambda + \mu)m$ versus p with n/p (dashed) for Comparison

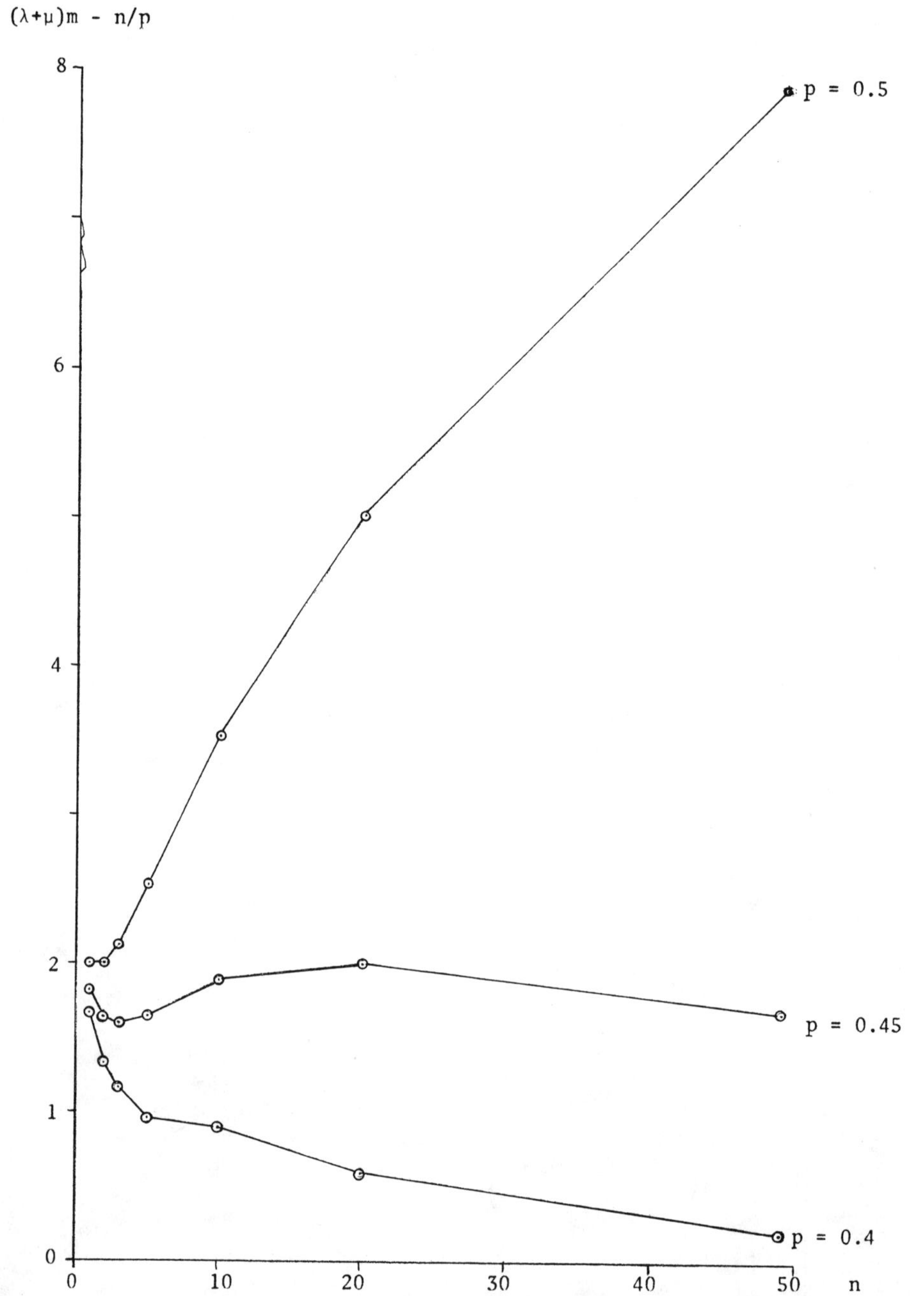

Figure 3

$(\lambda + \mu)m - n/p$ versus n

References

1. Feller, W. (1957), An Introduction to Probability Theory and its Applications, Vol. 1, John Wiley, N. Y., Chapter III.

2. Milch, P. R. and M. H. Waggoner (1970), "A Random Walk Approach to a Shutdown Queueing System," SIAM J. Appl. Math., 19, 103-115.

3. Pearson, K. (1956) ed., Tables of the Incomplete Beta-Function, Cambridge.

OPTIMAL SOLUTION OF SCHEDULING PROBLEMS USING LAGRANGE MULTIPLIERS: PART II

Marshall L. Fisher

University of Chicago, Chicago, Illinois, USA

ABSTRACT

Lagrange multipliers may be used to obtain a strong lower bound on the cost of an optimal solution for the resource-contrained network scheduling problem. A branch-and-bound algorithm designed to complement this procedure for obtaining bounds is presented here. We show how the concept of an active schedule may be used to eliminate nodes in the tree and how to update the multipliers when we branch.

1. Introduction

This paper extends the results of a previous paper [4] in which the author used Lagrange multipliers to develop an algorithm for the resource-constrained network scheduling problem. This problem or special cases of it has received considerable attention recently [1, 2, 3, 4, 5, 7, 10, 11, 12, 13, 14, 15]. The version of the problem we will consider is described by the following six conditions.

1. A set of I jobs must be performed. The ith job consists of n_i tasks numbered 1 to n_i. The time to perform each task is a known integer represented by p_{ij} for the jth task of the ith job.
2. A set of K different resources is available. R_{kt} is the amount of the kth resource available in any time period. r_{ijk} is the amount of the kth resource required by task ij during its processing. For example, in the job shop scheduling problem, the resources correspond to the machines of a job shop, and each task requires only a single machine during the interval of its processing.

3. No preemption of task performance is allowed. Once task ij is started it must be processed until completed in p_{ij} time units. If t_{ij} and f_{ij} represent the start and finish times for task ij then this condition is tantamount to requiring

$$f_{ij} = t_{ij} + p_{ij}, \quad j = 1, \ldots, n_i, \quad i = 1, \ldots, I \quad .$$

4. The start times of the tasks on a given job are constrained by a cycle-free precedence network of the CPM-PERT type. This network is determined by a set of constraints for each job of the form

$$t_{ij} \geq f_{i\ell} = t_{i\ell} + p_{i\ell}, \quad \ell \in P(ij)$$

where $P(ij)$ is the predecessor set for task ij and contains all tasks of job i which must be completed before ij is started.

5. Define $f_i = \max_{j=1,\ldots,n_i} (f_{ij})$ to be the completion time of job i. There is an integer upper bound T on the number of time periods until all jobs are completed, i.e.,

$$f_i \leq T, \quad i = 1, \ldots, I \quad .$$

6. We are required to find values for t_{ij}, $j = 1, \ldots, n_i$, $i = 1, \ldots, I$ which satisfy conditions 1-5 and which minimize

$$Z = \sum_i g_i(f_i)$$

where g_i is some specified nondecreasing function of f_i. For example, if d_i is a due date for job i and $g_i(f_i) = \max(0, f_i - d_i)$ then the objective is to minimize total job tardiness. More complicated functions of tardiness such as minimizing the square of job tardiness could also be used.

In reference [4] Lagrange multipliers are used to develop an algorithm for the problem defined by conditions 1-6. The important constraints of this problem are the network constraints (condition 4) and the resoucre constraints (condition 2). Lagrange multipliers are used to dualize the resource constraints, forming a Lagrangian problem in which the network constraints appear explicitly while the resource constraints appear only in the Lagrangian function. Because the network constraints do not interact among jobs, the problem of minimizing the Lagrangian separates into a subproblem for each job. Due to this separation and the absence of explicit resource constraints, the Lagrangian problem is generally much easier to solve than the original problem. Minimization of the Lagrangian with fixed multiplier values yields a lower bound on the objective function value of an optimal solution to the scheduling problem. The multipliers may be iteratively adjusted to strengthen the value of this bound. Computational experience has shown that the bound obtained in this manner is extremely strong. The bound may be used in a branch-and-bound algorithm for solving the scheduling problem. This approach is similar to that used by Held and Karp for the traveling salesman problem [8, 9] and by Fisher and Shapiro for integer programming [6].

In developing an algorithm for the scheduling problem using Lagrange multipliers we must deal with three problems: (1) Solution of the Lagrangian problem with multiplier values fixed. (2) Adjustment of multiplier values to obtain a sharp bound. (3) Development of a branch-and-bound algorithm which uses the lower bound obtained from the Lagrangian problem. Reference [4] treats problems (1) and (2) extensively and briefly considers problem (3). The purpose of this paper is to consider problem (3) more fully.

The plan of the paper is as follows. In section 2 the major results of [4] are reviewed, showing how Lagrange multipliers may be used to obtain a lower bound. In section 3 we show that it is desirable that the restrictions corresponding to a node of the branch-and-bound tree be separable by job. A basic enumeration scheme is presented which has this property and a theorem is proven which allows us to easily obtain bounds on new nodes from the solution to the Lagrangian problem at the preceding node. In section 4, the concept of an active schedule is used to show that some nodes in the branch-and-bound tree may be eliminated. In section 5 we discuss the adjustment of multiplier values obtained at one node so that they may be used when we wish to bound another node.

2. A Lower Bound on the Cost of an Optimal Schedule

In this section we review the procedure developed in [4] for using Lagrange multipliers to obtain a lower bound on the cost of an optimal solution to the scheduling problem. This development proceeds in several steps. First, an integer programming (IP) formulation of the problem described by conditions 1-6 is developed. Then we associate multipliers with those rows of this IP formulation which correspond to the condition 2 resource constraints and form a Lagrangian problem whose optimal solution yields a lower bound for the scheduling problem. This problem may be solved by solution of a subproblem for each job. Finally, we show how the multipliers may be adjusted to obtain a strong bound.

In developing an IP formulation of the scheduling problem we require the concept of a job schedule.

Definition 1. A job schedule is any schedule for the tasks of a particular job (i.e., a set of values for t_{ij}, $j = 1, \ldots, n_i$) which satisfies conditions 1, 3, 4, and 5.

The fact that all p_{ij} specified in condition 1 are integer may be used to show that all t_{ij} are integer for at least one optimal solution to the scheduling problem [5] so that we may restrict consideration to integer values of t_{ij}. This fact and the upper bound of T on all f_i implies that the number of job schedules for the ith job is finite. Denote this number by Q_i.

Further, let f_i^q be the completion time of job i under schedule q, and define $c_{iq} = g_i(f_i^q)$. We let α_{iqkt} be the amount of resource k used during time period t by job i under schedule q, i.e., $\alpha_{iqkt} = \sum_{j \epsilon J(iqt)} r_{ijk}$ where $J(iqt)$ is the set of all tasks on job i in process at time t. All of these quantities can be readily calculated for a given job schedule. The following 0-1 IP problem is equivalent to the scheduling problem defined in conditions 1-6.

$$\min \sum_{iq} c_{iq} x_{iq} \tag{1a}$$

subject to:

$$\sum_q x_{iq} = 1 \ , \qquad i = 1, 2, \ldots, I \tag{1b}$$

$$\sum_{iq} \alpha_{iqkt} x_{iq} \leq R_{kt} \ , \qquad \begin{array}{l} t = 1, 2, \ldots, T \\ k = 1, 2, \ldots, K \end{array} \tag{1c}$$

$$x_{iq} = 0 \text{ or } 1 \ , \qquad \begin{array}{l} q = 1, 2, \ldots, Q_i \\ i = 1, 2, \ldots, I \ . \end{array} \tag{1d}$$

In this formulation $x_{iq} = 1$ implies that we are using the qth job schedule for job i. The number of constraints in (1) is $I + KT$. The number of columns, $\sum_i Q_i$, will be quite large, but in the use we make of (1), only a subset of these columns will be explicitly needed and they will be generated as required.

We next associate multipliers with equations (1c) and form a Lagrangian problem. Let u_{kt}, $t = 1, \ldots, T$, $k = 1, \ldots, K$ denote a nonnegative multiplier associated with the kt^{th} constraint in (1c), and let $u = (u_{11}, u_{12}, \ldots, u_{1T}, \ldots, u_{Kt})$ be a vector of these multipliers. The following Lagrangian problem is formed.

$$w(u) = \min_{x_{iq}} \left(\sum_{iq} c_{iq} x_{iq} + \sum_{kt} u_{kt} \left(\sum_{iq} \alpha_{iqkt} x_{iq} - R_{kt} \right) \right) \tag{2a}$$

$$\text{s.t.} \quad \sum_{q} x_{iq} = 1 \qquad i = 1, \ldots, I \tag{2b}$$

$$x_{iq} = 0 \text{ or } 1 \quad q = 1, \ldots, Q_i, \quad i = 1, \ldots, I \; . \tag{2c}$$

The motivation for defining (2) is that positive values of u_{kt} will tend to impose the constraints (1c) by penalizing solutions which violate these constraints. It can be shown that $w(u)$ is a lower bound on the cost of an optimal solution in (1) for any $u \geq 0$.

To solve (2) when a nonnegative value for u has been specified we define $\pi_{iq}(u) = c_{iq} + \sum_{kt} \alpha_{iqkt} u_{kt}$ and rewrite (2) as

$$w(u) = \min_{x_{iq}} \sum_{iq} \pi_{iq}(u) x_{iq} - \sum_{kt} u_{kt} R_{kt}$$

$$\sum_{q} x_{iq} = 1 \qquad i = 1, \ldots, I$$

$$x_{iq} = 0 \text{ or } 1 \quad q = 1, \ldots, Q_i, \quad i = 1, \ldots, I \; .$$

Let $\pi_i(u) = \min_q \pi_{iq}(u)$. From the above reformulation of (2) it is clear that the optimal solution to (2) is obtained by setting an x_{iq} for which $\pi_{iq}(u) = \pi_i(u)$ to 1 for each i and all other x_{iq} to zero. The associated value of $w(u)$ is given by $w(u) = \sum_i \pi_i(u) - \sum_{kt} R_{kt} u_{kt}$.

Definition 2. Given a value for u, a job schedule q^* for job i which satisfies $\pi_{iq^*}(u) \leq \pi_{iq}(u)$, $q = 1, \ldots, Q_i$ is called an optimal job schedule. Finding such a schedule is called the job scheduling problem.

In [4] various algorithms are presented for solving the job scheduling problem. Computation experience with one of these algorithms indicates that very little work is required to find an optimal job schedule.

To obtain a value for u such that $w(u)$ is large, we consider the problem

$$\max w(u) \tag{3a}$$

$$\text{s.t. } u \geq 0 \tag{3b}$$

which is dual to (1). However, we do not necessarily require the exact optimum of (3) to obtain a usable lower bound, and in fact it may be desirable to seek only a good solution to (3) depending on the work involved in finding the exact optimum.

It can be shown that (3) is equivalent to the following linear programming problem.

$$\max w = \sum_i \pi_i - \sum_{kt} R_{kt} u_{kt} \tag{4a}$$

$$\text{s.t. } \pi_i \leq c_{iq} + \sum_{kt} \alpha_{iqkt} u_{kt} \qquad \begin{array}{l} q = 1, \ldots, Q_i \\ i = 1, \ldots, I \end{array} \tag{4b}$$

$$u_{kt} \geq 0 \qquad \begin{array}{l} k = 1, \ldots, K \\ t = 1, \ldots, T \end{array} \tag{4c}$$

$$w \text{ and } \pi_i,\ i = 1, \ldots, I \text{ unrestricted in sign .} \tag{4d}$$

Problem (4) is exactly the LP dual of (1) with integrality relaxed. While the constraints of (4b) are too numerous to even be explicitly written down, most of these constraints will be nonbinding at a particular extreme point solution. This fact is used in [4] to develop an algorithm for solving (4) in which we apply the simplex method, dealing explicitly with only a subset of the binding constraints. An algorithm for finding an optimal job schedule is used to generate other constraints of (4) as they become binding. When we solve (4) with this approach we may terminate before optimality and still obtain a usable lower bound.

The procedure for obtaining a lower bound on the cost of an optimal schedule by solving (4) in this manner was applied to a number of problems and the results reported in [4]. These results show the bound obtained to be extremely strong, in several cases actually equal to the optimal objective value. The bound currently available for problems of this type is obtained by simply ignoring resource constraints (see reference [15]). The bound from Lagrange multipliers was always as large as this bound, and in most cases considerably larger.

3. The Branch-and-Bound Algorithm

In this section we develop a branch-and-bound algorithm which uses the bound $w(u)$. Typically in a branch-and-bound algorithm we derive new problems by adding restrictions to the original problem such that the solution sets of the derived problems form a partition of the original solution set. Then we determine lower bounds on the optimal objective value of each of the derived problems.

In the problem considered here, when restrictions to define a derived problem are added we would make corresponding changes to problems (1)-(4). Then a lower bound is obtained from these modified problems in the manner discussed in section 2. In developing a branch-and-bound algorith, we must recognize that the ease of solving (2) depends in part on the separable structure which allows us to independently solve a job scheduling problem for each job. It is highly desirable that the restrictions added to form derived problems preserve this separability. This requirement forces us to exclude a commonly used enumeration scheme discussed by Schrage [15] in which the derived problems contain restrictions which impose a partial ordering on the sequence in which tasks are performed. These restrictions have the form $t_{ij} \geq t_{\ell m} + p_{\ell m}$ and if $i \neq \ell$, they destroy the separability of (2).

We present here a natural enumeration scheme for which the separability of (2) is preserved. In the usual fashion we represent the various derived problems by nodes in a tree. New nodes are created from an existing node by appending additional restrictions to the derived problem for that node. New nodes are joined to the nodes from which they were created by "branches" and the process of creating new nodes is called branching. The basic branching mechanism used is to select a resource k and a time period t and allocate the available resource R_{kt} to the various resource-feasible subsets of the tasks which may use resource k during time period t. Let S_{kt} be the set of tasks which conditions 4 and 5 allow to be in process during period t and for which $r_{ijk} > 0$. Let $T_{kt} \subseteq S_{kt}$ be any subset of S_{kt} which satisfies $\sum_{ij \epsilon T_{kt}} r_{ijk} \leq R_{kt}$. When we branch we create a node for every subset T_{kt}. At a node corresponding to a subset T_{kt}, the derived problem is formed by restricting all tasks contained in T_{kt} to be in process during period t and all tasks not in T_{kt} but in S_{kt} to not be in process during period t. The pair kt may be selected in a variety of ways, including the kt for which u_{kt} is maximum or the kt for which resource infeasibility is greatest for the current value of u.

The restrictions which define the derived problem at a node are exactly equivalent to restricting the allowable job schedules to be a certain subset of all job schedules. These restrictions will be represented by letting $A_i \subseteq [1, \ldots, Q_i]$ denote the set of allowable schedules for job i for a general derived problem. $A = A_1 \times A_2 \times \ldots \times A_I$ will denote the external product of these sets. The LP equivalent of the dual for this derived problem will be (4) with $q = 1, \ldots, Q_i$ in (4b) replaced by $q \in A_i$. We denote this problem by (4_A).

A step in the algorithm begins by applying the bounding procedure to the derived problem at a node. To obtain a lower bound for a derived problem we apply the algorithm referred to in section 2 to problem (4_A). The restriction of allowable job schedules for job i to the set A_i can be effected by direct modification of the algorithms in [4] for finding an optimal job schedule. We will also assume that any u_{kt} for which kt has been allocated has been removed from (4_A) and u correspondingly redefined.

Let Z^* denote the objective value of a known solution to the scheduling problem. When the bounding procedure is applied at a node, one of three outcomes is possible: (1) An optimal solution to the derived schedulng problem is obtained with objective value less than Z^*. In this case, Z^* is updated and the node is discarded. (2) The lower bound at the node is greater than or equal to Z^*, in which case the node is discarded. (3) The lower bound is less than Z^* but an optimal solution is not obtained. In this case, we create new nodes using the branching mechanism described.

In selecting a node at which to apply the bounding procedure, if new nodes have been created we select any of these with least bound. Otherwise, any node with least bound is selected.

When we create new nodes it is convenient to obtain a lower bound for each of these nodes before we actually apply the bounding procedure. The following theorem shows how a bound may be obtained for a node from the information available at the time it is created.

Theorem 1. Suppose we have applied the bounding procedure at a node for which the set of allowable job schedules is given by $\bar{A}$ and obtained a solution to $(4_{\bar{A}})$ which we denote by $\bar{\pi}_i$, $\bar{u}_{kt}$ and $\bar{w}$ with $\bar{w} < Z^*$. A lower bound for a new node created from this node by selecting a resource time period $\overline{kt}$ and a set of tasks $T_{\overline{kt}}$ which must be in process during period $\bar{t}$ and requiring all other tasks with $r_{ij\bar{k}} > 0$ not to be in process is given by $\bar{w} + \bar{u}_{\overline{kt}}(R_{\overline{kt}} - \sum_{ij\epsilon T_{\overline{kt}}} r_{ij\bar{k}})$.

Proof. Let $\bar{\bar{A}}$ denote the set of allowable job schedules for the new node and note that $\bar{\bar{A}}_i \subseteq \bar{A}_i$, i = 1, ..., I. Let $\hat{\pi}_i = \min_{q\epsilon\bar{\bar{A}}_i} \pi_{iq}(\hat{u})$,

where $\hat{u}$ is given by $\hat{u}_{kt} = \bar{u}_{kt}$, $kt \neq \overline{kt}$, $\hat{u}_{\overline{kt}} = 0$. Note that $w(\hat{u}) = \sum_i \hat{\pi}_i - \sum_{kt\neq\overline{kt}} R_{kt}\, \bar{u}_{kt}$ is a lower bound for the new node. Let $T_{i\overline{kt}}$ denote the set of tasks for job i contained in $T_{\overline{kt}}$.

Let $q^*\epsilon\bar{\bar{A}}_i$ denote a job schedule for which $\hat{\pi}_i = \pi_{iq^*}(\hat{u})$. This job schedule is feasible in the job scheduling problem for the node we are branching from (since $q^*\epsilon\bar{\bar{A}}_i \subseteq \bar{A}_i$) and with multiplier values $\bar{u}$, the value of this job schedule is $\pi_{iq^*}(\bar{u}) = \hat{\pi}_i + \sum_{ij\epsilon T_{i\overline{kt}}} r_{ij\bar{k}}\, \bar{u}_{\overline{kt}}$.

$\bar{\pi}_i = \min_{q\epsilon\bar{A}_i} \pi_{iq}(\bar{u})$ implies

$\bar{\pi}_i \leq \pi_{iq^*}(\bar{u}) = \hat{\pi}_i + \sum_{ij\epsilon T_{i\overline{kt}}} r_{ij\bar{k}}\, u_{\overline{kt}}$ which gives

$$w(\hat{u}) = \sum_i \hat{\pi}_i - \sum_{kt \neq \overline{kt}} R_{kt}\bar{u}_{kt} \geq \sum_i \bar{\pi}_i - \sum_{ij \in T_{\overline{kt}}} r_{ij\bar{k}}\, \bar{u}_{\overline{kt}} - \sum_{kt \neq \overline{kt}} R_{kt}\bar{u}_{kt}$$

$$= \bar{w} + \bar{u}_{\overline{kt}} \left(R_{\overline{kt}} - \sum_{ij \in T_{\overline{kt}}} r_{ij\bar{k}}\right) \quad . ||$$

Note that we also have the option of actually determining $\hat{\pi}_i$ by solving job scheduling problems for the new node with the multiplier $\hat{u}$. Computational experience has shown that the increase in the bound obtained by doing this does not warrant the additional computation. We also mention that Theorem 1 provides a rationale for selecting a kt to branch on for which u_{kt} is maximum.

We conclude this section by showing that the restrictions added to form the derived problem at a node may imply other restrictions of the same type. Including these other restrictions when we apply the bounding procedure will generally reduce the work required to solve (4). Another representation for the restrictions which have been added at a particular node is given by sets $A(ij)$ and $B(ij)$ for each task ij where $A(ij)$ contains those time periods during which task ij must be in process and $B(ij)$ contains those time periods during which ij cannot be in process. We first show that these restrictions imply lower and upper bounds on the start times t_{ij} for all tasks.

Let a_{ij} and b_{ij} denote lower and upper bounds on t_{ij}. Initial values for a_{ij} and b_{ij} are obtained by noting that conditions 4 and 5 imply lower and upper bounds on t_{ij} which may be calculated by the standard critical path method algorithm. These bounds are then strengthened by considering $A(ij)$ and $B(ij)$. The set $A(ij)$ implies the conditions

$$\left.\begin{array}{l} a_{ij} \geq t - p_{ij} \\ b_{ij} \leq t - 1 \end{array}\right\} \text{for all } t \in A(ij) \quad .$$

while if $B(ij)$ contains a t in the interval $[a_{ij}, a_{ij} + p_{ij}]$, then we may increase a_{ij} to the largest such t. We may successively apply this test to increase a_{ij} until we obtain an a_{ij} such that $B(ij)$ contains no t in the interval $[a_{ij}, a_{ij} + p_{ij}]$. An analogous procedure may be used to decrease b_{ij} using the interval $[b_{ij} + 1, b_{ij} + p_{ij}]$.

The following 2 lemmas show how additional restrictions can be derived using a_{ij} and b_{ij}.

Lemma 1. For $t \in A(ij)$, every integer τ in the interval $[t^1, t^2]$ given by $t^1 = \min(t, b_{ij} + 1)$, $t^2 = \max(t, a_{ij} + p_{ij})$ can be appended to $A(ij)$.

Proof. a_{ij}, b_{ij} and $t \in A(ij)$ gives us the following constraints on t_{ij}.

$$t_{ij} \geq a_{ij} \tag{5a}$$

$$t_{ij} \leq b_{ij} \tag{5b}$$

$$t_{ij} \leq t - 1 \tag{5c}$$

$$t_{ij} \geq t - p_{ij} \quad . \tag{5d}$$

If $a_{ij} + p_{ij} \geq t$, then (5a) and (5c) imply that task ij must be in process during periods $t, \ldots, a_{ij} + p_{ij}$. If $b_{ij} + 1 \leq t$, then (5b) and (5d) imply that task ij must be in process during periods $b_{ij} + 1, \ldots, t$. Together these establish the lemma.||

Lemma 2. Let $A'(ij)$ denote additional assignments made by applying lemma 1 at a node and define $\bar{R}_{kt} = R_{kt} - \sum_{\substack{\text{all } ij \text{ such that} \\ t \in A'(ij)}} r_{ijk}$. Then for any ij for which $t \notin A'(ij)$ and $r_{ijk} > \bar{R}_{kt}$, we may append t to $B(ij)$.

Proof. This restriction is directly implied by the restrictions $A'(ij)$.||

These lemmas could be used when new nodes are created to strengthen the restrictions added to form derived problems.

4. Elimination of Inactive Schedules

Conway, Maxwell, and Miller [2] point out that for most objective functions (including the one considered here) it is necessary to consider only a subset of the possible schedules, the so-called active schedules. In this section we present the definition of an active schedule in the context of our problem and prove 2 theorems which indicate for the important special case of the general scheduling problem in which $r_{ijk} = 0$ or 1 for all ijk, $\sum_k r_{ijk} = 1$ and all $R_{kt} = 1$, when a node of the branch-and-bound tree cannot lead to an active schedule. These nodes, of course, need not be considered.

Definition 3. An active schedule is a set of values $\bar{t}_{ij}$ for the start times of all tasks which satisfy conditions 1-5 and the further condition that there exist no other values $\hat{t}_{ij}$ which satisfy conditions 1-5 with $\hat{t}_{ij} \leq \bar{t}_{ij}$ for all ij and $\hat{t}_{ij} < \bar{t}_{ij}$ for at least one ij.

Given the requirement in condition 6 that $g_i(f_i)$ be a non-decreasing function, there must be at least one active schedule which is optimal.

The following theorem describes a class of nodes which cannot lead to an active schedule.

Theorem 2. We assume that $r_{ijk} = 0$ or 1 for all ijk, $\sum_k r_{ijk} = 1$ and $R_{kt} = 1$. Let S_k be the set of tasks with $r_{ijk} = 1$ which have no predecessors (i.e., $P(ij) = \phi$ for $ij \in S_k$) and let $\tau_k = \min_{\substack{ij \notin S_k \text{ and} \\ r_{ijk} = 1}} a_{ij}$. Let $S_{k\ell}$ be any subset of S_k which is maximal in the sense that $\sum_{ij \in S_{k\ell}} p_{ij} = \tau_{k\ell} \leq \tau_k$ but for any ij in S_k not contained in $S_{k\ell}$, $\tau_{k\ell} + p_{ij} > \tau_k$. Let $t_k = \min_{\ell} \tau_{k\ell}$. If we are creating nodes by allocating resource-time period kt with $t \leq t_k$, then a node requiring that no task be in process during time period t cannot lead to an active schedule and can be deleted.

Proof. We will show that any feasible schedule obtained from the node described must be inactive. Hence the node can be eliminated without eliminating any active schedules.

Consider any feasible schedule obtained from such a node. This schedule will have no task in process during period t. Let $\bar{S}_k$ denote the set of tasks with $r_{ijk} = 1$ in process before or during period τ_k. If there is an $ij \in \bar{S}_k$ with $t_{ij} \geq t$, let $t_{\overline{ij}}$ denote the smallest such value. Then reducing $t_{\overline{ij}}$ to $t - 1$ produces a feasible schedule, indicating that the current schedule is inactive. If there is no $ij \in \bar{S}_k$ with $t_{ij} \geq t$, then we must have $\bar{S}_k \subseteq S_{k\ell}$ for some ℓ. By adding tasks to $\bar{S}_k$ to obtain some $S_{k\ell}$ and reducing one or more t_{ij}, $ij \in \bar{S}_k$ while preserving the sequence of these tasks, we can obtain a new schedule in which at least one t_{ij} is reduced and none is increased.||

Computational experience has shown that elimination of nodes covered by the theorem can substantially reduce solution time.

In particular, Table 1 gives the reduction in solution time for four 5-job, 20-task problems reported in [4] as a result of implementing Theorem 2. Column (2) of the table gives the times originally reported in reference [4]. The problem numbers are those used in reference [4].

Table 1

Effect of Theorem 2

Problem	(1) Solution Time (360-65 seconds) Without Theorem 2	(2) Solution Time With Theorem 2
5	45.14	30.14[a]
6	108.24	62.34
7	29.80	30.14[a]
8	48.17	30.14[a]

[a] Only the total solution time for problems 1, 3, and 4 is known. This time is 90.42 seconds.

The following theorem indicates how elimination of inactive schedules may be used to strengthen the restrictions defining a node.

<u>Theorem 3</u>. Assume $r_{ijk} = 0$ or 1 for all ijk, $\sum_k r_{ijk} = 1$, and $R_{kt} = 1$ and that we are branching by allocating resource time period kt. Consider a node in which kt is allocated to a task $\overline{ij}$ which has no predecessors. Let S be the set of tasks other than $\overline{ij}$ with $r_{ijk} = 1$ which satisfy $a_{ij} < t$. Let $t^1 = \max\limits_{ij\varepsilon \overline{S}} b_{ij}$. Then any schedule with $t_{\overline{ij}} > t^1$ is inactive. If $t^1 < t$, $\overline{ij}$ must be in process during periods $t^1 + 1, \ldots, t$.

Proof. If $t_{\overline{ij}} > t^1$ for some schedule we can decrease $t_{\overline{ij}}$ to t^1 since task $\overline{ij}$ has no predecessors and no task can be in process in the interval $[t^1, t_{ij}]$.||

There are undoubtedly other situations in which nodes can be shown to produce only inactive schedules. The problem, of course, is to obtain conditions describing such nodes which are not so intricate that their application requires more computation than they save. Even with this requirement, it seems likely that the results of this section could be further extended, especially in the direction of relaxing the restrictions on the form of the resource constraints considered.

5. Updating the Multipliers When Branching

When we obtain a bound at a node by solving problem (4) modified to correspond to the derived problem at that node, the procedure suggested in [4] is to begin with the basic feasible solution corresponding to $u = 0$. One would expect, however, that the multiplier values at the node previously bounded could be adjusted to obtain a starting solution when we bound a new node. In this section we show exactly how this is done.

We consider first the case where we have applied the bounding procedure to a node for which the set of allowable job schedules is $\overline{A}$, and then branch by creating new nodes. We would like to apply the bounding procedure to one of these nodes defined by $\overline{\overline{A}}$ with $\overline{\overline{A}} \subseteqq \overline{A}$. To do this we must solve $(4_{\overline{\overline{A}}})$. We have available a basic feasible solution to $(4_{\overline{A}})$. Let $w_{\overline{A}}$ denote the bound for this solution. We will show how to obtain a basic feasible solution to $(4_{\overline{\overline{A}}})$ from the solution to $(4_{\overline{A}})$ by a sequence of pivots in which the objective value for $(4_{\overline{\overline{A}}})$ increases monotonically from

$w_{\bar{A}}$. This is much preferred to starting $(4_{\bar{\bar{A}}})$ with the solution $u = 0$ since computational experience has shown that $w_{\bar{A}}$ is generally much larger than the bound for $u = 0$ in $(4_{\bar{\bar{A}}})$.

We let s denote a vector of slacks for constraints $(4_{\bar{A}}\, b)$ and partition s and u into s_B, s_N, u_B and u_N corresponding to basic and nonbasic variables. We let $\pi = (\pi_1, \ldots, \pi_I)$. In obtaining the current multiplier values we have transformed $(4_{\bar{A}})$ to obtain the following expression of basics in terms of nonbasics.

$$w = \bar{w} \quad + c_1 u_N + c_2 s_N \tag{6a}$$

$$\pi = \bar{\pi} \quad + A_{11} u_N + A_{12} s_N \tag{6b}$$

$$u_B = \bar{u}_B \quad + A_{21} u_N + A_{22} s_N \tag{6c}$$

$$s_B = \bar{s}_B \quad + A_{31} u_N + A_{32} s_N \tag{6d}$$

$\bar{w}$, $\bar{\pi}$, $\bar{u}_B$, and $\bar{s}_B$ represent the current values of the basic variables. c_1, c_2 and the matrices A_{ij} denote the coefficients obtained when we transform. They have the dimensions required to conform with w, π, u_B, and s_B. The procedure in the algorithm for solving $(4_{\bar{A}})$ given in [4] is to store explicitly only the rows (6a), (6b) and (6c). Constraints (6d) are represented implicitly as feasible solutions to job scheduling problems and are "available" through the use of an algorithm for the job scheduling problem.

Problem $(4_{\bar{\bar{A}}})$ may be obtained from problem $(4_{\bar{A}})$ by removing rows which correspond to job schedules contained in $\bar{A}$ but not in $\bar{\bar{A}}$. Rows corresponding to (6d) may be removed simply by modifying the algorithm used for the job scheduling problem to rule out the job schedules corresponding to these rows. When we remove a row for (6d) we also remove the slack associated with that row. Hence the resulting solution is basic and feasible in the new system of equations.

If we remove one or more rows corresponding to (6b) and (6c), however, the current solution given by $u_B = \bar{u}_B$, $\pi = \bar{\pi}$, while feasible is no longer basic in the new system of equations. We will show how to adjust the current solution to obtain a basic feasible solution with objective value at least as great as $w_{\bar{A}}$.

A basic step in this process is a procedure for recovering a basic solution when a single row is removed. To simplify notation this procedure will be stated in terms of the following general problem which is written in a partitioned form which will facilitate our analysis. It will be assumed that we wish to remove row (7c) from (7).

$$w = d_1x_1 + d_2x_2 + d_3x_3 \tag{7a}$$

$$B_{11}x_1 + B_{12}x_2 + B_{13}x_2 = b_1 \tag{7b}$$

$$B_{21}x_1 + B_{22}x_2 + B_{23}x_3 = b_2 \quad . \tag{7c}$$

In (7) w is 1×1, x_1 is $m - 1 \times 1$, x_2 is 1×1, x_3 is $n - m \times 1$, b_1 is $m - 1 \times 1$, and b_2 is 1×1. All other matrices have dimensions to conform with these. Let

$$B = \begin{pmatrix} B_{11} & B_{12} \\ B_{21} & B_{22} \end{pmatrix} \quad \text{and} \quad b = \begin{pmatrix} b_1 \\ b_2 \end{pmatrix} .$$

We assume that B^{-1} is known along with a basic feasible solution given by

$$\begin{pmatrix} x_1 \\ x_2 \end{pmatrix} = B^{-1}b = \begin{pmatrix} \bar{x}_1 \\ \bar{x}_2 \end{pmatrix} .$$

We will show how to revmove row (7c) from (7) and obtain a basic feasible solution from this current solution.

The basic approach is to obtain a representation for x_1 and w in terms of x_2. x_2 is then either increased or decreased from $\bar{x}_2$ (depending on how w varies with x_2) until either x_2 or a component of x_1 reaches zero. This provides the required basic feasible solution.

The representation for w and x_1 is given by

$$w = d_1 B_{11}^{-1} b_1 + (d_2 - d_1 B_{11}^{-1} B_{12}) x_2 + (d_3 - d_1 B_{11}^{-1} B_{13}) x_3$$

$$x_1 = B_{11}^{-1} b_1 - B_{11}^{-1} B_{12} x_2 - B_{11}^{-1} B_{13} x_3 .$$

To obtain this representation we need to find B_{11}^{-1}. Let $B^{-1} = ||\bar{b}_{ij}||$. It is easy to show that B_{11}^{-1} exists if and only if $\bar{b}_{mm} \neq 0$ so we assume that the choice of which basic variable to let be x_2 was made so that $\bar{b}_{mm} \neq 0$. In this case B_{11}^{-1} may be obtained from B^{-1} as follows. First we obtain the inverse

$$\begin{pmatrix} B_{11} & 0 \\ B_{21} & 1 \end{pmatrix}^{-1}$$

from B^{-1} by the usual procedure of the revised simplex method for updating an inverse when one column of a matrix has been changed. Then we obtain

B_{11}^{-1} by noting that

$$\begin{pmatrix} B_{11} & 0 \\ B_{21} & 1 \end{pmatrix}^{-1} = \begin{pmatrix} B_{11}^{-1} & 0 \\ -B_{21}B_{11}^{-1} & 1 \end{pmatrix} .$$

If $d_2 - d_1B_{11}^{-1}B_{12} > 0$ then x_2 is increased from $\bar{x}_2$, otherwise x_2 is decreased. The value of x_2 at which a variable first reaches zero is determined in the usual manner of the simplex method from ratios of appropriate components of $B_{11}^{-1}b_1$ and $B_{11}^{-1}B_{12}$. This value of x_2 and the corresponding values of x_1 and w are a basic feasible solution for the new system.

The algorithm for updating the dual solution is now clear. Once we ascertain which rows must be removed from (6b) and (6c) we use the procedure just presented to successively remove these rows and adjust the current solution. As the solution is adjusted, we also implicitly check constraints (6d) as modified and add those which become binding as described in [4]. If we are branching on kt, we give preference to the variable u_{kt} when selecting a variable to remove from the basis. When this process is complete we have a basic feasible solution to $(4_{\bar{\bar{A}}})$ and may continue to apply the simplex method to $(4_{\bar{\bar{A}}})$ as described in [4].

Now suppose the node we wish to bound is not a descendant of the node for which multiplier values are known. Let $\bar{u}$ denote the multiplier values and $\bar{\bar{A}}$ and $\bar{A}$ denote the allowable job schedules for the node we wish to bound and the node for which multipliers are known respectively. In this case, we first solve a job scheduling problem for each job using $\bar{\bar{A}}$ as the set of allowable job schedules to obtain values for π and w

which we denote by $\bar{\pi}$ and $\bar{w}(\bar{u})$. We also determine $w(o)$ and if $w(o) \geq \bar{w}(\bar{u})$, we use the starting solution $u = 0$. Otherwise, we have a feasible but nonbasic solution to $(4_{\bar{A}})$ given by $\bar{u}$, $\bar{\pi}$, and $\bar{w}(\bar{u})$. A basic feasible solution may be recovered from this solution in a manner similar to that just discussed. Although the situation is now more complicated, the basic approach can be outlined. As a result of solving $(4_{\bar{A}})$ a set of rows of $(4_{\bar{A}})$ are known explicitly. From this set we remove those constraints which are either no longer valid in $(4_{\bar{A}})$ or are no longer binding at the current solution. The resulting system of equations will generally be insufficient in number to determine the components of u and π at nonzero values in the current solution. Hence, we determine a subset of these variables to act as basic variables and determine an expression for these in terms of the remaining variables which act as nonbasics. The nonbasics are then systematically adjusted up or down from their current values, while the constraints not explicitly known are checked, until a basic feasible solution is obtained.

ACKNOWLEDGMENT

The author would like to express his appreciation to Linus Schrage for his encouragement and suggestions while this paper was being written.

REFERENCES

[1] E. Balas, "Machine Sequencing via Disjunctive Graphs: An Implicit Enumeration Algorithm." Operations Research, XVII, No. 6, (November-December, 1969).

[2] R. W. Conway, W. L. Maxwell, and L. W. Miller, Theory of Scheduling, Addison-Wesley, 1967.

[3] E. W. Davis and G. E. Heidorn, "An Algorithm for Optimal Project Scheduling Under Multiple Resource Constraints." Management Science, XVII,(1971), B-803 - B-816.

[4] M. L. Fisher, "Optimal Solution of Scheduling Problems Using Lagrange Multipliers: Part I." Forthcoming in Operations Research.

[5] ___________, "Optimal Solution of Resource Constrained Network Scheduling Problems." Technical Report No. 56, Operations Research Center, Massachusetts Institute of Technology, September, 1970.

[6] ___________ and Jeremy F. Shapiro, "Constructive Duality in Integer Programming." Report 7224, Center for Mathematical Studies in Business and Economics, University of Chicago, May, 1972.

[7] B. Giffler and G. L. Thompson, "Algorithms for Solving Production Scheduling Problems." Operations Research, VIII, (1960), 487-503.

[8] M. Held and R. M. Karp, "The Traveling Salesman Problem and Minimum Spanning Trees." Operations Research, XVIII, No. 6, (1970), 1138-1162.

[9] ______ and __________, "The Traveling Salesman Problem and Minimum Spanning Trees: Part II." Mathematical Programming, I, (1971), 6-25.

[10] T. J. R. Johnson, "An Algorithm for the Resource Constrained Project Scheduling Problem." Unpublished Ph.D. thesis, Sloan School of Management, MIT, (August, 1967).

[11] J. F. Muth and G. L. Thompson, Industrial Scheduling, Englewood Cliffs, N. J.: Prentice-Hall, Inc., 1963.

[12] G. T. Patton, "Optimal Scheduling of Resource Constrained Projects." Ph.D. dissertation, Stanford University, August, 1968.

[13] A. B. Pritsker, L. J. Watters, and P. M. Wolfe, "Multiproject Scheduling with Limited Resources: A Zero-One Programming Approach." Management Science, XVI, (1969), 93-108.

[14] L. E. Schrage, "Solving Resource-Constrained Network Problems by Implicit Enumeration--Nonpreemptive Case." Operations Research, XVIII, No. 2 (March-April, 1970), 273-278.

[15] ___________, "Obtaining Optimal Solutions to Resource Constrained Network Scheduling Problems." Paper presented at the American Institute of Industrial Engineers Systems Engineering Conference, Phoenix, Arizona, February 11-13, 1971.

DISCUSSION OF

" OPTIMAL SOLUTION OF SCHEDULING PROBLEMS USING LAGRANGE MULTIPLIERS: PART II"

Victor A. Zaloom

Auburn University, Auburn, Alabama, USA

General Comments

I would like to congratulate Dr. Fisher on the quality of his research paper. This research shows much promise for advancing the resource scheduling problem from one for which only theoretical solution procedures exist to one which can be solved in practice. Branch and bound algorithms appear to present the greatest hope in the near term for obtaining practical solutions to research scheduling problems. Hence the branch and bound algorithms presented in Part II of Dr. Fisher's paper coupled with and complimenting the bounding procedures presented in Part I appear to represent a significant improvement over existing methods of solving resource scheduling problems.

There are several key ideas appearing in these papers which are worthy of noting. The first is the idea of making use of the separability or independence of tasks belonging to one job from those belonging to all other jobs. For example, a general network requiring ten tasks or activities might appear as in Figure 1. In Figure 2, there are ten tasks but they are distributed over two jobs, each task in any job being independent from all tasks in the other job. This fact has been recognized by other researchers but to my knowledge has not been capitalized on until this paper.

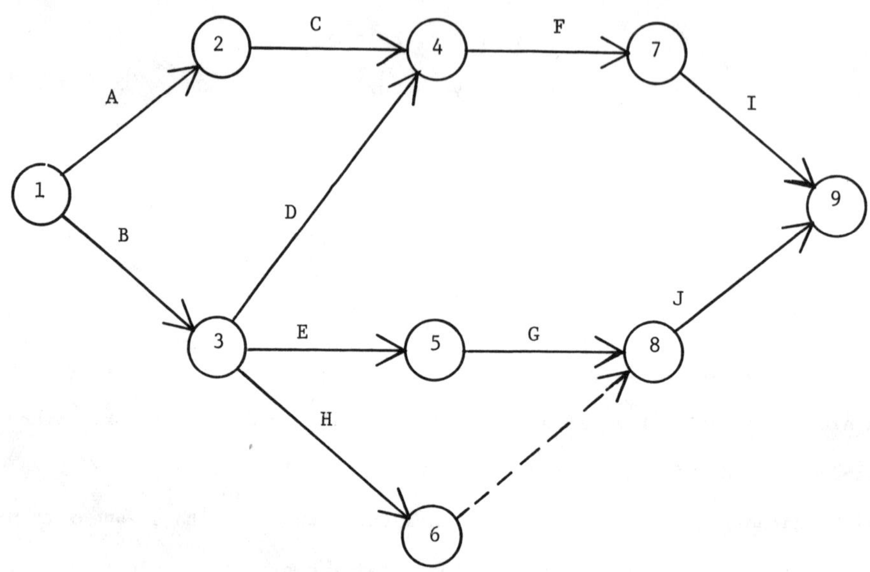

Figure 1. General Project Network

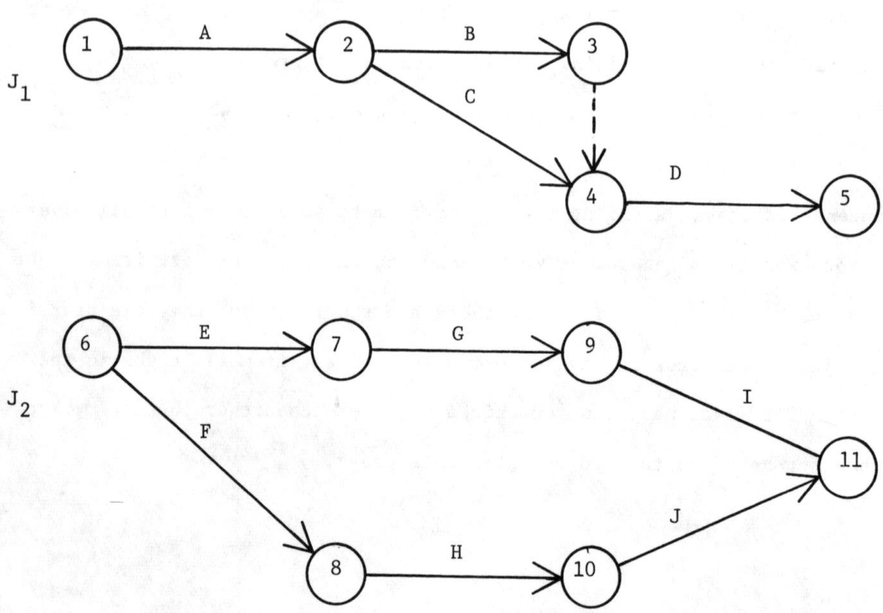

Figure 2. General Job Shop Network

Another idea presented in this research is that of defining a Lagrangian problem so that a sub-problem may be solved for each job independent of the other jobs. To quote the author: "Minimization of the Lagrangian with fixed multiplier values yields a lower bound on the objective value of the optimum solution to the scheduling problem".

The branch and bound algorithm which is the main theme of Part II appears to have great promise especially in light of the fact that constraints corresponding to a node of the branch-on-bound tree are separable by job.

Specific Comments and Questions

It is my purpose in this section of the discussion to bring to light certain specific areas of the paper. It is hoped that the comments and questions raised will provide a platform and a catalyst for a more complete discussion of these specific areas of the paper.

In describing the branch and bound algorithm (Section 3, page 10), the set S_{KL} is defined as the set of tasks which conditions 4 and 5 allow to be in process during period t and for which $r_{ij} > 0$. Condition 4 restricts the starting time of task ij to conform to the precedence restrictions and condition 5 defines $f_i = \max_{j=1,\ldots,n_i} (f_{ij})$, to be the completion time of job i. T is defined to be an integer upper bound on the number of time periods until all jobs are complete $f_i \leq T$ for $i = 1,\ldots,I$.

How is T initially obtained and is S_{KT} based on T? On page 13 A(ij) is defined to be the time periods during which task ij must be in process and B(ij) is the set of periods during which ij must not be in process. Are these sets also based on T?

The efficiency of the algorithms is not presented and so the paper ends without a strong conclusion. I hope for the sake of all of us interested in this research that the question of the algorithm's efficiency will be answered in a subsequent paper. Do you have any preliminary efficiency comparisons with other available algorithms?

Section 4 of this paper deals with improving the efficiency of the algorithm by eliminating inactive schedules. It is noted that this section deals exclusively with the restricted job shop scheduling problem depicted in Figure 3. Each job consists of an ordered set of tasks (as opposed to the partially ordered set Figure 2.) requiring unit resource of a single type. It is also assumed that one unit of each type of resource is available in each time period. When this assumption is relaxed, what is the estimated decrease in efficiency? Or what is the relative efficiency of the two different algorithms presented in 3.2 of Part I?

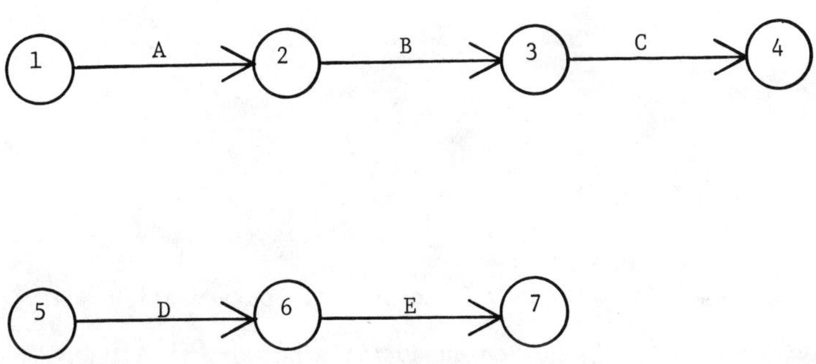

Figure 3. Restricted Job Shop Network

ON PROJECT COST-DURATION ANALYSIS PROBLEM WITH QUADRATIC AND CONVEX COST FUNCTIONS

K. C. Kapur

Department of Industrial Engineering and Operations Research
Wayne State University, Detroit, Michigan, USA

Optimality conditions for convex programming problem are applied to the project cost-duration analysis problem for project network with convex cost. The optimality conditions give an optimality curve or an out-of-kilter type diagram for this problem. A solution is optimal if and only if when the values for activities are plotted on their optimality diagram, the values lie on the optimality curve. An algorithm is developed for the case when the cost is quadratic and convex. The different steps of the algorithm generate the solutions in such a way that it always lies on the optimality curve. Thus, problems with convex cost functions can be approximated by quadratic functions. This is better than linear approximation as is generally discussed in literature.

The mathematical formulation of the project cost-duration analysis problem is as follows (2,3). Given an activity network $[\mathcal{N};A]$, where $\mathcal{N}$ is the set of N nodes of the network numbered as 1, 2, .., i, .., j, ..., N, and where 1 is the starting node and N is the terminal node, and A is the set of all arcs (i, j) representing the activites of the project.

For each activity (i, j), the lower and upper limits on the durations of the activity are given. The lower limit is given by l_{ij} and the upper limit is given by u_{ij}, i.e., $l_{ij} \leq t_{ij} \leq u_{ij}$, where t_{ij} is the duration of the activity (i, j). The cost of reducing the duration of the activity below the lower limit l_{ij} is infinite and thus l_{ij} represents the crashed duration of the activity (i, j). u_{ij} represents the tight or the normal duration of the activity and to reduce the duration of the activity from the level of u_{ij}, a certain cost is incurred. This cost is a certain function of the level of the reduction in the duration of the activity. This is an arbitrary nonlinear function of the reduction in the duration of the activity given by $C_{ij}(r_{ij})$, where $r_{ij} = u_{ij} - t_{ij}$. It is clear that $C_{ij}(0) = 0$. In general, $C_{ij}(r_{ij})$ is a monotone increasing function of r_{ij}, i.e., the higher the reduction, the more does it cost to reduce the duration of the activity. Given a certain vector of the reductions in the various activities $R = \{r_{ij}, (i, j) \varepsilon A\}$, we can uniquely determine the vector $T = \{t_{ij} = u_{ij} - r_{ij}, (i, j) \varepsilon A\}$ for the durations of the different activities. The total project duration or the length of the project is a function of T and may be denoted by L(T). It should be observed that, given a certain network, for every T, there is one L, but the same L may be obtained from more than one T. The length of the project is given by the critical path and for a given duration of the network, there may be many critical paths. The total project cost PC is a function of T, and therefore a function of L(T). Hence we have

$$PC(T) = PC\ \{L(T)\} = \sum_{(i,j)} C_{ij}\ (u_{ij} - t_{ij}) \qquad [1]$$

Let L_l and L_u be the lower and upper limits on the duration of the project given by L.

$$L_\ell = L\ \{(T) | t_{ij} = l_{ij},\ \text{for all}\ (i,j) \in A\}$$

$$L_u = L\ \{(T) | t_{ij} = u_{ij},\ \text{for all}\ (i,j) \in A\}$$

The problem in cost-duration analysis is to find the optimum value or the minimum cost of the total project for each possible duration of the project L, given by $L_\ell \leq L \leq L_u$. Hence

$$PC^*(L) = \min_{T} PC\ \{L(T)\},\ L_\ell \leq L \leq L_u,\ (\text{all}\ T{:}L(T) = L) \qquad [2]$$

Generally, the objective is to find a curve for PC* versus L.

Let t_i denote the realization time of event i. For the activity network, we can determine t_i for all $i \in \mathcal{N}$ given a certain T. We can also write [2] as a nonlinear programming problem which is given below:

$$PC^*(L) = \text{Min} \sum_{(i,j) \in A} C_{ij}\ (u_{ij} - t_{ij})$$

subject to

$$\begin{aligned} & t_j - t_i \geqq t_{ij} && \text{for all}\ (i,j) \\ & t_N - t_1 \leqq L && \qquad [3] \\ & l_{ij} \leq t_{ij} \leq u_{ij} && \text{for all}\ (i,j) \in A \end{aligned}$$

Generally, $t_1 = 0$. In the above formulation, we are saying that given a duration L for the activity network, find the minimum additional cost for realizing this duration.

NATURE OF THE COST FUNCTION

The cost function $C_{ij}(r_{ij})$, $0 \leq r_{ij} \leq (u_{ij} - l_{ij})$ could be any arbitrary nonlinear function. In general, it will be non-decreasing. Problem [3] as such is a nonlinear programming problem and hence, theoretically could be solved by using the techniques available in nonlinear programming. From a computational point of view, these techniques as such are not very efficient. Hence, we use the special structure of these problems which is given by the underlying network. These problems have been shown equivalent to nonlinear network flow problems (2,3). Thus, we could use the techniques available for nonlinear network flow problems. So far, not many computational techniques are available to solve these nonlinear problems exactly. Most of the techniques available approximate the nonlinear function by piecewise linear function and then solve these approximate problems by the techniques available in network flows and mathematical programming (7). Zangwill (8) considers the nonlinear network problem where a concave objective function is minimized. In this case, we get extremal solutions (8) because a concave function achieves its local minimum at the boundary point of the convex feasible region. This characterization is exploited in (5) to develop some algorithms for the nonlinear concave minimization network problem. In this case, an optimal solution can be obtained by some type of search procedure of the extreme points of the convex polyhedron set. For practical problems, the feasible region given by [3] will be bounded and hence will have a finite number of extreme points.

The problem as formulated and given by [3] does not always require integer solutions. Thus, consider the variables as real numbers. This means that durations of the activities is given by certain time units and any real value of the units is possible. In the literature, generally integer values have been assumed when the network approach is used. Let us assume that the function $C_{ij}(r_{ij})$ is any arbitrary convex function defined on the region $0 \leq r_{ij} \leq (u_{ij} - l_{ij})$. In this case, the feasible region is convex. Thus, we are minimizing a convex function on a convex set, and hence the optimal solution could be anywhere in the feasible region, both the boundary and the interior of the feasible region. Hence, there are infinite number of possible solutions which could be optimal. Due to convexity, one advantage is that a local optimal solution will be a global optimal solution. We will exploit this property in terms of necessary and sufficient conditions for optimality for convex programming to develop an approach towards the solution aspects for nonlinear project-cost duration analysis problem.

It is well known (1,6) that a convex function f(x) defined on a convex set $X \subset R$ i.e., an interval in R given by $\ell \leq x \leq u$, where ℓ and u are two real numbers, is continuous at every point in the interval and has both right-hand and left-hand derivatives. Let $f'^{(-)}(x)$ denote the left-hand derivative and $f'^{(+)}(x)$ denote the right-hand derivative. Then. $f'^{(+)}(x)$ as well as $f'^{(-)}(x)$ is monotone non-decreasing function of x and at every interior point $x \varepsilon X$, we have

$$f'^{(+)}(x) \geqq f'^{(-)}(x)$$

Also, $f'^{(-)}(x_1) \geqq f'^{(+)}(x_2)$ if $x_1 > x_2$.

The characteristic curve of a convex function f(x) is the set of points (x, x*) such that x is an interior point of X and $f'^{(-)}(x) \leqq x^* \leqq f'^{(+)}(x)$, and if x is the left end of X, then $-\infty < x^* \leqq f'^{(+)}(x)$; and if x is the right end of X, then $f'^{(-)}(x) \leqq x^* < \infty$. Also, we have

$$f(x_1) = \int_{(x_0,\ x_0^*)}^{(x_1,\ x_1^*)} x^* d\ x + \text{an integration constant}$$

where x_0 is the left hand end point of the interval. Thus, we can represent a convex function by a curve, known as the characteristic curve and given by

$$f'^{(-)}(x) \leqq x^* \leqq f'^{(+)}(x)$$

on the plane with abscissa x^* and ordinate x.

Hence the objective function for [3] for the convex case can be written as

$$\text{Min } PC^*(L) = \text{Min } \Sigma \int_0^{r_{ij}} c_{ij}(r)dr \qquad [4]$$

where $c_{ij}(r_{ij})$ is given by

$$C_{ij}'^{(-)}(r_{ij}) \leqq c_{ij}(r_{ij}) \leqq C'^{(+)}(r_{ij}) \qquad [5]$$

See Figure 1. Then $c_{ij}(r_{ij})$ defines the characteristic curve for the convex function $C_{ij}(r_{ij})$. Hence, we have

$$C_{ij}(r_{ij}) = \int_0^{r_{ij}} c_{ij}(r)d\ r$$

(shown by the shaded area in Figure 1)

The constant of integration is zero because $C_{ij}(0) = 0$. Thus, the curve $c_{ij}(r_{ij})$ is monotone and non-decreasing and satisfies other properties as stated before.

In many cases, the cost functions $C_{ij}(r_{ij})$ are quadratic and convex. In this case, we have

$$C_{ij}(r_{ij}) = \frac{1}{2} a_{ij}\ r_{ij}^2 + b_{ij}\ r_{ij} + d_{ij}$$

Then, $c_{ij}(r_{ij}) = \dfrac{dC_{ij}(r_{ij})}{dr_{ij}} = a_{ij}\ r_{ij} + b_{ij}$

Now, C_{ij} is convex if and only if $a_{ij} \geqq 0$.

In order to write the optimization problem, it may be convenient to express the objective function in terms of the variables t_{ij}. The

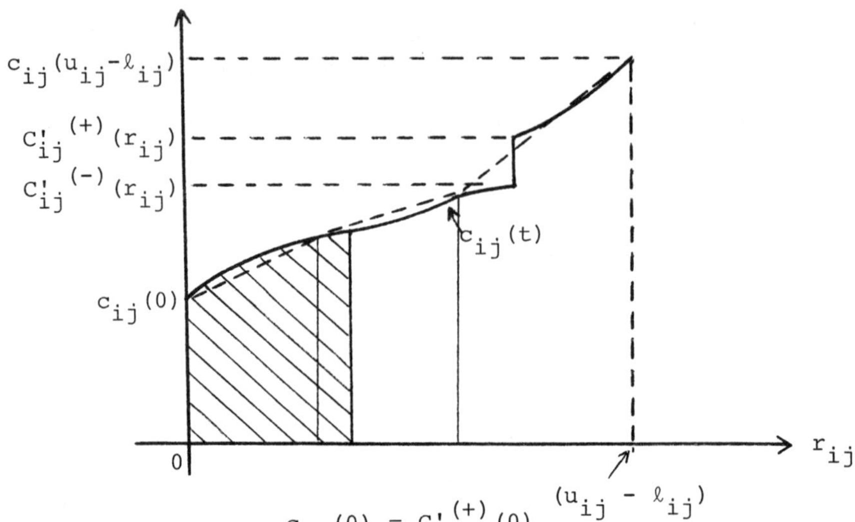

$$c_{ij}(0) = C'^{(+)}(0)$$

$$c_{ij}(u_{ij}-\ell_{ij}) = C'^{(-)}(u_{ij}-\ell_{ij})$$

FIGURE 1

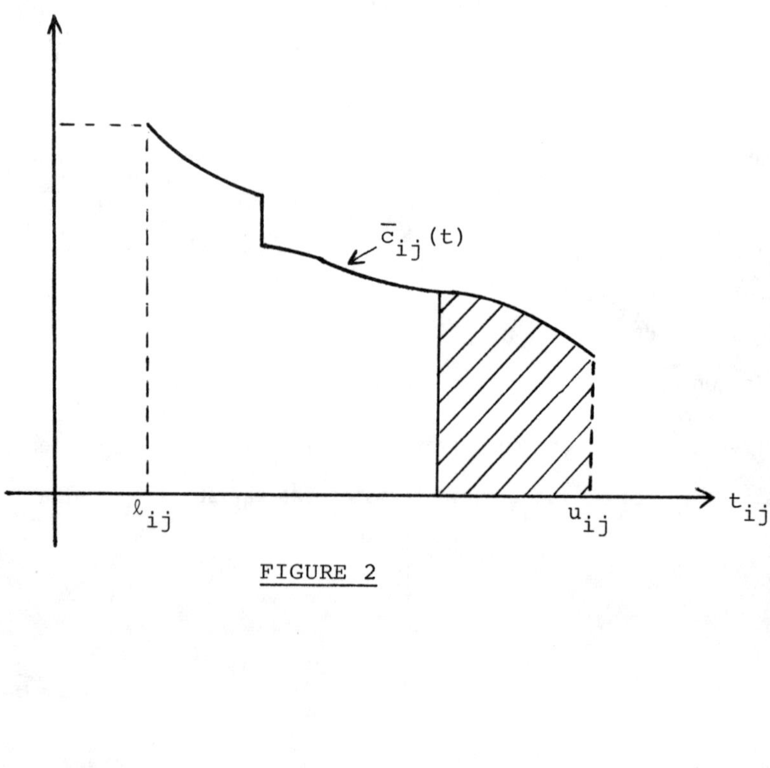

FIGURE 2

curve in Figure 2 is obtained after proper translation of the curve in Figure 1. Let the curve in Figure 2 be denoted by $\bar{c}_{ij}(t_{ij})$, defined on the interval $\ell_{ij} \leq t_{ij} \leq u_{ij}$. Then $\bar{c}_{ij}(t_{ij}) = c_{ij}(u_{ij}-t_{ij})$.

Also,

$$\int_0^{r_{ij}} c_{ij}(r)\, dr = \int_{\ell_{ij}}^{u_{ij}} \bar{c}_{ij}(t)dt - \int_{\ell_{ij}}^{t_{ij}} \bar{c}_{ij}(t)dt$$

The first integral on the right hand side is a constant, i.e., area under the curve $\bar{c}_{ij}$. Hence, instead of minimizing the objective function as given by [4], we should

$$\text{Maximize} \quad \sum_{(i,j)\varepsilon A} \int_{\ell_{ij}}^{t_{ij}} \bar{c}_{ij}(t)dt \qquad [6]$$

subject to

$$t_i + t_{ij} - t_j \leqq 0 \qquad \text{for all } (i,j)\varepsilon A \qquad [7]$$

$$-t_1 + t_n \leqq L \qquad [8]$$

$$t_{ij} \leqq u_{ij} \qquad \text{for all } (i,j)\varepsilon A \qquad [9]$$

$$-t_{ij} \leqq -\ell_{ij} \qquad \text{for all } (i,j)\varepsilon A \qquad [10]$$

It should be clear that $\bar{c}_{ij}(t)$, $\ell_{ij} \leq t_{ij} \leq u_{ij}$ is a monotone and non-increasing function of t_{ij}. Thus, the objective function in [6] is a concave function. Thus, we are maximizing a concave function and again we can use Kuhn-Tucker necessary and sufficient conditions as discussed in the next section.

OPTIMALITY CONDITIONS

The problem given by [6] to [10] is a concave programming problem. We can write easily Kuhn-Tucker necessary and sufficient conditions (9) for optimality. Associate the Lagrange multipliers f_{ij} with [7], Q with [8], g_{ij} with [9] and h_{ij} with [10]. Then, the Lagrangian function is

$$\sum_{(i,j)\varepsilon A} \int_{\ell_{ij}}^{t_{ij}} \bar{c}_{ij}(t)dt - \sum_{(i,j)} f_{ij}(t_i + t_{ij} - t_j) - Q(-t_1 + t_N - L)$$

$$-\sum_{(i,j)} (t_{ij}-u_{ij})g_{ij} - \sum_{(i,j)} (-t_{ij} + \ell_{ij})h_{ij}$$

$f_{ij} \geqq 0$, $g_{ij} \geqq 0$ and $h_{ij} \geqq 0$ for all $(i,j)\varepsilon A$, and $Q \geqq 0$.

A solution t_{ij}^*, t_i^* is optimal to problem [6] to [10] if and only if there exist Lagrange multipliers f_{ij}^*, Q^*, g_{ij}^* and h_{ij}^* such that

$$\bar{c}_{ij}(t^*_{ij}) - f^*_{ij} - g^*_{ij} + h^*_{ij} = 0 \text{ for all } (i,j)\varepsilon A \qquad [11]$$

$$(t_i^* + t^*_{ij} - t_j^*)\, f_{ij}^* = 0 \text{ for all } (i,j)\varepsilon A \qquad [12]$$

$$(-t^*_1 + t^*_N - L)\, Q^* = 0 \qquad [13]$$

$$(t_{ij}^* - u_{ij})\, g_{ij}^* = 0 \text{ for all } (i,j)\varepsilon A \qquad [14]$$

$$(-t^*_{ij} + \ell_{ij})\, h^*_{ij} = 0 \text{ for all } (i,j)\varepsilon A \qquad [15]$$

Equation [11] is obtained by differentiating the Lagrangian with respect to t_{ij} and equating it equal to zero. Equations [12] to [15] are the well-known complementary slackness type conditions for this problem. Thus, when $t^*_{ij} = u_{ij}$, then $h^*_{ij} = 0$ and hence,

$$\bar{c}_{ij}(u_{ij}) - f^*_{ij} = g^*_{ij} \geqq 0$$

$$\text{or } \bar{c}_{ij}(u_{ij}) \geqq f^*_{ij} \qquad [16]$$

And when $t^*_{ij} = \ell_{ij}$, then $g^*_{ij} = 0$

and hence

$$\bar{c}_{ij}(\ell_{ij}) - f^*_{ij} = -h^*_{ij} \leqq 0$$

$$\text{or } \bar{c}_{ij}(\ell_{ij}) \leqq f^*_{ij} \qquad [17]$$

And when $\ell_{ij} < t^*_{ij} < u_{ij}$, then $h^*_{ij} = g^*_{ij} = 0$

$$\text{and hence } \bar{c}_{ij}(t^*_{ij}) = f^*_{ij} \qquad [18]$$

Combining these equations, we get the optimality curve or out-of-kilter type diagram for the nonlinear problem as shown in Figure 3. This type of diagram is well known for the linear case. A solution to problems [6] to [10] is optimal if and only if when the values for different arcs or activities are plotted on their optimality or out-of-kilter diagram, the values lie on the dark curve shown in Figure 3 for each arc or activity.

Now, in order to find the optimum project cost for a given duration of the project L, we know that the length L is given by the critical path for the project network. Thus, $\{t_i\}$ can be considered as potentials on different nodes and $t^*_N - t^*_1 = L$. $\{f_{ij}\}$ can be considered as a flow function on the network and it satisfies the following conditions:

$$\sum_j (f_{ij} - f_{ji}) = \begin{cases} Q & i = 1 \\ 0 & i \neq 1, N \\ -Q & i = N \end{cases} \qquad [19]$$

Also, equation [12] shows us that flow will be positive if $t_{ij} = t_j - t_i$, i.e. there is no slack in the activity (i,j). All the activities with slack will have zero flow.

In Figure 3, an activity could be in any of the four states A, B, C or D. In state A, there is a slack in the activity. For optimality, due to [12], the flow f_{ij} in this arc should be zero. In state B, the activity has the normal duration and we have to send a flow of $\overline{c}_{ij}(u_{ij})$ through this arc before we can start decreasing the duration of this activity. Activities in state B are also sometimes called crucial because only after a flow of $\overline{c}_{ij}(u_{ij})$ through these activities is sent, can these activities be used for a reduction in the duration of the project. As given by eq. [18], when $l_{ij} < t_i{}^*{}_j < u_{ij}$, then $\overline{c}_{ij}(t_i{}^*{}_j) = f^*{}_{ij}$. Activities which satisfy this condition are in state C. It is this condition that makes the nonlinear problem harder to solve. In the linear case, the activities in state C would have been represented by a horizontal line and hence duration of the activities or the potentials could be changed without changing the flow through the corresponding arc (3,4). For the nonlinear case, in order to remain optimal, a simultaneous change in the flows and potentials is needed because we have to remain on the curve C shown in Figure 3 for optimality. This simultaneous change affects the whole activity network. In the next section, an algorithm is developed which accomplishes this simultaneous change in the flows and the potentials. The steps of the algorithm are illustrated by an example. Activities in state D are saturated and their duration cannot be reduced any further. In this state, we can send as much flow as we want and thus the capacity is infinite. Thus, the flow capacity of arcs in state A is zero, in state B is $\bar{c}_{ij}(u_{ij})$, in state C is $\bar{c}_{ij}(\ell ij)-\bar{c}_{ij}(u_{ij})$ and in state D is infinite.

The algorithm developed here is for quadratic and convex cost functions The optimality curve for these functions is shown in Figure 4. In this case, the curve C is replaced by a straight line with a constant slope of $(-a_{ij})$. In this case, to change flows and potentials simultaneously, a system of linear equations and inequalities has to be solved as shown later.

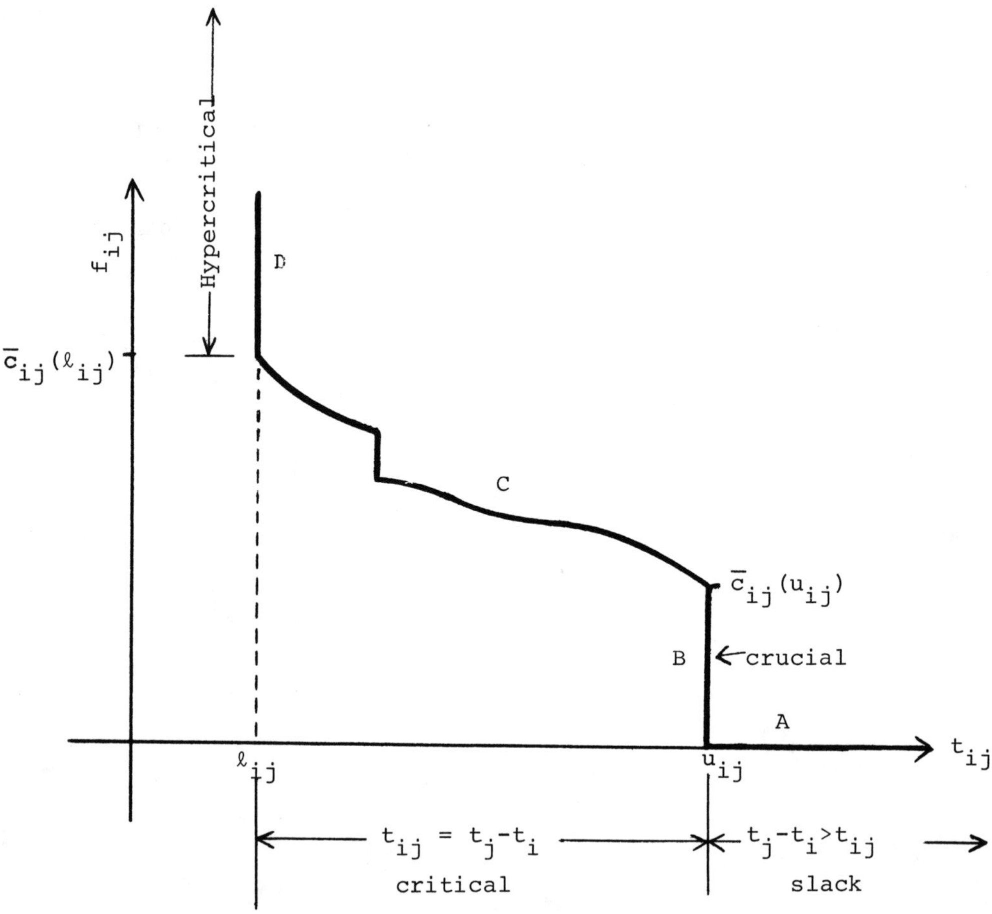

Hypercritical
f_{ij}
$\bar{c}_{ij}(\ell_{ij})$
D
C
$\bar{c}_{ij}(u_{ij})$
B
crucial
A
t_{ij}
ℓ_{ij}
u_{ij}
$t_{ij} = t_j - t_i$
critical
$t_j - t_i > t_{ij}$
slack

FIGURE 3

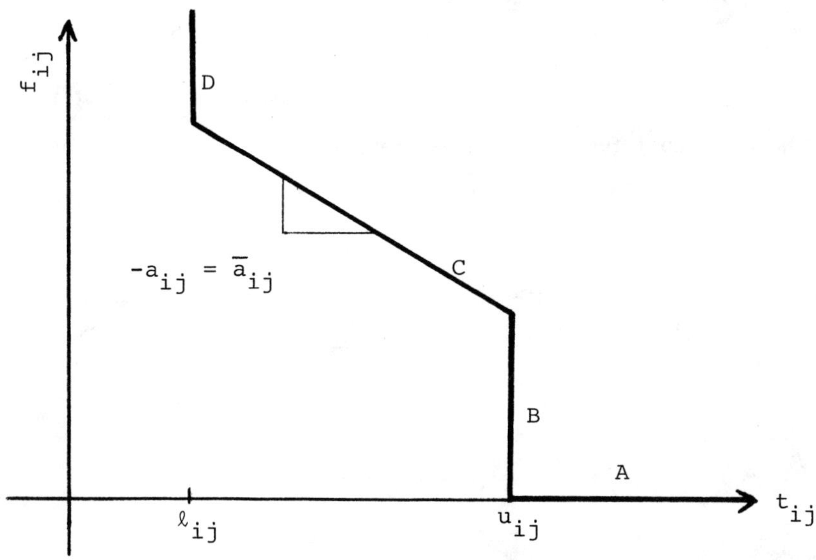

FIGURE 4

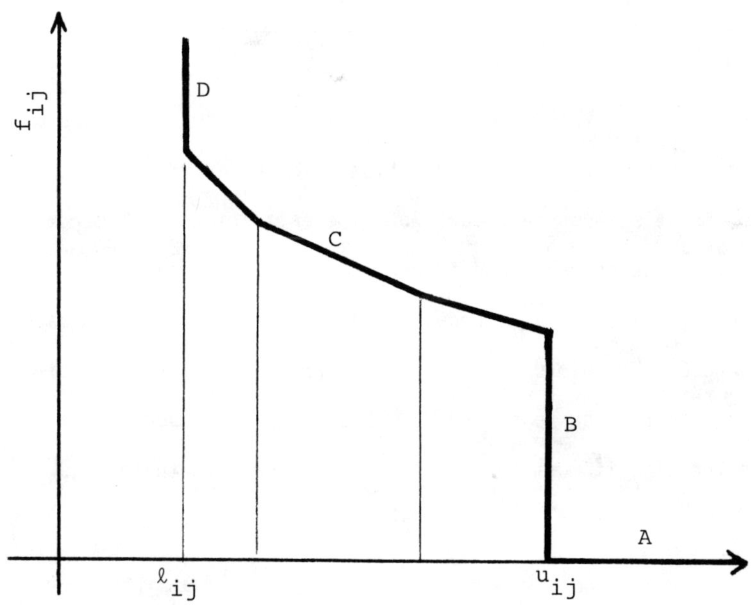

FIGURE 5

EXAMPLE

Now, we give a numerical example which illustrates the algorithm for convex quadratic cost functions. This example is a generalization of the example given in (3) where only linear cost function is considered. The activity network is given in Figure 7a with the following data:

Arc	ℓ_{ij}	u_{ij}	$\bar{c}_{ij}(t_{ij})$	$\bar{a}_{ij}$
1,2	4	6	$-2t_{12} + 18$	-2
1,3	4	8	$-t_{13} + 15$	-1
1,4	3	5	$-2t_{14} + 11$	-2
2,4	3	3	∞	---
2,5	3	5	$-t_{25} + 8$	-1
3,6	8	12	$-2t_{36} + 40$	-2
4,6	5	8	$-4/3t_{46} + 41/3$	-4/3
5,6	6	6	∞	---

Diagrams similar to Figure 4 for each arc are shown in Figure 6. The duration for the project when duration of each activity is u_{ij} is 20. This is shown in Figure 7b. The entries in each circled node i are the node number i and its earliest realization time t_i. Let $f_{ij} = 0$ for all $(i,j) \varepsilon A$. Hence, we can plot f_{ij} and t_{ij} for each arc in Figure 6. The plots for iteration 1 are shown by a small circle with number 1 in the square brackets. This example is used to illustrate the algorithm given in the next section. In these figures, the state of each activity is given by letters A, B, C or D for a particular iteration. The lower limit and the upper limit for the duration of the activity is shown by a column vector $\binom{u_{ij}}{\ell_{ij}}$ on each arc in the activity network. There is a row vector associated with each arc also. The first element of the row vector is the present flow through that arc and the second element is the remaining capacity of that arc in the present state. The capacities for the different states have been given before. For example, in Figure 7c, for the arc (1,3), the vectors $\binom{8}{4}$ - (7, 4) represent that $u_{13} = 8$, $\ell_{13} = 4$, the present flow through arc (1,3) is 7 units and the remaining flow capacity for arc (1,3) in state C is 4 units. The two dimensional row vector associated with each row represent the labels for that node. Details for this representation can be found in reference (3).

A chain for the activity network is a sequence of nodes and directed arcs (in one direction) from the origin node 1 to the terminal node N. Thus, $1,(1,j_1), j_1,(j_1,j_2)\ldots,(j_n,N),N$, where all $(1,j_1)$, $(j_1,j_2),\ldots,(j_n,N)$ belong to A is a chain. Many times a chain is just denoted by a sequence of nodes such as $1\text{-}j_1\text{-}j_2,\ldots,\text{-}j_N\text{-}N$.

ALGORITHM FOR MINIMAL COST-DURATION FUNCTION WITH CONVEX QUADRATIC COST

The algorithm given below is a generalization of the method known for linear cost functions and as given by Elmagrably (3).

Step 1: Determining $L_u = L\{(T) | t_{ij} = u_{ij}, \text{ for all } (i,j) \varepsilon A\}$

Put $t_{ij} = u_{ij}$ and determine $t_i(E)$, the earliest completion time of all nodes i=1,--, N by any known method (3). This is shown in Figure 7b. Let the initial flow through the network be zero, i.e. $f_{ij} = 0$ for all (i,j). Using $\{t_i\}$ and f_{ij}, plot them in Figure 6 for each arc. This is shown by small circle numbered [1] in Figure 6. Only a path, whose arcs are in state B, C or D, will be a critical path. Thus, chain 1-3-6 is a critical path with project duration of 20 units.

Step 2: Determining feasibility of L(T) less than L_u

Label the origin with (∞,0). For any mode i which is labelled with ∞, check all nodes j connected with it for

$$S_{ij} = t_i + \ell_{ij} - t_j = 0 \qquad \text{(State D)}$$

If the above equation holds for j, label it with (∞,i). Continue this procedure. If the node N cannot be labelled with ∞, it is possible to reduce L further. If the node N can be labelled, we cannot reduce L further and terminate. If the node N cannot be labelled, go to step 3. The above is also equivalent to finding a path, all of whose arcs are in state D. If such a path exists, we cannot reduce L further.

In our case, we can reduce the project duration further, as is clear from Figure 7b.

Step 3: Sending flows at minimum cost

Flow can only be sent through an arc in state B, C or D. Find all the paths whose arcs are in state B or C or D. These are the critical paths.

Step 3a: Sending flow without change of potentials

Start with the origin, node 1, which is always labelled $(\infty,0)$. For any labelled node i, check all nodes j connected with it so that we have the following four conditions:

(i) $S_{ij}^{(1)} = t_i + u_{ij} - t_j < 0$ and $f_{ij} = 0$ State A

(ii) $S_{ij}^{(2)} = t_i + u_{ij} - t_j = 0$ and $f_{ij} < \bar{c}_{ij}(u_{ij})$ State B

(iii) $S_{ij}^{(3)} = t_i + t_{ij} - t_j = 0$ and $f_{ij} < \bar{c}_{ij}(\ell_{ij})$ State C

(iv) $S_{ij}^{(4)} = t_i + \ell_{ij} - t_j = 0$ State D

In case (ii), label node j with (q_j, i) where

$$q_j = \min \{q_i, \bar{c}_{ij}(u_{ij}) - f_{ij}\}$$

and in the fourth case, label it with (q_i, i).

Continue the labelling until one of the following two conditions obtains: (a) the terminal node N cannot be labelled and this is a nonbreakthrough situation, or (b) the terminal node N is labelled. In case (b), use the flow change routine which is given next. Start with node N. For any labelled node j with label (q_j, i), increase the flow f_{ij} by q_j and reduce the residual capacity in the arc (i,j) by q_j, and erase the labeling of j. Continue until node 1 is reached.

As shown in Figure 7b, we can label the node 6 with (7,3). Hence a breakthrough occurs and we can send a flow of 7 units along the path 1-3-6. This is shown in Figure 7c. Now, if we try to label again using step 3a, we get a cut-set given by arcs (1,3), (4,6) and (5,6). Now, go to step 3b because we obtain the case (a) discussed above.

Step 3b:

In case (a), in order to increase the flow, arcs in state C have to be used. To use these arcs, simultaneous change in flow and potential is needed. Our objective is to send flow along the arcs with the least cost. The flow capacities of the arcs are variable. The arc in state A has zero capacity, in state B has $\bar{c}_{ij}(u_{ij})$ capacity, in state C has a total capacity of $\bar{c}_{ij}(\ell_{ij}) - \bar{c}_{ij}(u_{ij})$. In state D, the capacity of the arc is infinite.

Critical path is the longest chain in the activity network. Generally, there are many critical paths. For the existing network, we can find all the second longest paths. For example, for Figure 7c, there are two second longest paths 1-2-5-6 and 1-2-4-6 with durations of

17 units each. This means that we can reduce the present critical path by at most 3 units, because then other paths will also become critical. Another restriction which we may have to consider to determine the maximum possible reduction of the project duration from the present iteration is given next.

In the case of a nonbreakthrough, let Z be the set of arcs (i,j) such that one node is labelled and the other is not. Then Z is a cut set. Then Z consists of activities either in state A or in state C.

$$Z_A = \{(i,j) \mid i \text{ labelled, } j \text{ unlabelled and } S_{ij}^{(1)} < 0\}$$

$$Z_C = \{(i,j) \mid i \text{ labelled, } j \text{ unlabelled and } t_{ij} > \ell_{ij}\}$$

Define $\delta_1 = \min\limits_{Z_A} \{-S_{ij}^{(1)}\}$

$$\delta_2 = \min_{Z_C} \{t_{ij} - \ell_{ij}\}$$

and $\delta = \min(\delta_1, \delta_2)$

Then, the maximum reduction possible in the duration of the project is $\Delta L_{max} = \delta$. As we will see later on, there are other constraints which we have to consider, which may reduce the maximum reduction possible in the project still further.

In our example, using the above criteria, maximum reduction denoted by ΔL_{max} in the duration of the project is 3 units.

Let P be a critical path.

P has arcs in state B, C or D

P^B - subset of arcs in state B

P^C - subset of arcs in state C

P^D - subset of arcs in state D

Then, $P = P^B \cup P^C \cup P^D$

L(T) is the present duration of the path.

Let ΔL be the maximum reduction possible in the duration of the critical path. Then

$$\sum_{(i,j)\in P^C} (t_{ij} - \Delta t_{ij}) = L - \Delta L$$

$$\text{or} \quad \sum_{(i,j)\in P^C} \Delta t_{ij} = \Delta L \qquad [20]$$

It is clear that we reduce durations of only those activities which are in state C.

The relationship between the flow and the durations can be developed. A reduction Δt_{ij} of the duration of the activity (i,j) in state C is equal to an increment $\Delta f_{ij} = a_{ij}\Delta t_{ij}$ of the flow in the arc (i,j) in state C. Now, we have to decrease the durations and simultaneously increase the flows in the arcs in such a way that we satisfy all the constraints. The constraints are

$$\sum_{(i,j)\varepsilon P^C} \Delta t_{ij} \leqq \Delta L$$

$$f_{ij} + \Delta f_{ij} \begin{cases} \leqq \bar{c}_{ij}(u_{ij}) & \text{if } (i,j)\varepsilon P^B \\ \leqq \bar{c}_{ij}\ (\ell_{ij}) & \text{if } (i,j)\varepsilon P^C \end{cases}$$

and $t_{ij} - \Delta t_{ij} \geqq u_{ij}$ for all (i,j)

Thus, for a decrease ΔL in the duration of the critical path, the equations are

$$\sum_{(i,j)\varepsilon P^C} \Delta t_{ij} = \Delta L$$

$$\text{or} \quad \sum_{(i,j)\varepsilon P^C} \frac{\Delta f_{ij}}{a_{ij}} = \Delta L$$

Let there be p critical paths, $P_1, P_2, \ldots, P_k \ldots, P_p$.

Along a certain path P_k, the incremental flow is given by Δf_k. Then

$$\sum_{(i,j)\varepsilon P_k^C} \frac{1}{(a_{ij})} \Delta f_k = \Delta L$$

Then, we have to reduce the lengths of all the critical paths simultaneously. This reduction must satisfy the following p simultaneous equations.

$$\sum_{(i,j)\varepsilon P_1^C} (1/a_{ij})\ \Delta f_1 + \sum_{(i,j)\varepsilon P_1^C \cap \bigcup_{k=2}^{p} P_k^C} 1/a_{ij} \Big(\sum_{k\varepsilon I_{ij}^C} \Delta f_k\Big) = \Delta L$$

and

$$(i,j) \notin \bigcup_{k=2}^{p} P_k^C \qquad [21]$$

$$\vdots \qquad \vdots \qquad \vdots \qquad \vdots$$

$$\vdots \qquad \vdots \qquad \vdots \qquad \vdots$$

$$\sum_{(i,j)\varepsilon P_p^C} (1/a_{ij})\Delta f_p + \sum_{(i,j)\varepsilon P_p^C \cap \bigcup_{k=1}^{p-1} P_k^C} 1/a_{ij}\Big(\sum_{k\varepsilon I_{ij}^C} \Delta f_k\Big) = \Delta L$$

and

$$(i,j) \notin \bigcup_{k=1}^{p-1} P_k^C$$

where $I^C_{ij} = \{k|(i,j)\varepsilon P_k^C,\ k=1,\ldots,p\}$

and $I^B_{ij} = \{k|(i,j)\varepsilon P_k^B,\ k=1,\ldots,p\}$

The Δf_k, k=1,...,p determined by solving the above system of equations should be such that the resulting flows are feasible, i.e.

$$f_{ij} + \sum_{k\varepsilon I^C_{ij}} \Delta f_k \leqq \bar{c}_{ij}(\ell_{ij}) \qquad [22]$$

$$\text{and } f_{ij} + \sum_{k\varepsilon I^B_{ij}} \Delta f_k \leqq \bar{c}_{ij}(u_{ij}) \qquad [23]$$

Thus, for step 3b, we have to know all the critical paths, and solve equations [21] such that the resulting flows satisfy [22] and [23]. This is illustrated by the example later on. The right hand vector for equation [21] has the same co-efficients. Thus, from equation [21], we can obtain a relationship between the flows Δf_k, k=1,...,p i.e. each flow is a constant multiple of the other flows. Physically, this means that in order to reduce the duration of the project, we have to send an incremental flow $\Delta f = \sum_{k=1}^{p} \Delta f_k$, through the network. In order to maintain optimality, any incremental flow Δf is subdivided optimally into flows Δf_k through the different chains of the network. This optimality is maintained by [21], [22] and [23].

After step 3b, we return to step 2 and if we can reduce to duration any further, we return to step 3a or step 3b and iterate again.

In Figure 6, the iterations are denoted by circles numbered with numbers in square brackets. The numbers 1,2,3,4,5 denote the iterations. Numbers with primes like [1'] denote the situation after a breakthrough. Thus circles with [1] correspond to Figure 7b, [1'] corresponds to Figure 7c and so on.

Example Continued

In our example, $\Delta L_{max} = 3$ units. Now, there is only one critical path

$P = \{1,(1,3),3,(3,6),6\}$

$P^B = \{3,6\}$

$P^C = \{1,3\}$

Then, using eq. [21]

$(1/1)\Delta f_1 = \Delta L = 3$

and using equations [22] and [23], we find that

$7 + 3 \leqq \bar{c}_{13}(4) = 11$

Hence, $\Delta f_1 = 3$ is admissible.

This is illustrated in Figure 7d and by circles marked with [2] in Figure 6.

Now, there are three critical paths of length or duration of 17 units each

$P_1 = \{1,(1,3),3,(3,6),6\}$

$P_2 = \{1,(1,2),2,(2,5),5,(5,6),6\}$

$P_3 = \{1,(1,2),2,(2,4),4,(4,6),6\}$

$P_1^B = \{3,6\}$ $\qquad P_1^C = \{1,3\}$

$P_2^B = \{(1,2),(2,5),(5,6)\}$ $\qquad P_2^C = \{\emptyset\}$

$P_3^B = \{(1,2),(4,6)\}$, $P_3^C = \{\emptyset\}$, $P_3^D = \{2,4\}$

At this stage step 3a is applicable. We can label the node 6 by using the two paths 1-2-5-6 and 1-2-4-6. Hence, we send a flow along both of these paths. This leads us to a situation in Figure 7e and [2'] in Figure 6. Now, we cannot send any additional flow without changing potentials and flows simultaneously. The breakup of the activities involved in the critical paths are

$P_1^B = \{3,6\}$ $\qquad P_1^C = \{1,3\}$

$P_2^B = \{\emptyset\}$, $P_2^C = \{(1,2,(2,5)\}$ $\qquad P_2^D = \{(5,6)\}$

$P_3^B = \{\emptyset\}$, $P_3^C = \{(1,2),(4,6)\}$ $\qquad P_3^D = \{(2,4)\}$

Then, using eqs. [21], [22] and [23], we have

$$\Delta f_1 = \Delta L$$

$$\Delta f_2 + \tfrac{1}{2}(\Delta f_2 + \Delta f_3) = \Delta L$$

$$\tfrac{3}{4}\Delta f_3 + \tfrac{1}{2}(\Delta f_2 + \Delta f_3) = \Delta L$$

Also, $Z_A = \{1,4\}$

$Z_C = \{(1,2),(1,3)\}$

$\delta_1 = 4$

$\delta_2 = \min_{z_c}\{2,1\} = 1$

$\delta = \min\{1,4\} = 1$

Hence, $\Delta L_{max} \leqq 1$

Thus, we get $\Delta f_2 = \frac{3}{4}\Delta f_3$

and when $\Delta L = 1$, we have

$$\Delta f_2 = 6/13$$

$$\Delta f_3 = 8/13 \quad , \quad \Delta f_1 = 1$$

We can easily verify that this satisfies the constraints given by [22] and [23].

The values corresponding to this iteration are shown in Figure 6 by [3] and see Figure 7f. At this stage we will find that we can carry out the step 3a of the algorithm. By the labelling procedure, we can send a flow of 5 units along the chain 1-3-6. The new values are shown by [3'] in Figure 6 and see Figure 7g.

Again, there are the same three critical paths with duration of 16 units each. We will not be able to get a breakthrough by the labelling procedure. Hence, again using equations [21], [22] and [23], we have

$$1/2\Delta f_1 = \Delta L$$

$$\Delta f_2 + 1/2(\Delta f_2 + \Delta f_3) = \Delta L$$

$$\frac{3}{4}\Delta f_3 + 1/2(\Delta f_2 + \Delta f_3) = \Delta L$$

This again gives the relationships

$$\Delta f_2 = (3/4)\Delta f_3 \text{ and } \Delta f_1 = \Delta L/2$$

and we have the following constraints

$$\Delta f_2 \leqq 20/13$$

$$\Delta f_2 + \Delta f_3 = 38/13$$

$$\Delta f_3 \leqq 68/13$$

Thus,

$$\Delta f_2 + \frac{4}{3}\Delta f_2 \leqq 38/13 \quad \text{or} \quad 7/3\ \Delta f_2 \leqq 38/13$$

$$\text{or } \Delta f_2 \leqq \frac{38\text{x}3}{13\text{x}7} = \frac{114}{91}$$

Hence, $\Delta f_2 = \frac{114}{91}$

$$\Delta f_3 = \frac{114}{91} \text{ x } \frac{4}{3} = \frac{152}{91}$$

$$\Delta t_{25} = \frac{114}{91} \quad , \quad \Delta t_{46} = \frac{114}{91} \quad , \quad \Delta t_{12} = \frac{19}{13}$$

$$\Delta T = \frac{19}{13} + \frac{114}{91} = \frac{133+114}{91} = \frac{247}{91}$$

These values are plotted in Figure 6 by circles numbered [4]. Again, we have three critical paths with a duration of $\frac{1209}{91}$ units. Again, there is no breakthrough and hence using step 3b we can have additional flows through these paths and for the next iteration, we have

$$\Delta f_2 = \frac{26}{91} \quad , \Delta f_3 = \frac{26}{91} \text{ x } \frac{4}{3} = \frac{104}{273}$$

$$\Delta T = \frac{26}{91} \quad , \quad \Delta f_2 + \Delta f_3 = \frac{26}{91} + \frac{104}{273} = \frac{182}{273}$$

Again, we have the same three critical paths with a duration of 13 units. These results are shown in Figure 6 by [5] and in Figure 7i. Now, if we apply the step 2 of the algorithm, we find that we cannot reduce the duration of the project any more. Hence, the algorithm terminates.

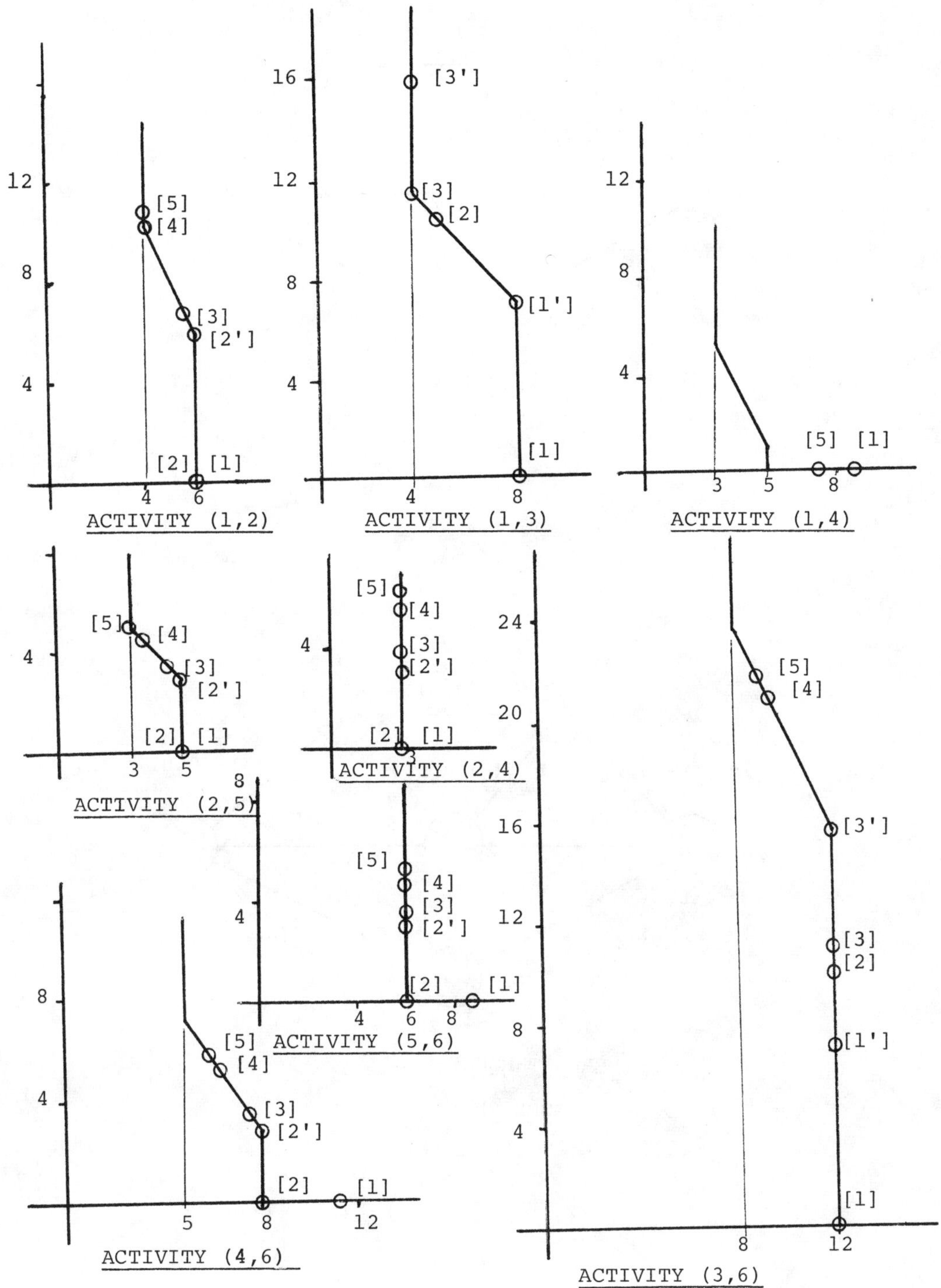

OPTIMALITY CURVES FOR VARIOUS ACTIVITIES

FIGURE 6

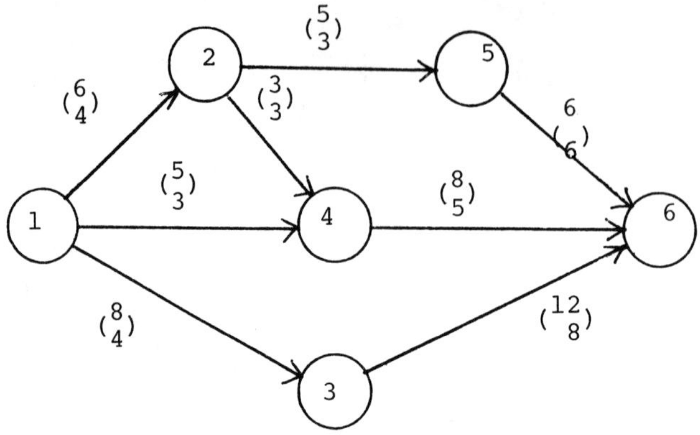

FIGURE 7a

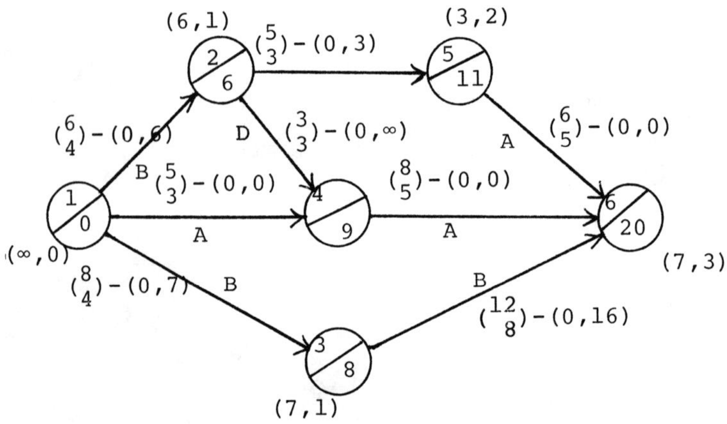

BREAKTHROUGH-STEP 3a

FIGURE 7b

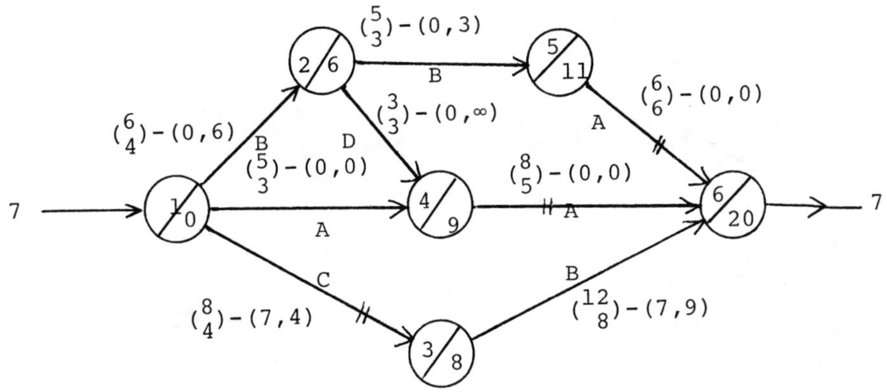

NON-BREAKTHROUGH

SIMULTANEOUS CHANGE OF FLOW AND POTENTIALS (Step 3b)

FIGURE 7c

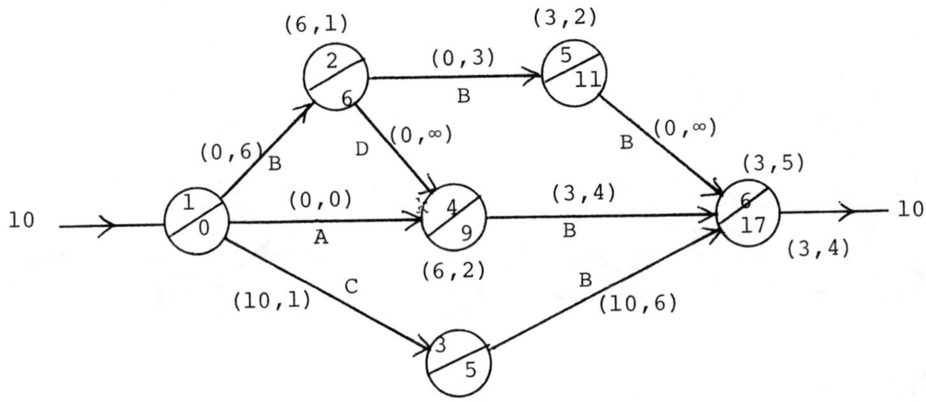

BREAKTHROUGH

Step 3a - Send a flow of 3 units along 1-2-5-6 and a flow of 3 units along 1-2-4-6

FIGURE 7d

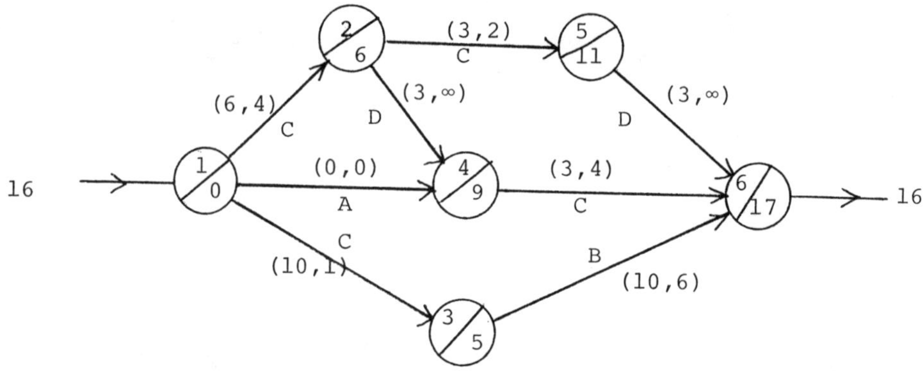

NONBREAKTHROUGH

SIMULTANEOUS CHANGE IN FLOWS AND POTENTIALS (Step 3b)

FIGURE 7e

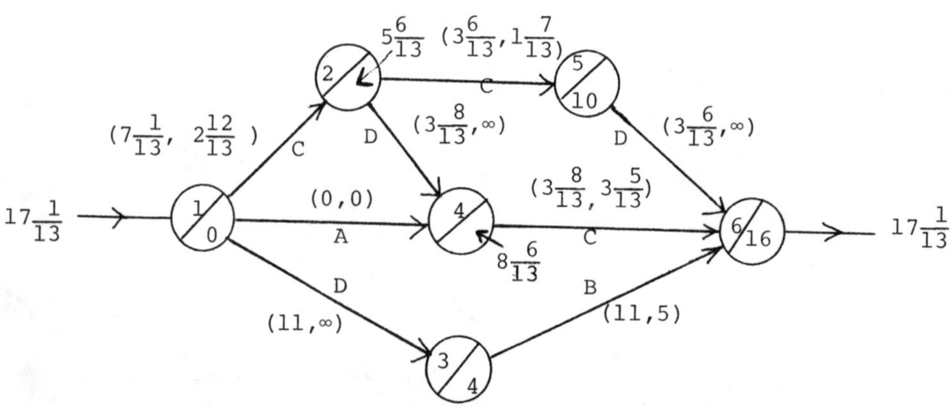

BREAKTHROUGH

STEP 3a - Send a Flow of 5 units along 1-3-6

FIGURE 7f

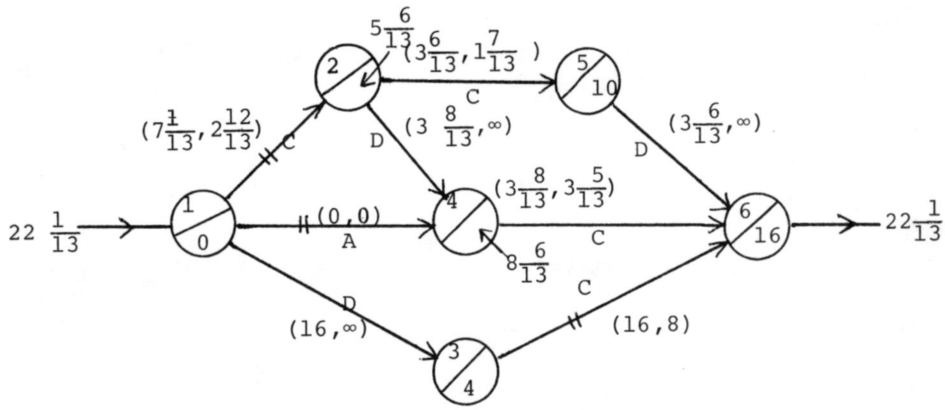

NONBREAKTHOUGH
STEP 3b

FIGURE 7g

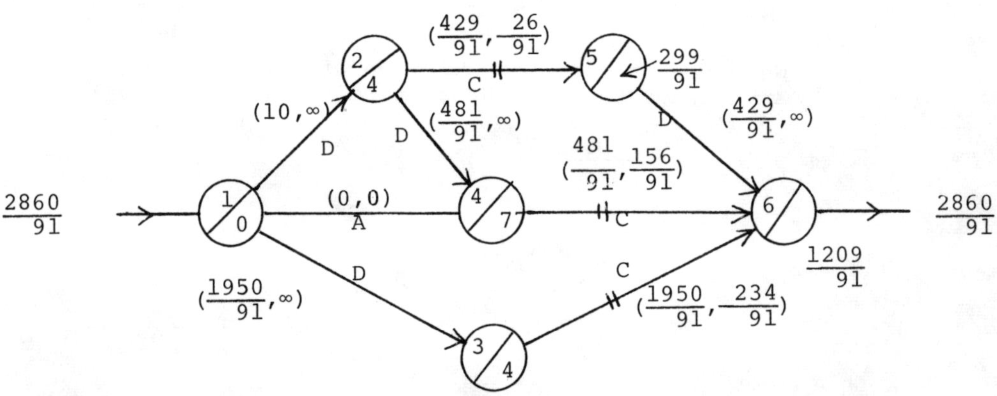

NONBREAKTHROUGH
STEP 3b

FIGURE 7h

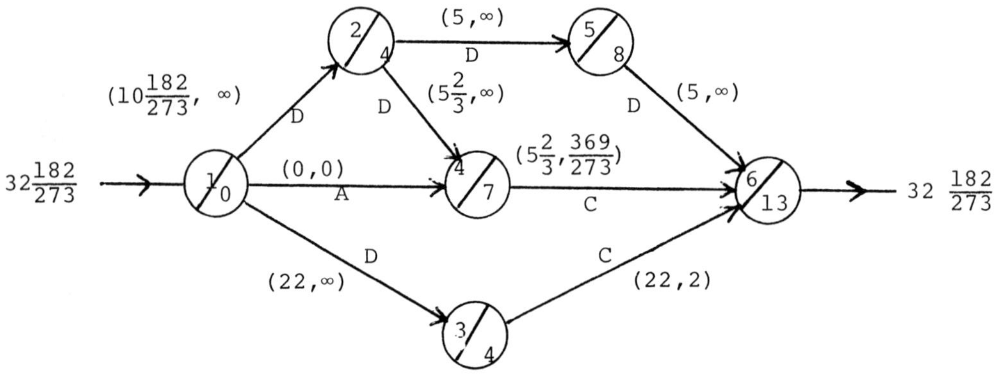

We cannot reduce the duration of the project any more (Step 2). Hence terminate.

FIGURE 7i

QUADRATIC APPROXIMATION OF THE CONVEX COST

When the cost function is a convex function of the duration of the activity, we will get a curve as shown in Figure 2. The quadratic approximation of the convex cost is shown by dotted lines in Figure 1. Here, the domain is subdivided, and on each interval, the function $c_{ij}(r_{ij})$ given by [5] has a constant slope on each subinterval. The algorithm given before could be generalized to handle this kind of problem. For this approximation, the equivalent of Figure 4 is shown in Figure 5. The author is working on these generalizations and other refinements of the algorithm given here. The computational aspects for large scale problems is also being studied.

REFERENCES

1) Berge, C., and Ghouila-Houri, A., Programming, Games and Transportation Networks, John Wiley & Sons, Inc., 1962.

2) Dessouky, M. I., and Dunne, E. J., "Cost-Duration Analysis with the Cut Network," AIIE Transactions, Vol. III, No. 2, pp. 150-163, 1971.

3) Elmagraby, S. E., The Design of Production Systems, Reinhold Publishing Corp., 1966.

4) Fulkerson, D. R., "A Network Flow Computation for Project Cost Curves," Management Science, 7, No. 3, pp. 167-178, Jan. 1961.

5) Florian, M. and Robillard P., "An Implicit Enumeration Algorithm for the Concave Cost Network Flow Problem," Management Science, Vol. 18, No. 3, pp. 184-193, Nov. 1971.

6) Masad, Iri, Network Flow, Transportation and Scheduling Theory and Algorithms, Academic Press, 1969.

7) Minty, G.J., "Monotone Networks ," Proceedings of the Royal Society of London, Ser. A, pp 194-212, 1960.

8) Zangwill, W.I., "Minimum Concave Cost flows in Certain Networks," Management Science, Vol. 14, No. 7, pp 429-450, 1968.

9) Zangwill, W.I., Non-Linear Programming: A Unified Approach, Prentice-Hall, Inc.,Englewood Cliffs, New Jersey, 1969.

A PROBLEM IN SINGLE FACILITY SCHEDULING WITH SEQUENCE INDEPENDENT CHANGEOVER COSTS†

Henry L. W. Nuttle and Adel A. Aly

North Carolina State University, Raleigh, North Caroline, USA

This paper treats a problem of optimally scheduling a number of products on a single facility when changeover costs are sequence independent. After some observations, a rather cumbersome, but straightforward, initial formulation is transformed into a zero-one knapsack problem with side constraints, for which an existing algorithm is modified to provide an efficient solution procedure. A numerical example and computational results are included.

1. INTRODUCTION

We consider a problem of optimally scheduling the production of n products on a single facility. For product i, i=1,...,n, the production per lot, R_i, and the lot processing time, w_i, are known. Whenever the facility is changed over from production of some other product to production of product i, a setup cost of c_i dollars is incurred. The net demand for product i, D_i, i=1,...,n, is specified and occurs at the end of a given finite planning horizon, T, i.e., production is against orders all of which are due T periods in the future. If demand for product i is not met, then a penalty is incurred which is proportional to the deficit. Inventory carrying charges are negligible and any initial inventories have already been netted out against demand.

In addition we make the following assumptions:

a) Only one product may be processed at a time.

b) Once production of a lot is started it must be completed.

c) No lot splitting is permitted, i.e., for product i only lots of size R_i may be scheduled.

This problem is an offshoot of a problem treated by Hoyte and Nuttle [7] in which lots consist of units of several of the n products are arranged in a particular sequence. For additional references on the scheduling of lots on a

† This research was partially supported by ONR Contract No. N00014-70-A-0120-0002

single facility see Elmaghraby et al [3],Elmaghraby and Mallik [4], Hoyte [6], Aly [1], and references cited therein.

2. MATHEMATICAL MODEL

Let x_{it} be an indicator variable such that $x_{it} = 1$ implies product i is scheduled during period t and $x_{it} = 0$ implies that product i is not scheduled during period t. In addition let

θ_i ≡ per unit penalty for unsatisfaction of demand of product i.

$S_i(T)$ ≡ total production of product i by the end of the horizon.

i_o ≡ the product on the facility at the end of the previous planning horizon, $i_o \in \{1,\ldots,n\}$.

We can then write the mathematical model for the problem of Section 1 as Problem I:

$$\text{minimize} \sum_{t=0}^{T-1} \sum_{j=1}^{n} \sum_{\substack{i=1 \\ i\neq j}}^{n} c_i\, x_{jt}\, x_{i,t+1} + \sum_{i=1}^{n} \theta_i\, [D_i - S_i(T)]^+$$

$$\text{subject to} \sum_{i=1}^{n} x_{it} \leq 1, \qquad t=1,\ldots,T \tag{1}$$

$$S_i(T) = (R_i/w_i) \sum_{t=1}^{T} x_{it}\,, \qquad i=1,\ldots,n$$

If $x_{it} = 1$ while $x_{i,t-1} = 0$, then

$$\sum_{j=t}^{t+pw_i-1} x_{ij} = pw_i\,, \qquad \text{for some } p,\ p=1,2,\ldots;\ i=1,\ldots,n_j\ t=1,\ldots,T \tag{2}$$

$$x_{ij} \in \{0,1\}\,, \qquad i=1,\ldots,n;\ t=1,\ldots,T$$

where $x_{i_o,0} = 1$; $x_{i,0} = 0$, $i \neq i_o$. We assume w_i and R_i, i=1,...,n, are positive integers, D_i, i=1,...,n are nonnegative and integer, and c_i and θ_i, i=1,..,n, are nonnegative and $[D_i - S_i(T)]^+ = \max(0; D_i - S_i(T))$. Equation (1) guarantees that at most one product is scheduled in any period, while Equation (2) assures that lots are not split.

At first glance Problem I appears to be a rather nasty nonlinear zero-one programming problem. However, by capitalizing on the model structure, we are able to transform it into a zero-one knapsack problem with side constraints which is amenable to efficient solution.

3. MODEL TRANSFORMATION

Since changeover costs are sequence independent, all demands occur at the end of the horizon, and carrying charges are negligible, there is really no sequencing element in our problem beyond the question of which product is scheduled first. (If product i_o is first then no changeover cost is incurred; if product i, $i \neq i_o$, then a cost of c_i is incurred). Thus we have,

Observation 1: There exists an optimal schedule such that if x_i, $x_i \geq 1$, lots of product i are scheduled, then these lots are scheduled consecutively.

Given Observation 1, we may answer the "initial product" question by,

Observation 2: If an optimal schedule includes product i_o, then i_o is scheduled first.

Capitalizing on Observation 1, let x_i, i=1,...,n represent the number of lots of product i scheduled in the horizon and replace c_{i_o} with zero.

Since for any product the question of the amount to schedule involves a trade-off between set-up (changeover) costs and demand penalty costs, and since, typically, overproduction of one product is at the expense of another, we may make the following:

Observation 3: There exists an optimal schedule such that if $x_i > 0$, then

(a) $x_i \geq x_i^*$ where $x_i^* = \left\lceil \frac{c_i}{\theta_i R_i} \right\rceil$† and (b) $x_i \leq \bar{x}_i$ where $\bar{x}_i = \left\lceil \frac{D_i}{R_i} \right\rceil$.

From Observation 3 we see that if $x_i^* > \bar{x}_i$ then product i should not be scheduled at all. In the future we shall assume that such products have been eliminated but retain n as denoting the number of schedulable products.

If none of the demand for product i were satisfied, the loss would be $\theta_i D_i$. Consequently, any production above the zero level represents a "gain" from this penalty. Thus if x_i lots, each of R_i units, are scheduled, the "gain" would be $\theta_i\ x_i\ R_i$. Since all the demand may be satisfied (by producing $\bar{x}_i$ lots), the "gain" in the objective function is equal to min $(\theta_i x_i R_i, \theta_i D_i)$, $x_i \varepsilon \{0, x_i^*, x_i^*+1, \ldots, \bar{x}_i\}$ recalling that there is no penalty for excess production over demand. But since there is a setup cost c_i incurred whenever $x_i > 0$ lots are scheduled; a "cost" of $c_i \delta(x_i)$ must be subtracted from the "gain" due to producing the x_i lots.

Hence, using Observations 1, 2, and 3, we may rewrite Problem I as Problem II:

$$\max \sum_{i=1}^{n} [\min (\theta_i x_i R_i, \theta_i D_i) - c_i \delta(x_i)] \tag{3}$$

$$\text{subject to } \sum_{i=1}^{n} w_i x_i \leq T$$

$$x_i \ \varepsilon \{0, x_i^*, x_i^* + 1, \ldots, \bar{x}_i\} \ , \qquad i=1,\ldots,n$$

where $\delta(x_i)$ is the Kronecker Delta.

† $\lceil A \rceil$ denotes smallest integer greater than or equal to A.

Although simpler in appearance than Problem I, Problem II does not appear to be easily solvable either. We shall have to make one further transformation.

From Observation 3, x_i is either 0 or is bounded from below by x_i^* and from above by $\bar{x}_i$. If product i is scheduled at its minimum schedulable amount, i.e., $x_i = x_i^*$, then the contribution to the objective function, Equation (3), is $\theta_i x_i^* R_i - c_i$, while $w_i x_i^*$ periods of the horizon are consumed. If $x_i^* + j$, $j \geq 1$; $x_i^* + j < \bar{x}_i$, lots are scheduled, then the contribution to the objective function is $\theta_i (x_i^* + j) R_i - c_i$ while $w_i (x_i^* + j)$ periods are consumed. Finally if $x_i = \bar{x}_i$, the contribution to the objective function is $\theta_i D_i - c_i$ and $w_i \bar{x}_i$ periods are consumed.

Define m_i according to

$$m_i = \bar{x}_i - x_i^* + 1 , \qquad i=1,\ldots,n$$

and let, for each i,

$$\begin{array}{llll} v_{i1} = \theta_i x_i^* R_i - c_i & ; \; a_{i1} = x_i^* w_i & & \\ \vdots & \quad \vdots & & \\ v_{ij} = \theta_i (x_i^*+j-1) R_i - c_i & ; \; a_{ij} = (x_i^*+j-1) w_i , & 1 < j < m_i & \qquad (4) \\ \vdots & \quad \vdots & & \\ v_{im_i} = \theta_i D_i - c_i & ; \; a_{im_i} = (x_i^*+m_i-1) w_i = \bar{x}_i w_i & & \end{array}$$

We replace the x_i's with m_i indicator variables x_{ij}, $j=1,\ldots,m_i$, where

$$x_{ij} = \begin{cases} 1, & \text{if exactly } x_i^*+j-1 \text{ lots of product i are scheduled} \\ 0, & \text{otherwise} \end{cases}$$

and rewrite Problem II as

Problem III:

$$\max \sum_{i=1}^{n} \sum_{j=1}^{m_i} v_{ij} x_{ij} \tag{5}$$

$$\text{subject to} \quad \sum_{i=1}^{n} \sum_{j=1}^{m_i} a_{ij} x_{ij} \leq T \tag{6}$$

$$\sum_{j=1}^{m_i} x_{ij} \leq 1 , \qquad i=1,\ldots,n \tag{7}$$

$$x_{ij} \;\varepsilon\{0,1\} , \qquad i=1,\ldots,n;\ j=1,\ldots,m_i \tag{8}$$

Although we have replaced an n-variable problem with a $\sum_{i=1}^{n} m_i$-variable problem, Problem III appears more amenable to efficient solution.

4. TOWARD AN ALGORITHM

With the exception of constraint (7) Problem III is of the form:

$$\max \sum_{i=1}^{n} d_i v_i \tag{9}$$

$$\text{subject to} \quad \sum_{i=1}^{n} b_i y_i \leq T$$

$$y_i \;\varepsilon\{0,1\} , \qquad i=1,\ldots,n,$$

i.e., a zero-one Knapsack problem.

It is well known that if in the problem of (9) the integer restrictions are replaced with $0 \leq y_i \leq 1$, an optimal solution may be obtained by simply ranking the ratios d_i/b_i in nonincreasing order and then loading the horizon (until exhausted) "product by product" following the ranked ordering. Furthermore if the solution so obtained is integer it is also optimal for the integer problem.

In 1967 Kolesar [8] presented a branch-and-bound algorithm for the zero-one knapsack problem which capitalizes on the fact that, while the easily obtained non-integers solution need not be optimal, it typically provides an excellent upper bound on the optimal value of the objective function for the integer problem. In 1970 Greenberg and Hegerich [5] presented an alternative branch-and-bound algorithm which has the same basis but uses a different branching scheme. The computational results reported in their paper appear quite good and, in many cases, significantly better than those reported by Kolesar. For this reason plus

a) the bounds provided by the non-integer problem are very tight, and

b) An optimal solution is typically obtained after having explicitly assigned values to very few variables, we have chosen to modify Greenberg & Hegerich's algorithm to accommodate the side constraints of Problem III.

In the spirit of Kolesar and Greenberg and Hegerich, suppose for Problem III we proceed as follows:

i) Compute the ratio v_{ij}/a_{ij} for all i,j.

ii) Form a list of the ratios ranked in nonincreasing order.

iii) Load the horizon T, until exhausted, starting with the associated number of lots of the product corresponding to the first ratio in the list and allowing positive only the variable corresponding to the maximum ratio for each product.

At first glance it would appear that this approach would yield an optimum if the associated solution were integer, or a good upper bound if the solution were fractional. Unfortunately this need not be the case.

Let $r_i = D_i - \lfloor D_i/R_i \rfloor < R_i$. Using Equation (4), after some algebra, we obtain

$$\frac{v_{ij}}{a_{ij}} = \theta_i \frac{R_i}{w_i} - \frac{c_i}{(x_i^*+j-1)w_i} , \qquad 1 \leq j < m_i ,$$

while

$$\frac{v_{i,m_i}}{a_{i,m_i}} = \theta_i \frac{R_i}{w_i} - \frac{c_i}{(x_i^*+j-1)w_i} - \frac{\theta_i(R_i-r_i)}{(x_i^*+m_i-1)w_i}$$

Thus while v_{ij}/a_{ij} is monotone increasing in j for $1 \leq j < m_i$, there may exist k', $1 \leq k' < m_i$ such that for all j, $k' \leq j < m_i$,

$$\frac{v_{i,m_i}}{a_{i,m_i}} < \frac{v_{ij}}{a_{ij}} \tag{10}$$

If this is the case, then the scheme described in steps (i) through (iii) may fail to schedule the last, or "fractional", lot of product i and include an additional lot(s) of some other product(s), when in fact the fractional lot of product i contributes more to the objective function.

To circumvent this difficulty while maintaining the basic scheme of steps (i) through (iii) we proceed as follows,

<u>Rule 1:</u> If $v_{i,m_i}/a_{i,m_i} \geq v_{i,m_i-1}/a_{i,m_i-1}$,

maintain v_{i,m_i} and a_{i,m_i} as originally defined.

<u>Rule 2:</u> If $v_{i,m_i}/a_{i,m_i} < v_{i,m_i-1}/a_{i,m_i-1}$,

replace v_{i,m_i} with $\theta_i r_i$ and a_{i,m_i} with w_i, yielding

a ratio of $v_{i,m_i}/a_{i,m_i} = \theta_i r_i/w_i$.

If Rule 2 applies for product i, then $x_{i,m_i} = 1$ implies the scheduling of only the fractional lot, while $x_{ik} = 1$, $k < m_i$, implies the scheduling of $x_i^* + k-1$ lots of product i. For simplicity we shall maintain the notation v_{i,m_i} and a_{i,m_i} when Rule 2 applies, distinguishing the affected products by grouping them in a set P.

Application of the definitions from Equation (4) plus some algebra yields the following additional two observations:

Observation 4: If Rule 2 has been applied for product i, then in an optimal solution, $x_{i,m_i} = 1$ only if $x_{i,m_i-1} = 1$.

Observation 5: If under the original definition (Equation 4)),

$v_{i,m_i}/a_{i,m_i} < v_{i,m_i-1}/a_{i,m_i-1}$, then $\theta_i r_i/w_i < v_{i,m_i-1}/a_{i,m_i-1}$.

Observation 4 allows us to rewrite Problem III as,

Problem IV:

$$\max \sum_{i=1}^{n} \sum_{j=1}^{m_i} v_{ij} x_{ij} \tag{11}$$

$$\text{subject to:} \quad \sum_{i=1}^{n} \sum_{j=1}^{m_i} a_{ij} x_{ij} \leq T \tag{12}$$

$$\sum_{j=1}^{m_i} x_{ij} \leq 1, \qquad \text{for all } i \notin P \tag{13}$$

$$\sum_{j=1}^{m_i-1} x_{ij} \leq 1, \qquad \text{for all } i \in P \tag{14}$$

$$x_{i,m_i-1} \geq x_{i,m_i}, \qquad \text{for all } i \in P \tag{15}$$

$$x_{ij} \in \{0,1\}, \quad i=1,\ldots,n;\ j=1,\ldots,m_i \tag{16}$$

If after Rules 1 and 2 are applied the modified set of ratios is ranked as in (ii), then Observation 5 states that any product i such that Rule 2 applies, $v_{i,m_i}/a_{i,m_i}$ always appears after $v_{i,m_i-1}/a_{i,m_i-1}$. Thus in scanning the list while loading the horizon to obtain an upper bound (or possible optimal), when $x_{i,m_i-1} = 1$, we need never "back up" in the list to locate $v_{i,m_i}/a_{i,m_i}$.

5. BOUNDING THE OBJECTIVE FUNCTION

Given the ranked list of the ratios v_{ij}/a_{ij} for Problem IV, suppose a subset of the variables have been assigned, in some sequence, values 0 or 1 consistent with (12)-(15). Let M' be a one-by- $\sum_{i=1}^{n} m_i$ component vector corresponding to the assignment for $(x_{11},\ldots,x_{n,m_n})$, where in M' the variables are ordered according to the ranked list of ratios. An entry of 0 implies that $x_{ij} = 0$, an entry of 1, $x_{ij} = 1$, and an entry of 2 implies that x_{ij} is unassigned. It should be understood that a zero entry in M' corresponds to a variable which has been specifically assigned a value of 0 as contrasted with a variable forced to 0 by (13)-(15). Variables in the latter class will have value of 2 in M'. Let a(k) denote the subscript pair of the kth assigned (0 or 1) variable in M', and let M denote the set of all such subscript pairs associated with M'.

An upper bound on the value of the objective function for Problem IV, conditional on the assignment M' may be obtained by solving Problem IV with Equation (16) replaced by $0 \leq x_{ij} \leq 1$ for all i and j, and with the additional constraints, $x_{ij} = 0,1$, as assigned, $(i,j) \in M$.

Specifically, let B(M') be the ordered set of subscript pairs obtained by scanning the ranked list of ratios (left to right) omitting all ratios corresponding to $(i,j) \in M$ and observing constraints (13)-(15). Let (r,j_r) be the first element in B(M'), scanning from left to right, such that

$$\sum_{\substack{(i,j)\varepsilon B(M') \\ (i,j)\leq(r,j_r)}} a_{ij} > T - \sum_{(i,j)\varepsilon M} a_{ij}\, x_{ij} \tag{17}$$

where $(i,j) \leq (r,j_r)$ should be interpreted as (i,j) "is or precedes" (r,j_r) in the ordered set $B(M')$.

Then, the conditional upper bound is given by

$$U(M') = \sum_{\substack{(i,j)\varepsilon B(M') \\ (i,j)<(r,j_r)}} v_{ij} + \sum_{(i,j)\varepsilon M} v_{ij}x_{ij} + v_{r,j_r}\, x_{r,j_r}\ , \tag{18}$$

and the corresponding fractional solution is

$x_{ij} = 1$, if $(i,j)\varepsilon B(M'),(i,j) < (r,j_r)$

$x_{ij} = 0$, if $(i,j)\varepsilon B(M'),(i,j) > (r,j_r)$

$$x_{r,j_r} = \Big(T - \sum_{(i,j)\varepsilon M} a_{ij}\, x_{ij} - \sum_{\substack{(i,j)\varepsilon B(M') \\ (i,j) < (r,j_r)}} a_{ij}\Big)\Big/ a_{r,j_r} \tag{19}$$

$x_{ij} = 0$, if $(i,j) \notin B(M')\cup M$

$x_{ij} = 0$ or 1, depending on the assignment, $(i,j)\ \varepsilon\ M$.

If (r,j_r) is undefined, we set $x_{ij} = 1$ for all $(i,j)\ \varepsilon\ B(M')$. In this case, or if $x_{r,j_r} = 0$, we have an optimal solution conditional on the assignment M'. If $M = \phi$, the solution is an optimal for Problem IV.

Note also that the solution of (19), modified such that $x_{r,j_r} = 0$, represents a feasible solution to Problem IV. Thus the associated objective function value,

$$L(M') = \sum_{\substack{(i,j)\varepsilon B(M') \\ (i,j)<(r,j_r)}} v_{ij} + \sum_{(i,j)\varepsilon M} v_{ij}\, x_{ij} \tag{20}$$

is a lower bound on the optomal value of the objective function for Problem IV.

6. THE ALGORITHM

We are now in a position to state an algorithm to generate optimal solutions for Problem IV.

To this end, let $X(L^*, \chi)$ correspond to the best lower bound solution thus far obtained. χ is a one-by-$\sum_{i=1}^{n} m_i$ component vector, where the elements are indexed as in M' and all elements are either 0 or 1 in accordance with the solution. (Note that any element of M' with a value of 2 has value 0 in χ). L* is the associated value of the objective function, i.e.,

$$L^* = \sum_{x_{ij} \varepsilon \chi} v_{ij} \, x_{ij}$$

The algorithm is as follows:

1. Compute x_i^* and $\bar{x}_i$ for all products and eliminate any product i such that $x_i^* > \bar{x}_i$. Renumber all remaining products and denote the new total number by n. Go to Step 2.
2. If $\sum_{i=1}^{n} \bar{x}_i w_i \leq T$, an optimal solution is achieved by setting $x_i = \bar{x}_i$, $i=1,\ldots,n$. Otherwise, go to Step 3.
3. Compute m_i, $i=1,\ldots,n$, and generate v_{ij}, a_{ij} for all i and j using (4). Compute v_{ij}/a_{ij} for all i and j, apply Rules 1 and 2, and modify v_{ij} and a_{ij} where appropriate, generating the set P. Rank v_{ij}/a_{ij} for all i and j in nonincreasing order, left to right. Go to Step 4.
4. At the initial node, set k=0, $M' = \{2,2,\ldots,2\}$, and $\chi = 0$. Compute B(M') and apply Equation (19) to obtain a fractional solution to Problem IV.[†] If $x_{r,j_r} = 0$ or if (r,j_r) is undefined, the associated solution is optimal. Stop. If $0 < x_{r,j_r} < 1$, compute L* using Equation (20), and in χ set $x_{ij} = 1$ for all $(i,j) \varepsilon B(M')$, $(i,j) < (r,j_r)$ to form $X(L^*,\chi)$.

[†] In actual computing this may be done implicitly, using Equation (17).

(a) If $(r,j_r) = (i,m_i)$ for some $i \in P$, set $a(0) = (i,m_{i-1})$ adding a node to the tree, and set the $(i,m_i-1)^{th}$ component of M' to zero. Go to Step 5.

(b) If $(r,j_r) \neq (i,m_i)$ for some $i \in P$ set $a(0) = (r,j_r)$ adding a node to the tree and set the $(r,j_r)^{th}$ component of M' to zero. Go to Step 5.

5. Compute B(M'), apply Equation (19) to obtain a fractional solution, and compute U(M') using Equation (18). If $U(M') \leq L^*$, go to Step 7. If $U(M') > L^*$ and the solution to (19) is integer, or (r,j_r) is not defined, set $L^* = U(M')$, update χ to obtain a new $X(L^*,\chi)$, and go to Step 7. If $U(M') > L^*$ but the solution to (19) is not integer, go to Step 6.

6. Compute L(M') using Equation (20). If $L(M') > L^*$, set $L^* = L(M')$ and update $X(L^*,\chi)$ with $x_{r,j_r} = 0$. In any case set $k = k+1$, and

   (a) If $(r,j_r) = (i,m_i)$ for some $i \in P$ and $(i,m_{i-1}) \in B(M')$, set $a(k) = (i,m_{i-1})$ adding a new node to the tree, set the $(i,m_{i-1})^{th}$ component of M' to zero, and go to Step 5.

   (b) Otherwise, set $a(k) = (r,j_r)$, adding a new node to the tree, set the $(r,j_r)^{th}$ element of M' to zero, and go to Step 5.

7. If the $a(k)^{th}$ element of M' equals 0, change it to 1, and go to Step 5. Otherwise go to Step 8.

8. If k=0, the optimal solution is the current $X(L^*,\chi)$. Stop. If $k > 0$ change the $a(k)^{th}$ element of M' to 2, set $k = k-1$, removing a node from the tree, and go to Step 7.

It should be noted that the only required permanent storage with respect to the tree is for $X(L^*,\chi)$, M', {a(k)} (or M), and a vector for B(M').

7. NUMERICAL EXAMPLE

To illustrate the application of the algorithm, we consider a numerical example involving the scheduling of 5 products over a horizon of length 40. The data is shown in Table 1. For simplicity we assume that $w_i = R_i \quad i=1,\ldots,5$.

Table 1

Product Number, i	1	2	3	4	5
Set-up cost, c_i	6	9	7	4	10
Demand penalty, θ_i	3	2	2	5	1
Lot size, R_i	5	4	2	3	3
Processing time, w_i	5	4	2	3	3
Demand, D_i	12	12	11	16	8

$\bar{x}_i$ and x_i^* for each product are given in Table 2. Note that $x_5^* > \bar{x}_5$ so that product 5 may be eliminated. m_i for each remaining product is also given in Table 2.

Table 2

i	1	2	3	4	5
$\bar{x}_i$	3	3	6	6	3
x_i^*	1	2	2	1	4
m_i	3	2	5	6	-

v_{ij}, a_{ij}, and v_{ij}/a_{ij} for all i and j are given in Table 3, where a slashed entry indicates the application of Rule 2.

Table 3

(i,j)	v_{ij}	a_{ij}	v_{ij}/a_{ij}
11	9	5	1.80
12	24	10	2.40
13	~~30~~ 6	~~15~~ 5	~~2.00~~ 1.20
21	7	8	0.88
22	15	12	1.25
31	1	4	0.25
32	5	6	0.84
33	9	8	1.12
34	13	10	1.20
35	~~15~~ 2	~~12~~ 2	~~1.25~~ 1.00
41	11	3	3.67
42	26	6	4.33
43	41	9	4.55
44	56	12	4.67
45	71	15	4.73
46	~~76~~ 5	~~18~~ 3	~~4.22~~ 1.67

The data in Table 3 implies the following ranked list: {45,44,43,42,41,12,11, 46,34,22,13,33,35,21,32,31} .

The tree generated in obtaining an optimal solution is shown in Figure 1. The relevant data at each node is given in Table 4.

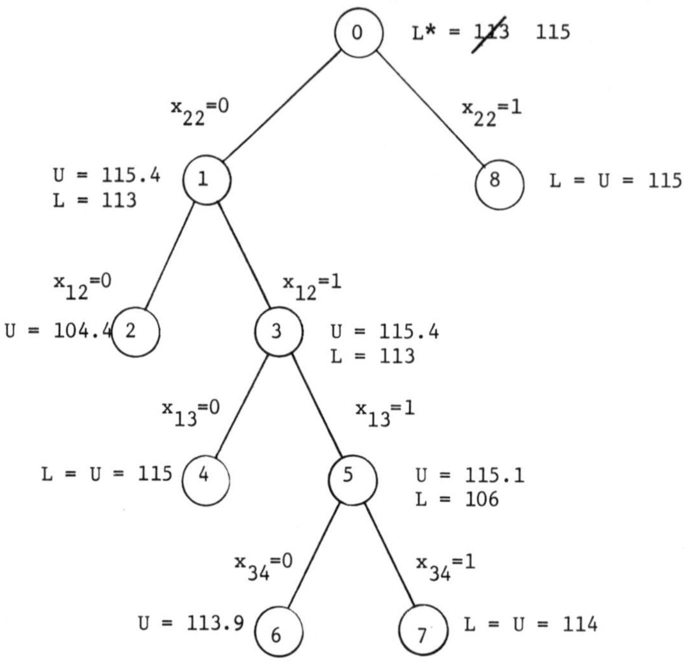

Figure 1

An optimal solution was obtained with 8 nodes. That solution is associated with node 4. The positive variables are $x_{45}= 1$, $x_{12}= 1$, $x_{46}= 1$, $x_{34}= 1$, and $x_{35}= 1$. The associated value is $L^* = 115$. Translated in terms of the original problem statement, the solution implies the scheduling of 2 lots of product 1, 0 lots of product 2, 5 lots of product 3, 6 lots of product 4, and, of course, 0 lots of product 5.

We note that there is an alternate optimum associated with node 8, for which the positive variables are $x_{45}= 1$, $x_{12}= 1$, $x_{46}= 1$, and $x_{22}= 1$, which translates the scheduling of 2 lots of product 1, 3 lots of product 2, 0 lots of product 3, 6 lots of product 4, and 0 lots of product 5.

Table 4

Node No.	B(M')	Positive variables in fractional solution	U(M')	L(M')	L*	k	a(k)[†]	Comments
0	{45,12,46,34,22, 13,35}	$x_{45}=1, x_{12}=1, x_{46}=1, x_{34}=1, x_{22}=1/6$		113	113	0	22^-	Initial branching
1	{45,12,46,34,13, 35,21}	$x_{45}=1, x_{12}=1, x_{46}=1, x_{34}=1, x_{13}=2/5$	115.4	113		1	12^-	U(M')>L*; continue branching per Step 6 (a)
2	{45,11,46,34, 35,21}	$x_{45}=1, x_{11}=1, x_{46}=1, x_{34}=1, x_{35}=1, x_{21}=5/8$	104.4			1	12^+	U(M')<L* backtrack
3	{45,46,34,13, 35,21}	$x_{45}=1, x_{12}=1, x_{46}=1, x_{34}=1, x_{13}=2/5$	115.4	113		2	13^-	U(M')>L*; continue branching
4	{45,46,34, 35,21}	$x_{45}=1, x_{12}=1, x_{46}=1, x_{34}=1, x_{35}=1$	115	115	115	2	13^+	Integer solution; L(M') = L*; set L*=L(M') and backtrack
5	{45,46,34,35, 21}	$x_{45}=1, x_{12}=1, x_{46}=1, x_{34}=7/10, x_{13}=1$	115.1	106		3	34^-	U(M')>L*; continue branching
6	{45,46,33,21}	$x_{45}=1, x_{12}=1, x_{46}=1, x_{13}=1, x_{33}=7/8$	113.9			3	34^+	U(M')<L*; backtrack
7	{45,46,35,21}	$x_{45}=1, x_{12}=1, x_{34}=1, x_{13}=1$	114	114		0	22^+	Integer solution; U(M')<L*; backtrack
8	{45,12,46,34, 13,35}	$x_{45}=1, x_{12}=1, x_{46}=1, x_{22}=1$	115	115				Integer solution; U(M')=L*; k=0, Stop.

[†] The "-" label indicates assigned value 0; "+" indicates assigned value 1.

8. COMPUTATIONAL RESULTS

We have programmed the algorithm in PL/I and ran approximately 275 test problems on an IBM 370/165. The parameter values for D_i, R_i, θ_i, c_i, and w_i were generated at random except that for subsets of problems,

$\bar{x} = \sum_{i=1}^{n} \frac{\bar{x}_i}{n_1}$ (i.e., the average number of lots per product) was held constant.

The results for number of nodes generated and solution time (given that node 0 is initialized) are given in Table 5. Each entry represents an average figure based on 10 problems.

Table 5

	$\bar{x} = 4$							
	Average No. of Nodes Generated				Average Time in Milliseconds			
T \ n	5	8	10	13	5	8	10	13
25	21.8	18.6	42.2	35.6	25	29	58	72
50	16.6	28	44.2	33	24	42	62	57
75	11.4	20	26.6	26.6	23	33	46	41
100	6	17.6	14.6	15.2	16	30	32	39

(a)

	n = 5							
	Average No. of Nodes Generated				Average Time in Milliseconds			
T \ $\bar{x}$	4	6	8	10	4	6	8	10
25	21.8	38.2	31.4	51.6	25	49	51	83
50	16.6	27	42	77.4	24	41	61	109
75	11.4	25.8	24.8	21.2	23	43	45	50
100	6	21.2	13.6	14.8	16	38	35	40

(b)

There appears to be a general trend toward more nodes generated and longer solution time as the number of products or the average number of lots per product increase, as would be expected. However there are many exceptions, especially with respect to number of nodes generated. Also there is no clear cut indication as to whether the number of products or number of lots per product has the more significant effect on number of nodes generated and solution time. It is interesting to note that more computation is typically required for the middle length horizon rather than either extreme. This phenomenon was also observed by Kolesar and Greenberg and Hegerich (G-H). A similar effect was noted when, for a given $(n,T,\bar{x})$, the w_i's were increased, yielding a larger $\sum_{i=1}^{n} \bar{x}_i w_i$.

As stated previously the times given in Table 5 were measured from the time of initial branching from node 0, i.e., they indicate the time required for the actual branching and bounding. Since there may be considerable computation required to prepare to branch from node 0, we also clocked the times to

a) Compute and rank ratios, and

b) Apply Rule 2, correct ranking, and determine the initial B(M') set.

For smaller problems the time to complete (b) outweighed the time for (a). However as the problem size increased, the ranking was reversed. Total initialization time resulted in an increase of solution time relative to the figures in Table 5 by a factor which averaged approximately 3.5. This is a significant increase. However total solution time still appears quite good.

In addition to the problems of Table 5 we have run several larger problems. The largest to date involved four problems with $n = 143$ and $\bar{x} = 7$, or an average of 1001 variables. For these problems the maximum number of nodes was 1,207, consuming 11.8 seconds, while the associated total solution time was 29 seconds.

In order to obtain a rough comparison with the computational results of Greenberg and Hegerich, Table 5 was generated such that $n\bar{x}$ is approximately equal to their N, i.e., the number of variables corresponding to each entry under both experiments is approximately the same. Comparing Tables 5(a) and 5(b) entry by entry with the Table of G-H shows in more than half of the cases our algorithm generated fewer nodes. The times in Table 5 are generally much less than the corresponding figures reported by G-H. However, if we inflate our figures to obtain total solution time (which is probably more relevant), the ranking is typically reversed with our solution times ranging from - 30% to 50% higher. This is not surprising since our problem requires more computation to initialize and to prepare to branch at each node. We hasten to add that these comparisons are at best tentative, since the algorithms were programmed by different people, in different languages, and run on different computers. However, we feel that they are encouraging.

9. DISCUSSION

From the computational experience to date it appears that the algorithm presented here is capable of solving large problems in a very reasonable amount of time. This, combined with the rather small permanent storage requirement, also suggests that it may be practical to run the algorithm on smaller and/or slower computers than the one used for the test problems.

It has been suggested[†] that a potential alternative approach to solving Problem IV is to follow the scheme described by Taha [9,10]. Let S_1 be the set of $x_{ij} \varepsilon [0,1]$ which satisfy constraint (12) and let S_2 be the set which satisfy all constraints, (12)-(15). The general idea is to generate nodes based only on S_1 while using constraints (13)-(15) as filters to fathom (or avoid generating) nodes for which the last assigned variable yields a solution infeasible with respect to (13)-(15). While this approach may prove very efficient for some classes of problems (See [10]) we feel that the algorithm presented in Section 6 is better suited to our particular problem.

† During an informal workshop at the Symposium on Theory of Scheduling and Its Applications, N. C. State University, May, 1972.

If we were to adopt the alternative scheme, using the G-H method to generate nodes and bounds based on S_1, and applying constraints (13)-(15) as filters when attempting to set $x_{r,j_r} = 1$, we would expect frequently to find that the upper bounds were overstated. This is because often many of the ratios associated with a given product occur in close proximity in the ranked list, so that in the absence of (13)-(15) their v_{ij}'s would simultaneously be included in the figure for the upper bound. It is also difficult, for the same reason, to see how one could, without considerable effort, determine if an integer solution generated for S_1 were feasible with respect to (13)-(15) (since variables associated with B(M') are not explicitly assigned) and thus potentially optimal.

To circumvent these difficulties we might adopt a Balasian approach for generating nodes based on S_1 by sequentially assigning variables a value of 1 (on the basis, say, of max v_{ij}/a_{ij} for eligible unassigned variables) using constraints (13)-(15) as filters at each node. However in this case we would lose the desirable property of the G-H approach, namely, the frequently realized potential to obtain an optimal solution having explicitly assigned values to very few variables.

We mention in passing that the latter approach would appear viable for problems of the form of Problem III in which, for a given i, the v_{ij}'s and a_{ij}'s are such that G-H bounding procedure, (i)-(iii) of Section 4 is invalid and for which no recovery, such as Rule 2, can be made.

REFERENCES

1. Aly, A. A., "Algorithms for Three Single Facility Scheduling Problems", unpublished M.Sc. Thesis, N. C. State University, 1972.

2. Balas, E., "An Additive Algorithm for Solving Linear Programs with Zero-One Variables", Operations Research, Vol. 13, pp. 517-545, 1965.

3. Elmaghraby, S. E., A. Mallik, and H. L. W. Nuttle, "The Scheduling of Lots on a Single Facility", AIIE Transactions, Vol. 2, pp. 203-213, 1970.

4. ________, ________, "The Scheduling of a Multi-Product Facility", O.R. Report No. 76, N. C.State University, April, 1972.

5. Greenberg, H. and R. L. Hegerich, "A Branch Search Algorithm for the Knapsack Problem", Management Science, Vol. 16, pp. 327-332, 1970.

6. Hoyte, D. S., "Scheduling Multi-Product Lots on a Single Facility", unpublished M.Sc. Thesis, N. C.State University, 1971.

7. ________ and H.L.W. Nuttle, "The Scheduling of Multi-Product Lots on a Single Facility", O.R. Report No. 74, N. C. State University,Oct., 1971 (also appears in the Proceedings of the 7th Annual Meeting of the Southeastern Chapter of TIMS, Blacksburg, Va., Oct. 1971.)

8. Kolesar, P. J., "A Branch-and-Bound Algorithm for the Knapsack Problem", Management Science, Vol. 13, pp. 723-735, 1967.

9. Taha, H. A., "A Note on the Solution of Specialized Zero-One Problems by Using Balas' Algorithm", Technical Report No. 71-8, Department of Industrial Engineering, University of Arkansas, Nov. 1971.

10. ________, "A Balasian-Based Algorithm for Zero-One Polynomial Programming", Management Science, Vol. 18, pp. 328-343, 1972.

RANDOM PATROL SCHEDULING UNDER OPERATIONAL CONSTRAINTS

M. Rosenshine and J. Evans
The Pennsylvania State University, University Park, Pennsylvania, USA

Random patrol scheduling is somewhat unique as an area of scheduling because it introduces uncertainty rather than avoid it or merely live with it. The purpose of the randomness is to make the patrols more unpredictable and hopefully more effective. However, certain practical constraints related to sensible use of patrol equipment and the ways in which police patrols operate need to be considered before random patrol scheduling can be considered for actual use. This paper describes the utilization of simulation to determine how well the random patrol of a network can be effected in the absence or modification of some of the theoretical requisites for random patrol.

INTRODUCTION--PROBLEMS CONSIDERED AND APPROACH TAKEN

In "Contributions to a Theory of Patrol Scheduling,"[1] the notion of random patrol of a network was defined. It was shown that to obtain random patrol it is necessary that the number of patrol cars in the network be a Poisson random variable. This raised the question of economic practicality of random patrol. Furthermore, no interferences with patrol schedule were considered. This, in turn, aroused concern about relevance of the situation considered.

The two concerns cited above led to speculation concerning effects of practical constraints on random patrol. To investigate these effects, two questions were posed.

1. What is the effect on random patrol of keeping the number of patrol cars in the network constant?
2. What is the effect on random patrol if the constant number of patrol cars encounter the delays that patrol cars might reasonably be expected to encounter?

To answer these questions, the General Purpose Systems Simulator (GPSS 360) was used to simulate the patrol of a street network because the constant number of patrol cars poses some rather stubborn analytical problems when the techniques of reference 1 are applied. It was easy to simulate the patrol of a network. Furthermore, for a small network it was inexpensive to simulate large numbers of patrol cars (2000) passing through the network over long periods of simulated time so that there is no reason to expect that steady-state results would differ substantially from those that would be obtained if mathematical analyses were feasible.

In addition to answering the questions posed, the results obtained in validating the simulation experiment led to an additional result that had been previously suspected, but forgotten. Patrol that is "as random as possible" according to the definition given in reference 1 cannot, in theory, be obtained in networks as general as those presented in reference 1. Fortunately, however, this modification does not affect any of the results presented therein. It merely serves to point out that if the network is sufficiently "peculiar," the properties of the patrol generated may depart somewhat from those of a random patrol.

THE CASE STUDY NETWORK CONFIGURATION AND PARAMETERS

The street network used for this investigation covers a ten square block area, consisting of nine streets, three of them one-way, a fourth one-way for one block, and eighteen intersections. This street network is illustrated in Figure 1. Minimum required patrol rates (in patrols per 60 time unit period) are assigned to each street segment, reflecting the need for more patrols in one section of the system than in another, presumably due to a higher crime rate. A high patrol rate is assigned to the street segment containing the station, so that each patrol will not spend an excessive amount of time in the network before returning to the station. If the patrol rate on this segment is not set high, the patrol rate requirements on the other streets will be met by fewer patrol cars, each circulating for a longer period of time.

A graphical representation of the street network is shown in Figure 2. Actual patrol rates were determined by solution of the linear programming problem described in reference 1. This solution resulted in the patrol rates shown in the rectangles adjacent to or immediately below the arc numbers. Those arcs on which the actual patrol rates turned out to be zero (arcs 5, 8, 11, 19, 21, 40) can be eliminated from Figure 2.

The routes taken by patrols through the network are determined from a decision matrix developed from the network and patrol rates of Figure 2, and a

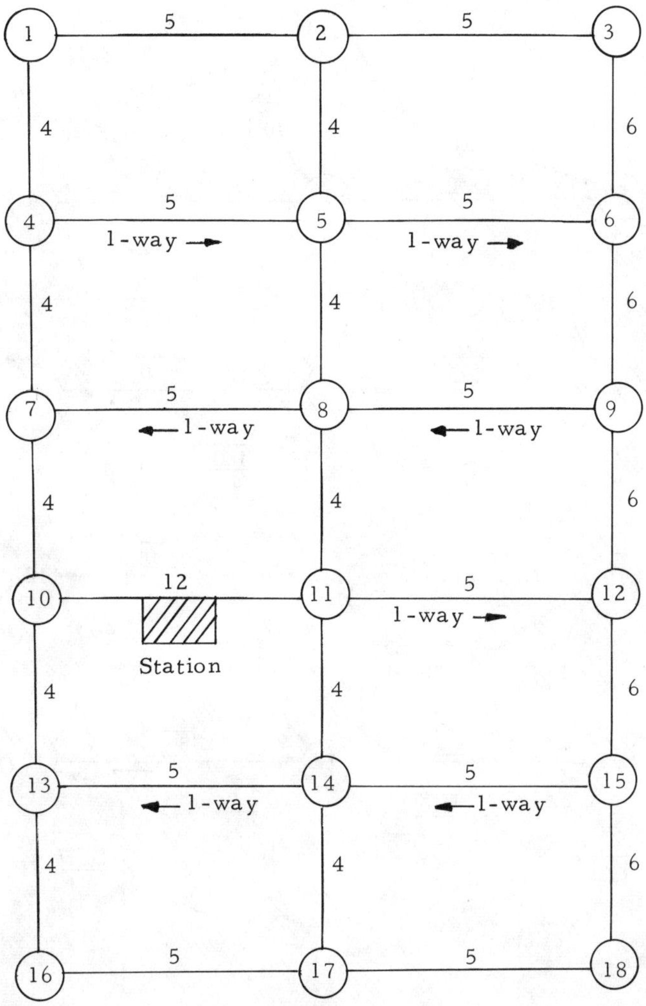

Intersection numbers are circled.
Numbers on streets are minimum number of patrols required per 60 time units

Fig. 1. Diagram of Street Network

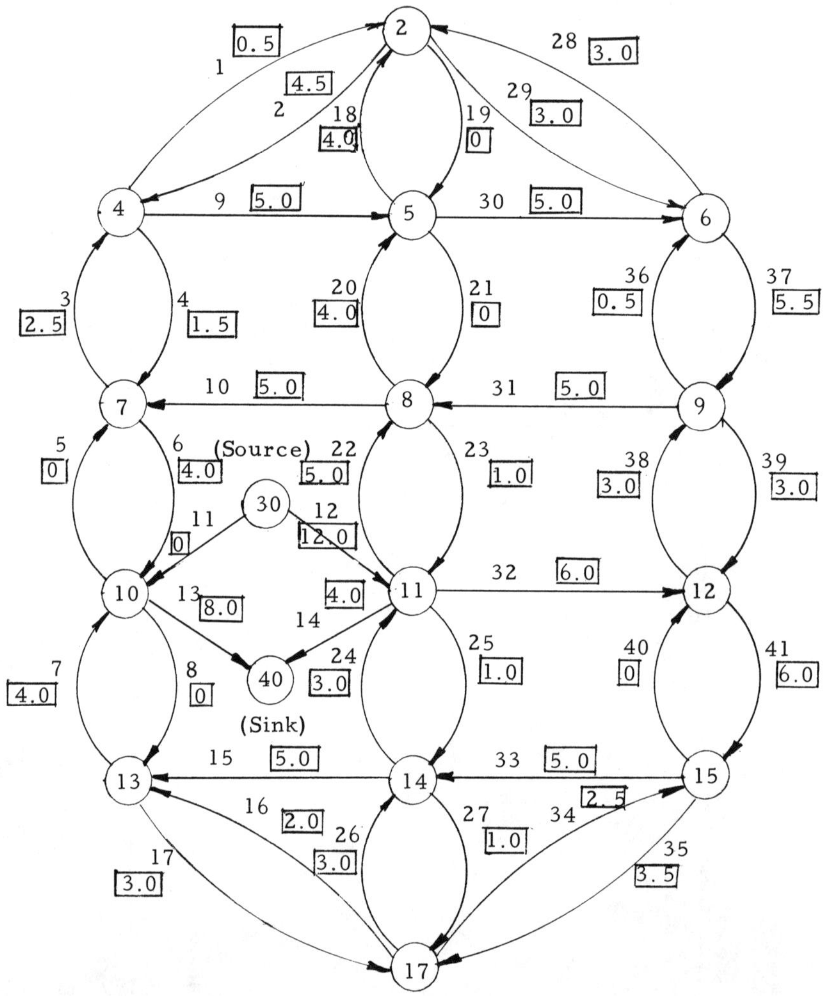

Node numbers are circles.

Numbers on arcs are arc numbers and actual patrol rates (in rectangles)

Fig. 2. Graphical Representation of Street Network

superimposed requirement prohibiting U turns. For example, consider that part of the network consisting of nodes 2, 4, 5, and 6 and the arcs connecting them. Since no U turns are permitted, we can see that if a patrol arrives at node 2 over arc 1 it must go to node 6 over arc 29. Similarly, if a patrol arrives at node 2 via arc 28, it must go to vertex 4 over arc 2. Because all of the patrols that traverse arc 28 must also traverse arc 2, to obtain a patrol rate of 4.5 on arc 2, we must get a rate of 1.5 from arc 18, in addition to the 3.0 from arc 28. Similarly, all of the patrols that traverse arc 1 must also traverse arc 29, so that to attain a rate of 3.0 on arc 29 we must get a rate of 2.5 from arc 18, in addition to the 0.5 from arc 1.

From this information a decision rule emerges for node 2. If a patrol is on arc 28, the probability that it will next traverse arc 2 is 1.0; similarly, if a patrol is on arc 1, the probability that it will next traverse arc 29 is 1.0. Reasoning that since 1.5 of the rate of 4.0 traversing arc 18 are required to traverse arc 2, the probability that a patrol traversing arc 18 will next traverse arc 2 can be calculated as

$$\text{Prob (Arc 18 to Arc 2)} = \frac{1.5}{4.0} = 0.375$$

Likewise, 2.5 of the mean rate of 4.0 traversing arc 18 are required to traverse arc 29, so that the probability that a patrol traversing arc 18 will next traverse arc 29 is

$$\text{Prob (Arc 18 to Arc 29)} = \frac{2.5}{4.0} = 0.625$$

These calculations yield the decision rule for node 2 which is summarized in Table 1.

TABLE 1. Decision Rule for Node 2

Arrive on Arc	Probability Next Arc Is	
	2	29
1	0	1.0
18	0.375	0.625
28	1.0	0

A similarly derived decision rule is used for each node. They are programmed into the simulation so that randomly selected patrol routes result with the desired patrol rates on each arc.

THE EXPERIMENTS

To determine the effect on random patrol of keeping the number of patrol cars in the network constant, the number of patrol cars in the network was constrained to remain equal to the expected number of patrol cars in the network.* This was accomplished by letting the Poisson stream of patrol cars introduced into the network build up to and then pass through the expected number of cars two times and assuming (somewhat arbitrarily) that the third time the expected number of cars is reached marks the beginning of steady state operation. In the transition period random patrol is followed, but once steady state is assumed the patrol generation mechanism is changed so that a new patrol car enters the street network only when one in the network returns to the station (sink). The time to traverse each arc was assumed constant at five time units.

To investigate the effect of random patrol if the constant number of patrol cars encounter delays, various probabilities (0.1, 0.2, 0.3, 0.4) of encountering delays were assigned to the various arcs. The delays were chosen from an Erlang-5 distribution with a mean of fifty time units. In addition, the time to traverse the various arcs in the absence of delays was no longer required to be the same for all arcs.**

---

* $E(N) = \sum_{i=1}^{35} \mu_i(t_i + p_i t_d)$ where μ_i, t_i, and p_i are, respectively, the patrol rate, the expected time to traverse without delay and the probability of a delay on arc i. t_d is the expected length of a delay.

** It appears from the description that everything was allowed to vary in this experiment without any reason. Actually, there were several controlled steps between the two experiments described in this paper. However, these intermediate experiments merely yielded "intermediate" results and are not reported upon here.

ANALYSIS AND RESULTS

An analysis of the experimental results was made to determine the effects of the various operational constraints simulated on the randomness of the patrols. This was accomplished by testing the distribution of inter-arrival times at selected points in the network for goodness of fit with an exponential distribution. The Kolmogorov-Smirnov (K-S) goodness of fit test was used for this purpose. It was performed on both experiments at a significance level of 0.10. This level was chosen rather than the usual 0.05 because acceptance of the null hypotheses (inter-arrival times are exponentially distributed) at a higher level of significance requires a closer fit, the power of the test being higher. Thus, with the higher significance level, the probability is higher of detecting that the hypothesis of an exponential distribution is incorrect.

Because different numbers of patrol cars pass over the various arcs in the network, the results of the K-S test make more sense if they are presented for each arc as the ratio of the absolute value of the maximum difference between the hypothesized and the observed cumulative distribution functions, d_m, and the critical value of this statistic at the 0.10 level of significance, $d_{.10}(N) = \frac{1.22}{N}$, where N is the number of cars which passed over the arc. Then, ratios less than 1.0 represent acceptance of an exponential fit while ratios greater than 1.0 represent rejection. Furthermore, since smaller values of the ratio tend to be associated with tighter fits, the sum of the ratios over all the arcs tested are used to provide a heuristic measure of the overall randomness of the patrol of the network.

The results obtained for the two experiments described in the previous section are shown in Table 2. For the first experiment these results show that six out of thirty arcs "fail" the K-S test. This is twice the expected number of rejections at a significance level of 0.10 if the distributions are exponential. Thus, it may be judged that the conditions for random patrol are not met. However, there are some rather interesting additional results available which give one cause to wonder whether it was not the network structure rather than the

TABLE 2. Goodness of Fit Ratios

Arc	Ratio d_M[a] · $d_{.10}(N)$[b]	
	Exper. 1	Exper. 2
2	1.140	0.547
3	0.492	0.411
4	0.438	0.500
6	0.600	0.628
7	0.633	0.560
9	0.875	0.300
10	0.575	0.487
12	1.480	0.518
14	0.625	0.396
15	1.700	0.837
16	0.631	0.409
17	2.580	0.683
18	0.932	0.356
20	0.848	0.333
23	0.256	0.275
24	0.842	0.483
26	0.582	0.448
27	0.917	0.265
28	0.885	0.404
29	0.706	0.442
30	1.085	0.375
31	0.930	0.524
33	0.978	0.391
34	1.890	0.514
35	0.541	0.200
36	0.401	0.500
37	0.692	0.289
38	0.930	0.268
39	0.423	0.615
41	0.920	0.237
Total	26.927	13.195

[a] d_M is the maximum difference between the observed cumulative distribution and the theoretical cumulative distribution.

[b] $d_{.10}(N)$ is the critical value at a level of significance 0.10 of the maximum difference between the observed cumulative distribution and the theoretical cumulative distribution.

scheduling technique which caused the hypothesis of randomness to be rejected. During the validation of the simulation program, an experiment with no constraints on the number of patrols utilizing the same network resulted in eight arcs "failing" the K-S test. It was found that three of these arcs carried flow among vertices 2, 4, 5, and 6 (the four vertices at the top of Figure 2), while the remaining five arcs carried flow among vertices 13, 14, 15, and 17 (the four vertices at the bottom of Figure 2). Calculations showed that, for example, the probability of a patrol on arc 2 returning to arc 2 in exactly fifteen time units is 0.111, and in exactly twenty time units is 0.185 when the time spent on each arc was exactly five time units. Similarly for arc 15, the probability of returning in exactly fifteen time units is found to be 0.154 and the probability of returning in exactly twenty time units is 0.216. One or more of these high "cycling" probabilities* also applied to all the other arcs among the eight for which exponentially distributed inter-arrival times were rejected. Since four of the six arcs which "failed" the K-S test in experiment 1 were among the arcs for which the exponential distribution was rejected in the validation run, the suggestion is that the network structure rather than the constant number of patrols is primarily responsible for not meeting the conditions for random patrol.

Regardless of the cause of the non-randomness of the patrol in experiment 1, the results obtained in experiment 2 indicate that neither cause will interfere with randomness of the patrols when delays are superimposed on the patrol schedule. The acceptance of exponential fits in all cases is somewhat surprising. However, the delays while on patrol remove almost all probability of returning to a given arc in any given number, say 15 or 20, of time units.** When one

---

* This is the phenomenon alluded to in the introduction which prevents patrols from being "as random as possible." However, if the probability of cycling is very low, or if cycle times are random and close to exponentially distributed, random patrol will be obtained.

** This reasoning can be justified from the results of an intermediate simulation run which differed from experiment 1 only in that an exponentially distributed delay time was assigned with probability 0.4 to a patrol on each arc it traversed. The exponential fit was accepted for all arcs in this case, too.

considers further than in experiment 2 the time spent on each arc is a random variable, any possible remaining non-random effects of introducing a new patrol every time one leaves the network should be eliminated.

CONCLUSIONS

A practical system of patrol scheduling which provides patrols that arrive as randomly as possible appears easily attainable. The requirement for a patrol system which operates with a nearly constant number of cars so as not to let cars and patrolmen sit idle introduces some non-random effects which are removed by the variable length delays that the patrols encounter as they travel the street network. While it is clear, for theoretical reasons, that delay distributions and probabilities can be chosen expressly to contradict the conclusions reached, it is not felt that these reflect the distributions of delay times that would be encountered in actual patrol situations.

REFERENCE

1. Rosenshine, M., "Contributions to a Theory of Patrol Scheduling," Operational Research Quarterly, Vol. 21, No. 2, March, 1970.

INTERACTION BETWEEN AGGREGATE AND DETAILLED SCHEDULING IN A JOB SHOP

Jeol Shwimer

52 Quai Louis Bleriot, Paris 16^{e}, France

The talk is an abstract of a Ph.D. dissertation submitted to the Alfred P. Sloan School of Management on May 4, 1972 in partial fulfillment of the requirements for the degree of Doctor of Philosophy.

The research is concerned with the different levels of decision-making in the scheduling of a production facility and how interaction between these levels can be used to improve the overall quality of the decisions being made. Although the particular models and modes of interaction are placed in the job shop context, the general methodology being employed can be useful in numerous other situations. In effect, this work is an attempt to synthesize and advance the work of many others who have studied only isolated portions of the job shop scheduling problem. Improved overall results can be achieved by treating the separate parts of this problem within a single framework, thereby considering the ways in which these parts are related.

A "total" model which incorporates both the aggregate (e.g., hiring, firing, use of overtime, and allocation of workers to machine groups) and detailed (e.g., what job is placed next on each machine, and at which machine each worker works next) decisions of job shop scheduling is constructed. The size and complexity of this model preclude its direction solution. However, the model delineates the basic structural aspects of the problem that must be considered in developing methods for its solution.

This model is then partitioned into an aggregate level model and a detailed level model. The aggregate model, although similar to some past work in this area, extends the applicability and usefulness of such a planning tool. A more realistic aggregate model of the shop is possible by considering a number of worker skill classes and machine groups, each with different capabilities. In addition, an important step has been taken towards the disaggregation of the solution, thus making implementation much easier. Although the aggregate model, as formulated, is a mixed-integer linear program, the linear approximation to this model can be solved quite efficiently. Using appropriate rounding techniques this becomes a good approximation to the mixed-integer model.

The detailed model is still of sufficient size and complexity to preclude the use of an optimal solution-seeking method. Therefore, heuristically based job dispatching (e.g., COVERT) and labor assignment rules within a simulation framework are used to obtain a good, but not necessarily optimal solution.

Iteration between these two submodels presents a method for obtaining a good solution to the "total" model. Each submodel requires certain information which is determined in the other submodel. A number of procedures for passing the necessary information and then utilizing it are proposed. This interaction between the two models, or levels of decision-making, has usually been ignored or been performed only by default in much of the job shop scheduling research.

A hypothetical job shop is used as the experimental environment within which a number of the proposed methods of interaction are tested. The results of this experimentation confirm that improved overall performance in a job shop can be obtained if the aggregate and detailed scheduling are performed jointly or by means of some reasonable coordination scheme between these two subproblems of the total job shop scheduling problem.

The complete dissertation is available as Technical Report No. 71 of the Operations Research Center, Massachusetts Institute of Technology, Cambridge, Massachusetts, May, 1972.

AN EXTENSION OF MOORE'S DUE DATE ALGOTITHM

Jeffrey B. Sidney

University of British Columbia, Vancouver, Canada

A one machine job shop is considered. Associated with each job is its processing time and due date. There exists a subset of jobs which must be completed on time in any feasible schedule. A feasible sequencing of the jobs is sought which minimizes the number of late jobs.

An algorithm is stated, and proved optimal. In addition, the resulting optimal set is further characterized. A second optimal algorithm is also presented.

Consider a job shop consisting of n jobs, 1,2,...,n to be processed on one machine. Associated with each job i is its deterministic processing time p_i and its due date d_i. A subset T of the set of jobs exists such that each job in T must be completed on time in any feasible schedule. One machine is continuously available from time t = 0, for processing the jobs, until they have all been completed. A feasible (i.e., jobs in T are on time) sequencing of the jobs is sought which minimizes the number of late jobs.

For the case in which T = ∅, Moore [2,3] has presented an elegant algorithm, which has since been modified by Hodgson [Moore, 3].

In this paper, two algorithms which generalize the Hodgson algorithm to cover the case of non-empty T are presented. The first is proven optimal, and further useful characteristics of schedules resulting from the algorithm are also presented.

The second algorithm is quite similar, and appears after the discussion of the first.

A subset S of the jobs is an on-time subset if all jobs in S can be completed on-time.

Lemma 1 (Jackson [1]): Let S be any on-time subset of the jobs. Then if the jobs in S are ordered by non-decreasing due dates, all jobs in S will be on time in the resulting schedule.

Note that, in the event there exists a due date d shared by several jobs, lemma 1 allows us to arbitrarily choose an order for these jobs.

A precedence relation R' is defined on the jobs such that if $d_i < d_j$, then $(i,j) \varepsilon$ R'. Based on lemma 1, it is clear that the search for an optimal schedule can be restricted to schedules such that all jobs which meet their due dates are ordered consistently with R'.

R' may not totally order the jobs, for there may be a due date d shared by several jobs. As previously mentioned, an arbitrary ordering of such jobs may be chosen. Hence, R' will be extended to a total ordering R of {1,...,n} by arbitrarily ordering the jobs which share the same due date, where we will require that this "arbitrary" ordering satisfy the one additional condition that if $d_i = d_j$ and $i \notin T$, $j \varepsilon T$ then $(i,j) \varepsilon R$. (In the second version of the algorithm, we shall require a totally arbitrary due date ordering. This will simplify the preliminaries to the algorithm, but complicate one step of the algorithm).

Lemma 2: There exists an optimal schedule such that the set S of jobs which are on-time are ordered consistently with R.

Proof: Discussion above.

Based upon lemma 2, we can view the original problem as that of finding a feasible on-time subset S of greatest cardinality, where S will be assumed ordered by R. Such a set S is called an optimal set.

From now on, it is assumed that the natural order (1,2,...,n) is the order given by R.

Now, consider any job i, and suppose k>i and $k \varepsilon T$. Since k must be in any feasible set S, it follows that, if job i is also to be on time, then job i must be completed by $d_k - p_k$. If the due date d_i were greater than $d_k - p_k$, it could therefore be reduced to $d_k - p_k$ without loss of feasibility of any set. Hence, we shall define new due dates d_i' in the following iterative manner, where it is assumed that m is the highest numbered job in T:

(1) If $i \geq m$, $d_i' = d_i$

(2) For i decreasing from m-1 to 1, suppose k is the first job following i in R which is also an element of T. Then $d_i' = \min\{d_i, d_k'-p_k\}$ (d_i may be used, rather than d_i'. The proof simplifies when d_i' is used).

From the above discussion, lemma 3 follows.

Lemma 3: There exists an optimal set S for the original problem such that S is also optimal for the problem with precedence relation R and due dates d_i'.

Let $N_k = \{1,2,\ldots,k\}$ and for any subset S of the job set, let $p(S) = \sum_S p_i$.

To simplify the algorithm, note that any job i such that $d_i' < p_i$ may be eliminated from the job set (if $i \varepsilon T$, there is no feasible solution). In addition, if the jobs in T, when ordered by non-decreasing due date, are not all on time, then again no feasible solution exists, and we stop.

Assuming the above checks have been made, the algorithm proceeds as follows:

1. Compute the d_i' and generate a precedence order R as described previously.

   Set $S_o = \emptyset$ and $k = 1$.

2. If $k = n+1$, stop. S_n is an optimal set.

3. If $p(S_{k-1}) + p_k \leq d_k'$ set $S_k = S_{k-1} \cup \{k\}$.

   Otherwise, let $p_r = \max\ \{p_j | j \varepsilon S_{k-1} \cup \{k\} - T\}$ and set $S_k = S_{k-1} \cup \{k\} - \{r\}$

4. Increment k by 1. Return to step 2.

How does this algorithm differ from the Hodgson version of Moore's algorithm? First, by the introduction of d_i' and the more specific definition of R.

Second, in step 3, it is required that the job r that is selected to become a late job cannot be a member of the set T.

Theorem 4: At the end of iteration k, the set S_k has the properties that

(1) S_k is feasible.

(2) S_k has maximal cardinality among all on-time feasible subsets of N_k.

(3) Suppose U is a q-job subset of S_k obtained by eliminating those $(|S_k|-q)$ jobs from S_k which have largest processing times amongst the jobs in

S_k-T. Then U is a feasible on-time set and has least processing time among all q-job feasible on-time subsets of N_k. (Note that $|S_k \cap T| \leq q \leq |S_k|$).

Proof: (induction on k):

The conclusions are true for k = 1.

Suppose the conclusions are true for $S_1, S_2, \ldots, S_{k-1}$.

Conclusion (1): If $S_k = S_{k-1} \cup \{k\}$, this is trivially true.

If not, then $p(S_{k-1}) + p_k > d'_k$. If r = k in step 3 of the algorithm, then $S_k = S_{k-1}$ and the result is true. Otherwise, $p(S_k) = p(S_{k-1}) + p_k - p_r \leq p(S_{k-1})$. Note that, by the definition of the d'_i, it must be true that $k \notin T$, since if $k \epsilon T$ and $j < k$, it follows that $d'_j \leq d'_k - p_k$. Every job in $S_{k-1} - \{r\}$ will be completed at least as early as before. Moreover, there exists $i \epsilon S_{k-1}$ such that $d'_i \geq p(S_{k-1})$. Hence, $d'_k \geq d'_i \geq p(S_{k-1}) \geq p(S_k)$ = completion time of job k after iteration k.

Hence all jobs in S_k are on-time.

Conclusion (2): The maximal cardinality among on-time feasible subsets of N_k will be either $|S_{k-1}|$ or $|S_{k-1}| + 1$, since at most $|S_{k-1}|$ of the jobs in N_{k-1} could be on-time.

If some maximal feasible subset V of N_k does in fact have cardinality $|S_{k-1}| + 1$, then V must contain k. Hence, a feasible subset V of N_k consisting of $|S_{k-1}| + 1$ elements exists if and only if there exists a feasible subset U of N_{k-1} consisting of $|S_{k-1}|$ elements having the property that $p(U) \leq d'_k - p_k$. By applying conclusion 3 to S_{k-1} with $q = |S_{k-1}|$, it is evident that such a U exists if and only if $p(S_{k-1}) \leq d'_k - p_k$ (In other words, if such a U exists, we can take $U = S_{k-1}$). Hence, step 3 of the algorithm will produce a maximum cardinality on-time feasible set.

Conclusion (3): Observe that for the case in which $p(S_{k-1}) + p_k > d'_k$, step 3 of the algorithm is equivalent to: S_k consists of all required (i.e., ϵT) jobs in N_k and the $|S_{k-1}| - |N_k \cap T|$ smallest jobs of $S_{k-1} \dot{\cup} \{k\} - T$. Otherwise, step 3 is simply: $S_k = S_{k-1} \dot{\cup} \{k\}$.

On the other hand, conclusion (3) of this theorem defines U to be that subset of S_k consisting of (all required jobs in N_k) $\cup$ (the $q-|N_k \cap T|$ smallest jobs in S_k-T).

These two observations yield the desired result. Q.E.D.

An additional result is also available. Suppose S_n has been determined, and suppose a new restriction on the problem occurs, namely that the machine is available only until time t_o. If $t_o \geq p(S_n)$ there is no problem. If $t_o < p(S_n)$, one could arrive at an optimal solution by using the algorithm with due dates $d_i^* = \min \{t_o, d_i'\}$. However, conclusion 3 implies that an optimal procedure is to simply eliminate jobs in S_n-T from S_n, in order of non-increasing processing times until the resulting set has total processing time $\leq t_o$.

An alternative algorithm follows (as before, it is assumed that all jobs having $p_i > d_i$ have been eliminated, and that the jobs in T, when ordered by non-decreasing due date, are all on time).

1. Renumber the jobs so that they are ordered by non-decreasing due date. Set $S_o = \emptyset$ and $k = 1$.
2. If $k = n+1$, stop. S_n is an optimal set.
3. If $p(S_{k-1}) + p_k \leq d_k$ set $S_k = S_{k-1} \cup \{k\}$. Otherwise, let $S_k^* = S_{k-1} \cup \{k\}$, and then eliminate from S_k^* in order of non-increasing processing time jobs in S_k^*-T, until all jobs in the remaining set are on time. This set will be the new S_k.
4. Increment k by 1. Return to 2.

Example: $p = (3,2,7)$, $d = (5,6,8)$, $T = \{3\}$.

Then in the third iteration, both jobs 1 and 2 will be eliminated in step 3, in order to accommodate job 3.

References

1. Jackson, J.R., "Scheduling a Production Line to Minimize Maximum Tardiness," Research Report 43, Management Sciences Research Project, UCLA, January, 1955.

2. Moore, J. Michael, "Sequencing Problems with Due Date Oriented Objective Functions," Ph.D. dissertation, Univ. of Michigan, 1967.

3. ________________, "An n Job, One Machine Sequencing Algorithm for Minimizing the Number of Late Jobs," Management Science, Vol. 15, No. 1 (Sept. 1968).

SOLVING SCHEDULING PROBLEMS BY APPLYING COST OPERATORS TO ASSIGNMENT MODELS*

V. Srinivasan and G.L. Thompson

The University of Rochester, Rochester, New York, USA

Carnegie-Mellon University, Pittsburgh, Pennsylvania, USA

ABSTRACT

The authors apply their previously developed operator theory of parametric programming for the transportation problem to subtour elimination methods for solving scheduling problems. The specific examples of the travelling salesman and decision critical path problems are discussed and illustrated with numerical examples. Computational tests of these methods is planned.

* This report was prepared as part of the activities of the Management Sciences Research Group, Carnegie-Mellon University, under Contract N00014-67-A-0314-0007 NR 047-048 with the U. S. Office of Naval Research. Reproduction in whole or in part is permitted for any purpose of the U. S. Government.

1. INTRODUCTION

Subtour elimination and related methods for solving scheduling and allocation problems such as the travelling salesman problem [1,2,4,6,10-13] and the time minimizing or bottleneck assignment problem [9,20] are well known. In such methods the scheduling problem is solved by solving a sequence of assignment problems. However each successive assignment problem in the sequence is usually only a minor modification of the preceding one, and hence a large part of these computations are repetitive. For this reason it seems clear that post optimization or parametric programming procedures can be used to eliminate most of the duplicate computations thereby improving the efficiency of these methods.

In [15,16] we have presented an operator theory of parametric programming for the upper bounded transportation problem which, of course, includes the assignment model as a special case. In the present paper we shall specialize the cost operator theory to assignment models and then apply it to solve two scheduling problems. In Section 3 we shall show how it can be used to accelerate the subtour elimination method of solving travelling salesman problems. And in Section 4 we show that it can be used to solve decision critical path problems introduced by Crowston and Thompson in [5].

Although a transportation problem is a special kind of linear programming problem its parametric programming theory is more useful. To see this note that the problem

$$\text{(I)}\qquad \begin{aligned} &\text{Maximize} \quad cx + g\delta \\ &\text{Subject to} \\ &\qquad Ax - \delta d = b \\ &\qquad x \geq 0 \end{aligned}$$

(where A is $(m\text{x}n)$, b is $(m\text{x}1)$, c is $(1\text{x}n)$, d is $(m\text{x}1)$, and g is a scalar defining the data and $x \in R^n$ and $\delta \in R^1$ are the variables) may be regarded as <u>both</u> a linear programming problem in its own right <u>and</u> as a parametric variation on the problem

$$\text{(II)} \quad \begin{array}{l} \text{Maximize} \quad cx \\ \text{Subject to} \\ \qquad Ax = b \\ \qquad x \geq 0 \end{array}$$

Thus instead of applying parametric programming on the linear programming problem II one can merely solve the somewhat larger problem I of the same kind.

The same result is not true for a transportation problem, since as can easily be seen, a variation of the data alone destroys the special transportation structure. Our "operator theory" of parametric programming [15,16] for the transportation problem permits the study of parametric variations of the costs and rim conditions of a transportation problem while staying within the transportation model framework. We have found the results to be surprisingly rich and capable of wide application, two such applications being illustrated in the present paper.

Other applications of the operator theory include (a) the optimal scheduling of warehouse capacity expansions [17], (b) optimal scheduling of cash flows to meet payment and liquidity constraints [14], and (c) optimal assignment of sources to uses [18].

Recent computational studies of the primal transportation algorithm by us [19] and by Glover, Karney, Klingman, and Napier [8] have demonstrated that large scale transportation and assignment problems can be solved in extremely short times. For instance, in [19] we showed that a 100 x 100 fully dense transportation <u>or</u> assignment problem could be solved in less than 2 seconds (on a UNIVAC 1108). This is about the same time as Kuhn's Hungarian method takes to solve only assignment problems of the same size, and is considerably faster than the Ford-Fulkerson's primal-dual method [7]. These extremely fast solution times make applications involving the transportation or assignment models extremely attractive.

The operator theory we developed in [15,16] begins with an optimal solution to a transportation problem. Then changes are made in the data of the problem: to change one or more cost coefficients, cost operators are applied; and to change rim coefficients, rim operators are applied. As the operators change the data of the problem they also change the primal and dual solutions as necessary in order to retain primal and dual feasibility. If the optimal solution after the application of an operator has the same set of basis cells as before then we call it a basis-preserving operator. A more complete description of cost operators for the assignment problem is given in the next section.

2. COST OPERATORS FOR THE ASSIGNMENT PROBLEM

Let $I = J = \{1,2,\ldots,n\}$; we shall write the classical $n \times n$ assignment problem A with costs c_{ij} assigning the ith source to the jth use as:

$$\text{(1)} \qquad \text{Minimize} \quad \sum_{i \in I} \sum_{j \in J} x_{ij} c_{ij} = Z$$

Subject to

$$\text{(2)} \qquad \sum_{i \in I} x_{ij} = 1$$

$$\text{(3)} \qquad \sum_{j \in J} x_{ij} = 1$$

$$\text{(4)} \qquad x_{ij} \geq 0 \ .$$

A set $X = \{x_{ij}\}$ of shipments satisfying (2), (3), (4) is said to be feasible. By a cell we shall mean a pair $(i,j) \in I \times J$. A basic solution consists of a set of B of $2n-1$ cells and a set X of feasible shipments such that (a) the graph $G = (I \cup J, B)$ is a tree (an acyclic connected graph); and (b) $x_{ij} = 0$ if $(i,j) \notin B$.

REMARK 1. It is well known that an assignment problem is a natural integer problem, that is, for any basic solution $x_{ij} = 1$ or 0 for all i and j. It follows that $x_{ij} = 1$ for exactly n cells of B, $x_{ij} = 0$ for the remaining n-1 cells of B, and the matrix of the x_{ij}'s is a permutation matrix.

Let the dual variables u_i be associated with row i and v_j with column j. The dual problem to (1)-(4) is:

(5) $$\text{Maximize} \quad \sum_{i\epsilon I} u_i + \sum_{j\epsilon J} v_j$$

Subject to

(6) $$u_i + v_j \leq c_{ij} \qquad \text{for } i\epsilon I \text{ and } j\epsilon J,$$

where the u_i and v_j variables are unrestricted in sign. By the well known complementary slackness theorem we can write

(7) $$u_i + v_j = c_{ij} \quad \text{for} \quad (i,j) \in B$$

for a basic solution with basis B. The 2n-1 equations (7) determine a one parameter family of dual solutions such that $d_{ij} = u_i + v_j$ is unique for all $i \in I$ and $j \in J$. A set of dual solutions $D = \{d_{ij} = u_i + v_j\}$ satisfying (6) is said to be <u>dual feasible</u>. Solutions X and D that are primal and dual feasible respectively are <u>optimal</u>.

In order to make this paper self-contained we shall now quote several definitions and theorems taken from [15] but specialized to assignment problems so that they may be applied to the solution of the problems in the next two sections. Proofs of quoted theorems will not be given here but can be found in [15,16].

DEFINITION 1. Let $\delta \geq 0$ be a number and (p,q) a given cell. A <u>(positive cell) cost operator</u> δC^+_{pq} transforms the optimum solution of problem A into that for problem A^+ with all data the same except for

(8) $$c^+_{pq} = c_{pq} + \delta.$$

In either case we shall denote by X^+, D^+, B^+, and Z^+ as the optimum primal and dual solutions, basis and objective function after the transformation. If

$B^+ = B$ the operator is said to be <u>basis preserving.</u> We denote by μ^+ the maximum <u>extent</u> for δ (i.e., $0 \leq \delta \leq \mu^+$) so that δC^+_{pq} is basis preserving.

The next two theorems are specializations of theorems 5 and 6 of [15].

THEOREM 1. (Primal solutions for basis preserving cost operators.) For the operator δC^+_{pq}, $X^+ = X$ and $Z^+ = Z + \delta x_{pq}$.

REMARK 2. By Remark 1, $x_{pq} = 0$ or 1 so that $Z^+ = Z$ or $Z^+ = Z + \delta$.

THEOREM 2. (Dual solutions for basis preserving operators δC^+_{pq} with $(p,q) \notin B$.) For the operator δC^+_{pq} with $(p,q) \notin B$, $D^+ = D$ and

$$\mu^+ = M \tag{9}$$

where M is a large positive number.

In order to study operators with $(p,q) \in B$ we need to find how the dual variables u_i and v_j change as c_{pq} is changed. For this we need the following remarks and definition:

REMARK 3. Row i is connected to row p in a set of cells Ω if there is a chain of distinct cells $(i_1,j_1), (i_2,j_2),\ldots,(i_k,j_k)$ in Ω with $i_1 = i$, $j_1 = j_2$, $i_2 = i_3,\ldots, j_{k-1} = j_k$ and $i_k = p$. The other definitions of a row connecting a column and columns connecting rows and other columns are similar.

DEFINITION 2. (See Def. 8 in [15]) Let (p,q) be a cell in B and let $\Omega = B - \{(p,q)\}$. Then we define the sets

$$\begin{aligned}
I_p &= \{p\} \cup \{i \in I \mid i \text{ is connected to } p \text{ in } \Omega\}\\
J_p &= \{j \in J \mid j \text{ is connected to } p \text{ in } \Omega\}\\
I_q &= \{i \in I \mid i \text{ is connected to } q \text{ in } \Omega\}\\
J_q &= \{q\} \cup \{j \in J \mid j \text{ is connected to } q \text{ in } \Omega\}.
\end{aligned} \tag{10}$$

The sets I_p and I_q partition I; i.e., $I = I_p \cup I_q$ and $I_p \cap I_q = \phi$. Similarly J_p and J_q partition J.

REMARK 4. The following <u>scanning routine</u> can be used for identifying the four sets just defined:

0. Label with 1 the columns of basis cells in row p except for basis cell (p,q); Label row p with 2.
1. If no new columns were labelled 1 on the preceding step go to 5. Otherwise go to 2.
2. For each column labelled with 1, label with 1 each row containing a basis cell in that column unless the row is labelled with 2; then change the label of the column to 2.
3. If no new rows were labelled 1 on the preceding step go to 5. Otherwise go to 4.
4. For each row labelled with 1, label with 1 each column containing a basis cell in that row unless the column is labelled with 2; then change the label of the row to 2. Go to 1.
5. Stop. The set I_p (J_p) consists of the rows (columns) labelled with 2; the set I_q (J_q) consists of the unlabelled rows (columns).

The label "1" in the above routine corresponds to the "labelled but not yet scanned" terminology of Ford and Fulkerson [7], while the label "2" corresponds to "labelled and scanned" rows or columns.

The next theorem is a specialization of Theorem 7 in [15].

THEOREM 3. (Dual solutions for basis preserving cost operators δC^{+}_{pq} with $(p,q) \in B$.)

For δC^{+}_{pq} with $(p,q) \in B$, the optimal dual solutions are given by

$$(11) \qquad u^{+}_{i} = \begin{cases} u_i + \delta & \text{for } i \in I_p, \\ u_i & \text{for } i \in I_q \end{cases} \qquad v^{+}_{j} = \begin{cases} v_j - \delta & \text{for } j \in J_p \\ v_j & \text{for } j \in J_q \end{cases} .$$

The maximum extent μ^{+} to which this basis preserving operator can be applied is given by

$$(12) \qquad \mu^{+} = \underset{(i,j) \in I_p \times J_q - \{(p,q)\}}{\text{Minimum}} (c_{ij} - u_i - v_j)$$

The <u>entering cell</u> (e,f) is a cell at which the minimum in (12) is taken on.

The next theorem will frequently be used in the next two sections for finding bounds for a branch and bound process.

THEOREM 4. Suppose $(p,q) \in B$, $x_{pq} = 1$ and c_{pq} is to be increased to +M so that it will leave the basis; let Z^M be the optimal value of the objective function for the new problem; then $Z^M \geq Z + \mu^+$ where μ^+ is given in (12).

PROOF. By Theorem 1, $Z^+ = Z + \delta x_{pq}$ for all δ in the range $0 \leq \delta \leq \mu^+$. When $\delta = \mu^+$ the entering cell (e,f) enters the basis B and some other cell leaves. If (p,q) is one of the minimum giver cells in the cycle it can be made to leave the basis and its cost increased to M (Theorem 2) so that $Z^M = Z^+ = Z + \mu^+$. Otherwise some other cell in the cycle determined by (e,f) and B leaves the basis and the above construction can can be repeated. In the latter case, after a finite number of applications of the above reasoning we end with $Z^M \geq Z + \mu^+$.

As a result of this theorem the value $(Z + \mu^+)$ is a lower bound on Z^M. The important thing about these lower bounds is that they can be computed from a tableau <u>without chaninging it</u> in any way. And we shall see in the subsequent examples this lower bound is often enough to eliminate a whole branch of a branch and bound search tree.

In fact it turns out that in both the travelling salesman and DCPM examples to be discussed next, only one optimum assignment problem tableau is needed to calculate all the bounds necessary to construct the complete branch and bound tree. Then, in each case, the further application of one or two cost operators suffices to produce the optimum solution.

Comparisons of methods on small examples may, of course, be misleading. The authors are presently engaged in further computational tests for much larger problems.

3. APPLICATION OF COST OPERATORS IN SOLVING TRAVELLING SALESMAN PROBLEMS

The Travelling Salesman Problem [2] corresponds to the situation in which a salesman must choose shortest <u>tour</u> that passes through each of n cities exactly once and returns to his point of origin. The inter-city distances c_{ij} from city i to city j are given. The problem of minimizing the total time of performing n jobs on a single machine with sequence-dependent setup times c_{ij} (the time to set up job j in any sequence in which it is immediately preceded by job i) can also be formulated as a travelling salesman problem [4]. This problem, although easy to state, turns out to be extremely difficult to solve. This is surprising because it is almost indistinguishable from the simple assignment problem (1)-(4) constructed by defining the cost c_{ij} to be the inter-city distance (or set-up time) when $i \neq j$ and setting $c_{ij} = M$ whenever $i = j$ (where M is a large positive number).

However a solution to the assignment problem may have subtours. For instance, consider the six-city travelling salesman example shown in Fig. 1 due to Little et. al [12]. (Note that the problem is asymmetric in the sense

CITIES	1	2	3	4	5	6		Duals u_i
1	M	27	43	(16)^1	30	26	1	0
2	(7)^1	M	16	(1)^0	30	25	1	-15
3	20	13	M	35	(5)^1	(0)^0	1	-15
4	21	(16)^1	25	M	(18)^0	18	1	-2
5	(12)^0	46	27	48	M	(5)^1	1	-10
6	23	[5]	(5)^1	9	(5)^0	M	1	-15
	1	1	1	1	1	1		
Duals v_j	22	18	20	16	20	15		

Subtours {(1,4), (4,2), (2,7)}, {(3,5), (5,6), (6,3)}

Optimal Solution to the Assignment Problems
Associated with the Travelling Salesman Problem
Figure 1

that c_{ij} is not necessarily equal to c_{ji} for $i \neq j$ which is usually the case in scheduling contexts.) The circled cells (i,j) denote an optimal basis to this problem with the optimal amounts x_{ij} displayed above the circles for basic cells and set equal to zero for nonbasic cells. The optimum solution has cost = 54 but involves two subtours {(1,4), (4,2), (2,1)} and {(3,5), (5,6), (6,3)} and hence is not a feasible tour for the travelling salesman problem.

After surveying a large number of algorithms, Bellmore and Nemhauser [2] concluded that the subtour elimination algorithm developed by Eastman [6] and extended by Shapiro [13] is probably the best one for solving large travelling salesman problems. Subsequent to their survey, Held and Karp's one-tree approach [10,11] has proved to be remarkably efficient for symmetric travelling salesman problems though the corresponding results for the asymmetric problems are yet to be published. We discuss below a procedure for accelerating the Eastman-Shapiro subtour elimination algorithm which is applicable to both symmetric and nonsymmetric problems. It can also be extended to the more recent subtour elimination algorithm due to Bellmore and Malone [1].

The Eastman-Shapiro algorithm is basically a branch-and-bound technique in which one divides the set of all tours into subsets (exhaustive but not necessarily disjoint) and calculates lower bounds on the cost of all tours in a subset. The initial bound is found by solving the associated assignment problem. If the optimal solution is infeasible because it contains subtours, then one branches into m subproblems where m is the number of arcs in <u>one</u> of the subtours. To continue with our example, we may take, for instance, the subtour {(1,4), (4,2), (2,1)} and branch into three subproblems. For subproblem 1, let $c_{14} = M$. The first subset then is the set of all tours in which (1,4) is prohibited. Similarly for subproblems 2 and 3 let $c_{42} = M$ and $c_{21} = M$ respectively. Note that this way of branching includes all tours at least once, but may contain some tours more than once.

The m new assignment problems are solved and determine the lower bounds for all tours in their respective subsets. If any of these m solutions are tours and if the cost of one of these tours is less than or equal to the lower bounds on the other subsets then that tour is optimal. If not, then one takes the subset with the lowest bound and branches again according to the subtours present in that solution. Eventually we are assured of finding an optimum tour.

The cell cost operators discussed in Section 2 are useful in two ways for speeding up this algorithm. First of all the new assignment problems can be easily solved by applying operators to older ones. For instance, in the example provided earlier, to obtain the solution to subproblem 1, one has to simply apply the cost operator δC^{+}_{14} to the original assignment problem in order to drive the cost c_{14} to M.

Second, as explained in Theorem 4 of Section 2, the new optimal cost Z^M after applying the operator δC^{+}_{pq} is such that

$$Z^M \geq Z + \mu^{+}_{1}$$

where μ^{+}_{1} is the maximum extent of δC^{+}_{pq} the first time it can be applied. Thus $(Z + \mu^{+}_{1})$ provides a lower bound on the cost Z^M. Consequently, while Z^M provides a <u>lower bound</u> (LB) for the cost of all tours in the subset in which (p,q) is prohibited, $(Z + \mu^{+}_{1})$ provides a <u>weaker lower bound</u> (WLB) for the same. The important thing to note is that this weaker lower bound can be calculated <u>without</u> actually applying the operator -- i.e., the current tableau is not affected.

Thus our modification to the Eastman-Shapiro solution procedure can be outlined as follows: We first solve, as before, the assignment problem associated with the travelling salesman problem. If the optimal solution is not feasible because it contains subtours, then we evaluate the WLB's corresponding to the prohibition of each arc in some m-city subtour. We choose the branch with the smallest WLB, (say, arc (p,q) is blocked in this branch) and apply the operator δC^{+}_{pq}. If the solution is a tour and if its cost is less than or equal to the WLB's or (LB's) of the other subsets, then that tour is optimal. Otherwise, we choose that subset with the smallest WLB (or LB) and branch again. During the procedure we maintain as the current best solution a tour with the lowest cost among the tours that have already been explored.

We illustrate the above procedure with the example in Fig. 1. As shown in Fig. 2 we branch into the three subproblems #2, 3, 4 with prohibited routes

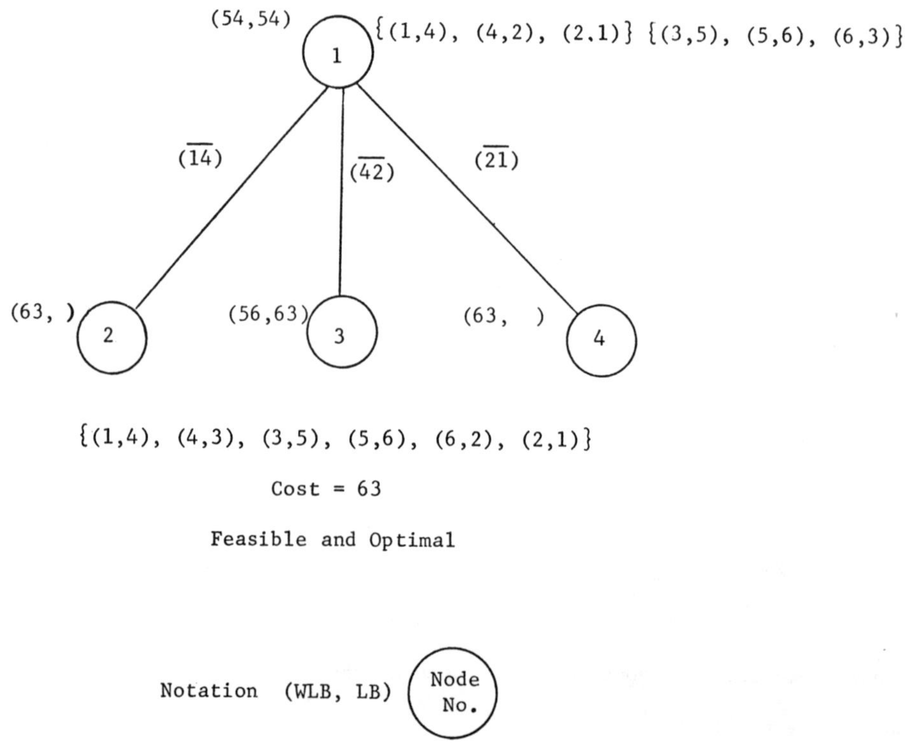

Branch-and-Bound Tree for Travelling Salesman Problem

Figure 2

(1,4), (4,2) and (2,1) respectively. For subproblem #2 the relevant cost operator is δC_{14}^{+}. Taking the cell (p,q) = (1,4) in Fig. 1 and applying the scanning routine of Remark 4 the set $I_p = \{1\}$ and $J_q = \{1,2,3,4,5,6\}$ so that

$$\mu_1^+ = \underset{(i,j)\epsilon[I_p xJ_q] - \{(1,4)\}}{\text{Min}} \{(c_{ij} - u_i - v_j)\}$$

$$= \{(M-0-22), (27-0-18), (43-0-20), (30-0-20), (26-0-15)\}$$

$$= 9$$

and the minimum is attained at cell (r,s) = (1,2). Consequently the WLB for subproblem 1 = 54 + 9 = 63 as shown in Fig. 2.

For the subproblem 3, (p,q) = (4,2) so that by Remark 4 $I_p = \{1,2,3,4,5,6\}$ and $J_q = \{2\}$ so that

$$\mu_1^+ = \underset{(i,j)\epsilon[I_p xJ_q]-\{(4,2)\}}{\text{Min}} \{(c_{ij} - u_i - v_j)\}$$

$$= \{(27-0-18), (M+15-18), (13+15-18), (46+10-18), (5+15-18)\}$$

$$= 2$$

and the minimum occurs at (r,s) = (6,2). For this subproblem WLB = 54 + 2 = 56.

For subproblem 4, (p,q) = (2,1) so that using Remark 4 the set $I_p = \{1,2\}$ and $J_q = \{1,2,3,5,6\}$ so that μ_1^+ can be verified to be equal to 9 with the minimum attained at (r,s) = (1,2). Consequently the WLB for subproblem 3 is also 54 + 9 = 63 as shown in Fig. 2. It should be noted that all the three WLB's were calculated using the original optimal tableau of Fig. 1 without changing it.

Of the nodes corresponding to subproblems 2, 3, and 4, #3 has the smallest WLB so that we proceed to apply the cost operator δC_{42}^{+} to the tableau of Fig. 1. From our previous calculations, it will be recalled that $\mu_1^+ = 2$ and (r,s) = (6,2). When (6,2) is added to the basis of Fig. 1 the cycle {(4,2), (4,5), (6,5), (6,2)} is obtained. Marking the alternate cells in this cycle as 'getter' and 'giver' starting with (6,2) as a 'getter' we see that the smallest giver cell is (6,5) with $x_{65} = 0$. Consequently, the new basis is as shown in Fig. 3 with amounts x_{ij} unchanged. Note that c_{42} has

CITIES	1	2	3	4	5	6		Duals u_i
1	M	27	43	(16)^{1}	30	26	1	2
2	(7)^{1}	M	16	(1)^{0}	30	25	1	-13
3	20	13	M	35	(5)^{1}	(0)^{0}	1	-13
4	21	(18)^{1}	[25]	M	(18)^{0}	18	1	0
5	(12)^{0}	46	27	48	M	(5)^{1}	1	-8
6	23	(5)^{0}	(5)^{1}	9	5	M	1	-13
	1	1	1	1	1	1		
Duals v_j	20	18	18	14	18	13		

Optimal Solution After Applying $2C^{+}_{42}$ to Fig. 1

Figure 3

been increased from 16 to18 and the duals u_i and v_j have been changed as per equation (11), i.e., $u_i' = u_i + 2$ for $i\varepsilon I_p = \{1,2,3,4,5,6\}$ and $v_j' = v_j - 2$ for $j\varepsilon J_p = \{1,3,4,5,6\}$ with $v_2' = v_2$.

To apply the cost operator δC^{+}_{42} further we find from Fig. 3 that corresponding to (p,q) = (4,2) the set $I_p = \{1,2,3,4,5\}$ and $J_q = \{2,3\}$ so that

$$\mu_2^{+} = \min_{(i,j)\varepsilon[I_p x J_q]-\{(4,2)\}} \{(c_{ij} - u_i - v_j)\}$$

$$= \{(27-2-18), (M+13-18), (13+13-18), (46+8-18),$$
$$(43-2-18), (16+13-18), (M+13-18), (25-0-18), (27+8-18)\}$$

$$= 7$$

and the minimum is attained at both (1,2) and (4,3). If we choose $(r,s) = (4,3)$ the cycle $\{(4,2), (4,3), (6,3), (6,2)\}$ is obtained as shown in Fig. 3. The minimum giver is (4,2) with $x_{42} = 1$ (whenever there is a tie for the minimum giver with (p,q) as one of them, it is always chosen to leave the basis). After adding 1 to the getters and subtracting 1 from the givers the new basis of Fig. 4 is obtained. Since (4,2) leaves the basis, it becomes nonbasic

CITIES	1	2	3	4	5	6		Duals u_i
1	M	27	43	(16)^{1}	30	26	1	9
2	(7)^{1}	M	16	(1)^{0}	30	25	1	-6
3	20	13	M	35	(5)^{1}	(0)^{0}	1	-6
4	21	M	(25)^{1}	M	(18)^{0}	18	1	7
5	(12)^{0}	46	27	48	M	(5)^{1}	1	-1
6	23	(5)^{1}	(5)^{0}	9	5	M	1	-13
	1	1	1	1	1	1		
Duals v_j	13	18	18	7	11	6		

Optimal Tour $\{(1,4), (4,3), (3,5), (5,6), (6,2), (2,1)\}$

Optimal Solution to Subproblem 3

Figure 4

and its cost can be increased to M directly using Theorem 2. The cost of this solution is $54 + 2 + 7 = 63$ and yields a tour $\{(1,4), (4,3), (3,5), (5,6), (6,2), (2,1)\}$. (If the cell (1,2) had been chosen as (r,s) then the solution obtained is not a tour and one more branching would be necessary.) Since the cost of this tour is no greater than the WLB corresponding to subproblems #2 and #4, we have an optimum for the travelling salesman problem with cost = 63. In other words further searching on branches #2 and #4 is unnecessary. It is interesting to note that our solution procedure for this example involved the application of only two basis-preserving operators and the branch-and-bound tree of Fig. 2 is considerably smaller than that of the Little et. al. solution to this problem [12].

Though the above procedure has not yet been coded for a computer, the results have been very encouraging on other sample problems solved using this method. In general the WLB's work out to be quite close to the LB's and the computational burden is considerably smaller than a straightforward application of the Eastman-Shapiro procedure. As discussed earlier the computational efficiency is due to (a) the relative ease in applying a cost operator rather than having to resolve the modified assignment problem and (b) the ability to calculate the WLB's without affecting the current tableau.

4. SOLUTION OF DCPM PROBLEMS

The Decision Critical Path Method (DCPM) was introduced by Crowston and Thompson [5]. It is an extension of the ordinary Critical Path Method (CPM) to permit the inclusion in the project graph of decision job sets with the requirement that some but not all the jobs in the decision set be performed. Thus part of the DCPM problem is to decide which jobs should be done. In this way the optimal interactions among project design, project scheduling, and project control can be simultaneously be obtained.

To define a DCPM problem we let $J = \{S_1, S_2, \ldots,\}$ be a set of job sets that must be done to complete a project. Some job sets are unit sets $S_i = \{S_{i1}\}$ and others called <u>decision job sets</u> have several members $S_i = \{S_{i1}, S_{i2}, \ldots, S_{i,k(i)}\}$. If S_i is a unit set it has a completion time t_i. With each job S_{ij} in a decision job set we associate a completion time t_{ij}

and a cost c_{ij}. We shall consider here only the mutually exclusive case in which exactly one job in the decision job set is to be performed. (More general cases are considered in [5]. The same techniques as we use here can be used to solve these more general cases but we shall not elaborate on them here.) We define a variable d_i which takes on one of the values in $\{1,2,\ldots,k(i)\}$ depending on which job in the set S_i is to be performed.

It is easy to draw a project graph for a DCPM problem by representing decision sets S_i as jobs coming out of a triangle with each job following

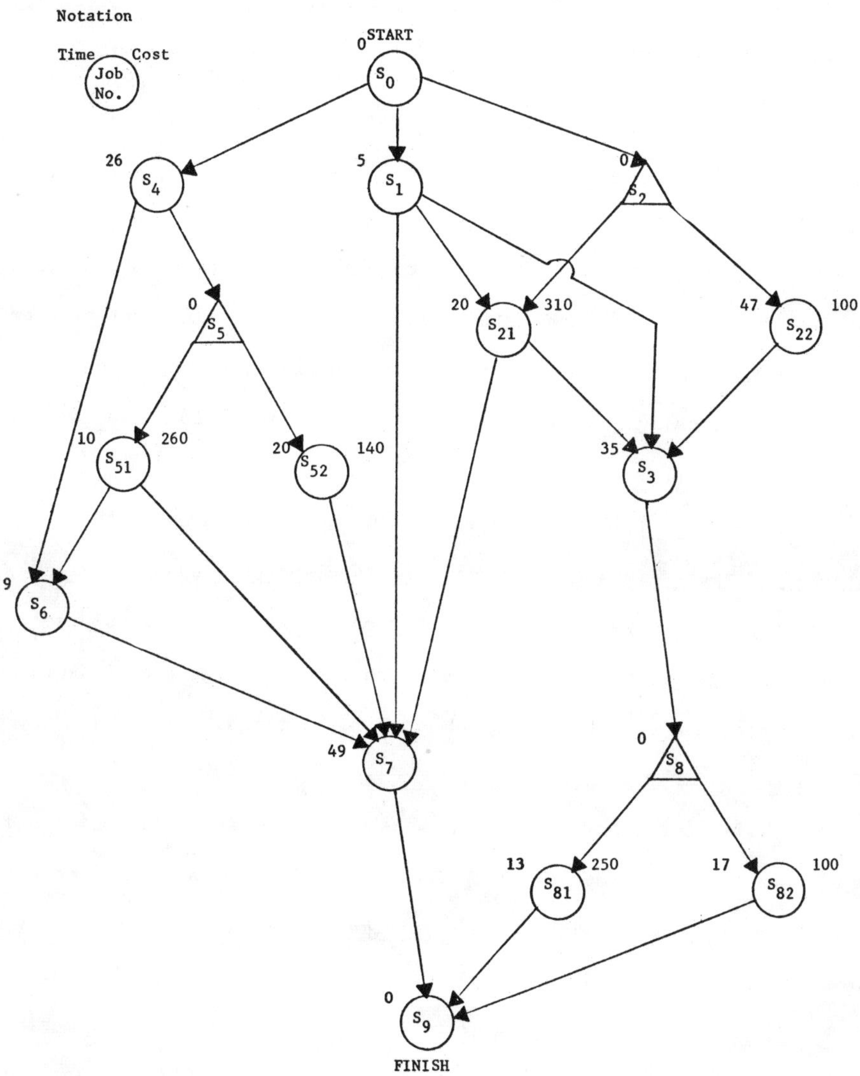

Figure 5

such a triangle being one of the jobs S_{ij}. Figure 5 shows an example which is a slightly more complicated version of the example shown in Figure 1 of [5] with three decisions sets S_2, S_5, and S_8 each having two jobs in it, and 5 unit sets S_1, S_3, S_4, S_6, and S_7 together with (dummy) jobs Start, S_0 and Finish, S_9. Job times appear above and to the left of the circle representing the job and job costs (for decision jobs) appear above and to the right.

In [5] the objective function was formulated in terms of a due date T, a reward function for project completion before T and a penalty function for completion after T. Here we consider the simpler problem of completing the project on or before the due date T at lowest possible cost.

To solve this problem by means of cost operators applied to assignment problem we first formulate the network problem as an assignment problem by using the method of Charnes and Cooper given in [3].

Figure 6 shows the assignment problem corresponding to the network in Figure 5. Job times are shown as negative because the assignment problem (1)-(4) is a minimizing problem and we want to find the longest path. All costs not specifically indicated are assumed to be equal to M, a large positive number.

In order to simplify the discussion we now make an assumption. <u>If any decision job appears in a longest path it is also the cheapest available job in the decision set</u>. We need this assumption so that we can simply solve the assignment problem without first blocking shorter and higher cost decisions. This assumption can easily be removed with only a slight increase in computational difficulty.

In Figure 6 the basis cells of an optimum solution are circled and a 1 or a 0 appears above each circle. The basis jobs that are on or above the main diagonal and marked with 1 are <u>critical</u>. Thus the optimum solution in Figure 6 has a critical (longest) path consisting of jobs S_0, S_{22}, S_3, S_{82} and S_9. (S_0 and S_9 are always critical.) The sum of the job times is 99. Since decision job S_5 does not appear on the critical path the best decision

Jobs		1	21	22	3	4	51	52	6	7	81	82	9		u_i ↓
	i \ j	1	2	3	4	5	6	7	8	9	10	11	12		
0	1	-5	-20	(-47) 1		(-26) 0								1	-95
1	2	(0) 1	(-20) 0		-35					-49				1	-69
21	3		(0) 1		-35					(-49) 0				1	-49
22	4			(0) 0	(-35) 1									1	-48
3	5				(0) 0						-13	(-17) 1		1	-13
4	6					(0) 1	-10	(-20) 0	-9					1	-69
51	7						(0) 1		(-9) 0	-49				1	-58
52	8							(0) 1		(-49) 0				1	-49
6	9								(0) 1	(-49) 0				1	-49
7	10									(0) 1			(0) 0	1	0
81	11										(0) 1		(0) 0	1	0
82	12											[0]	(0) 1	1	0
		1	1	1	1	1	1	1	1	1	1	1	1		
v_j →		69	49	48	13	69	58	49	49	0	0	-4	0		

Completion Time is 99. Critical Path is S_0, S_{22}, S_3, S_{82}, S_9. Costs not marked are = M.

Figure 6

is to choose the cheaper of the two alternatives which is S_{52}. The cost of the three decisions is 100 + 140 + 100 = 340. The optimal dual solution variables u_i and v_j are shown to the right and below the tableau.

Suppose now that $T = 96$ so that the solution shown in Figure 6 does not meet the due date and it will be necessary to make at least one more expensive decision. The only way we can shorten the critical path is to change one of the decisions S_2 or S_8 on the critical path. The difference in cost between S_{81} and S_{82} is 150 while the difference in cost between S_{21} and S_{22} is 210. We choose the smaller of these and want to change the decision at S_8 from S_{82} to S_{81}.

For this purpose we employ a cost operator to drive the time of S_{82} which is 17 to $-M$ so that job S_{82} can never be critical. In the assignment tableau of Figure 6 this means that we want to drive the cost -17 in cell (5,11) to M. Thus we want to apply the cost operator $\delta C^+_{5,11}$ to its fullest extent, and hence we must find the μ^+ of (12). To do this we need the sets I_p and J_q. Applying the scanning routine given in Remark 4 we easily find $I_p = I$ and $J_q = \{11\}$. Since all the costs in column 11 are M except for $c_{5,11} = -17$ and $c_{12,11} = 0$ we have to evaluate $c_{ij} - u_i - v_j$ only when $i = 12$ and $j = 11$ (since $(p,q) = (5,11)$). Hence in (12) we find $^+ = 0 - (-4) - 0 = 4$ and the incoming cell $(e,f) = (12,11)$. At this point we know there is a feasible solution with completion time at most 95 which is less than $T = 96$, which makes use of decision S_{81} and hence costs 340 + (250 - 100) = 490. We don't yet know what the critical path of this new solution is but we can find out by changing the cost $c_{5,11} = -17$ to $-17 + \delta$ and increasing δ to the full extent of 4. When $\delta = 4$ we will have the new solution value $z^+ = -99 + 4 = -95$ and the incoming cell $(e,f) = (12,11)$ can now enter the basis.

To find out which cell will leave the basis when (12,11) comes in we construct the cycle it determines in the basis as shown in Figure 6. Labelling (12,11) as a getter cell and going around the cycle labelling the cells alternately giver and getter cells we find the giver cells are (12,12), (10,9),

(8,7), (6,5), (1,3), (4,4), and (5,11). They can all give 1, so we alter shipments around the route. Now all givers have 0 and all getters have 1, and we can remove any giver cell from the basis. We choose to remove (5,11), since that was the cell we wanted to take out of the basis. It leaves at the cost of $c_{5,11} = -13$ and now, applying Theorem 2 we can increase its cost to M without altering the value of Z further. The resulting new solution is shown in Figure 7.

In Figure 7 notice that now $c_{5,11} = M$ and all dual variables are changed according to formula (11). The new critical path is now S_0, S_4, S_{52}, S_7, S_9 which is entirely different from the previous one (except, of course, for S_0 and S_9).

Let us now show how the above calculations can be incorporated into a branch and bound algorithm for finding the optimal DCPM solution. Figure 8 shows the branch and bound tree for the example. Inside the circle at each node of the tree the values of the decision variables are indicated; above and to the right the cost of these decisions is shown; and above and to the left are shown a pair of numbers, the first number being a weak upper bound (WUB), and the second number an actual upperbound (UB) for the project completion time given these decisions. The reason that we have upper bounds here instead of lower bounds as before is that we are trying to find the <u>longest</u> path instead of the <u>shortest</u> tour. Recall that we accomplished this by changing the signs of the costs.

Thus the uppermost node corresponds to the solution shown in Figure 6. The next two nodes below represent the branching on decision d_8 with $d_8 = 1$ to the left and $d_8 = 2$ to the right. The stars on these decisions mean that they are fixed for all lower nodes on the tree.

As outlined above, we can evaluate the left hand node with $*d_8 = 1$ to have a weak upper bound on its completion time of 95, which is less than the required due date of 96, and a cost of 490. This alternative is therefore feasible

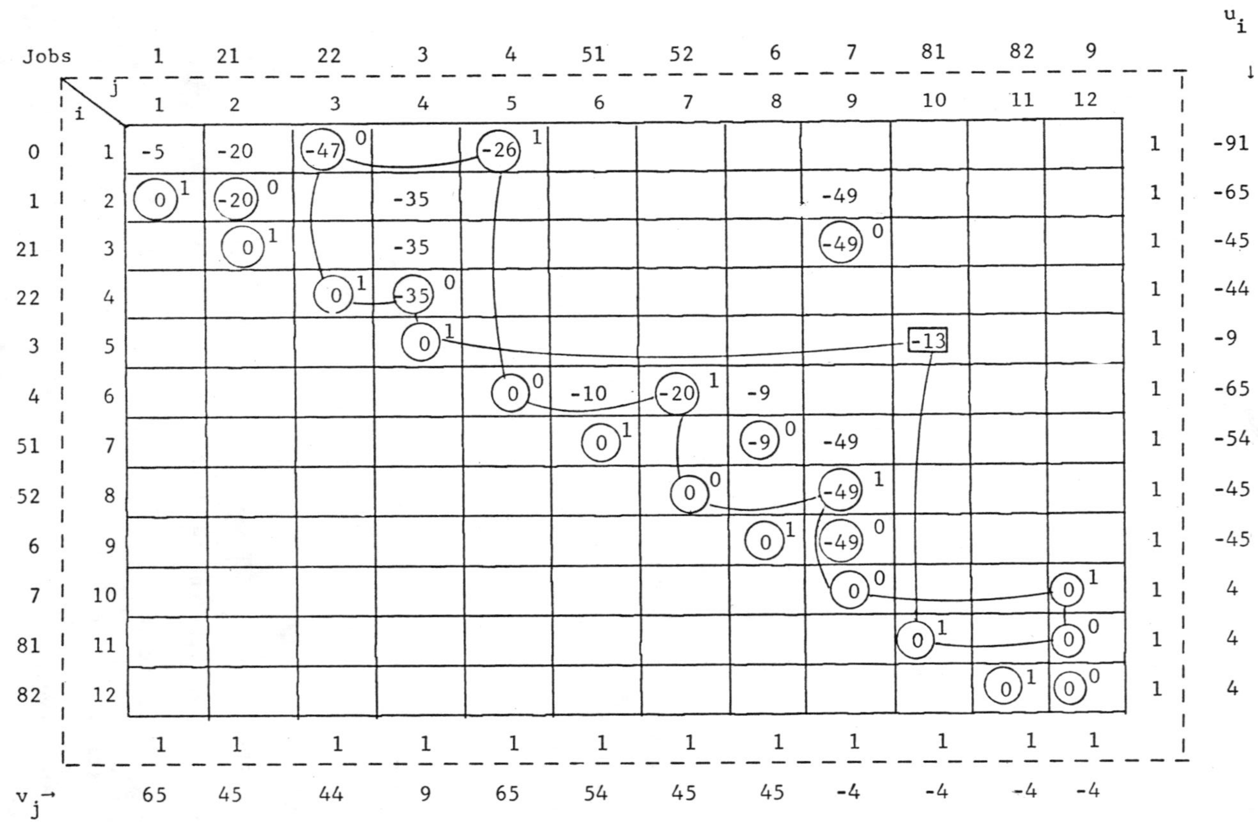

Completion time is 95. Critical path is S_0, S_4, S_{52}, S_7, S_9.

Figure 7

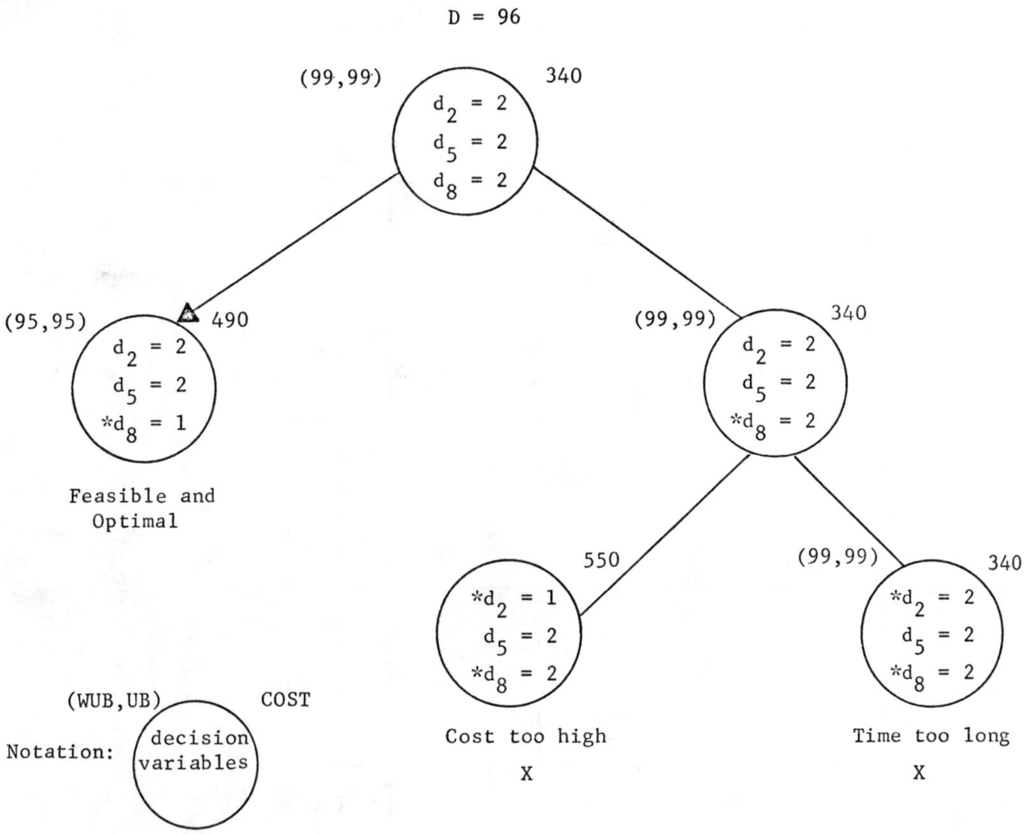

* Indicates decisions fixed at all lower nodes of this branch.

Branch and Bound Tree for

the DCPM Problem

Figure 8

Jobs		1	21	22	3	4	51	52	6	7	81	82	9		$u_i \downarrow$
	i \ j	1	2	3	4	5	6	7	8	9	10	11	12		
0	1	-5	-20	(-47) 1		(-26) 0								1	-91
1	2	(0) 1	(-20) 0		-35					-49				1	-65
21	3		(0) 1		-35					(-49) 0				1	-45
22	4			(0) 0	(-35) 1									1	-44
3	5				(0) 0						(-13) 1			1	-9
4	6					(0) 1	-10	(-20) 0	-9					1	-65
51	7						(0) 1		(-9) 0	-49				1	-54
52	8							(0) 1		(-49) 0				1	-45
6	9								(0) 1	(-49) 0				1	-45
7	10									(0) 1			0	1	4
81	11										(0) 0		(0) 1	1	4
82	12											(0) 1	(0) 0	1	4
		1	1	1	1	1	1	1	1	1	1	1	1		
$v_j \rightarrow$		65	45	44	9	65	54	45	45	-4	-4	-4	-4		

Completion time is 95. Alternate Critical Path is S_0, S_{22}, S_3, S_{81}, S_9.

Figure 9

(and we show it optimal later). The decision with $*d_8 = 2$, the right hand node is the same as the top node except that $d_8 = 2$ is fixed. So it has WUB = UB = 99 and cost 340.

Below it are two more nodes, corresponding to choices $*d_2 = 1$ to the left and $*d_2 = 2$ to the right. The right most node again has WUB = UB = 99, cost = 340 and it cannot be further shortened since the only remaining decision node, S_5, is not on the critical path. Hence this branch of the tree is terminated. We can also terminate the left hand node which has $*d_2 = 1$ without even evaluating it since its cost is 550 which is greater than the already achieved cost of 490.

Thus, using the optimal tableau in Figure 6 alone we were able to complete the branch and bound tree of Figure 8. Then applying cost operators to Figure 6 we constructed the optimal solution in Figure 7.

It is interesting to note in Figure 7 that $c_{5,10} - u_5 - v_5 = -13 + 9 + 4 = 0$ so that there is an alternate critical path that can be obtained by bringing cell (5,10) into the basis. The cycle it creates is shown in Figure 7. Again all givers can give 1 around thic cycle and any one of them can be removed. The alternate optimal solution after cell (10,12) is removed is shown in Figure 9. Its critical path is S_0, S_{22}, S_3, S_{81}, S_9. Thus Figure 9 gives an alternate critical path.

5. CONCLUSIONS

In this paper we have demonstrated the usefulness of the operator theory of parametric programming for assignment problems by using it in the branch and bound solution of travelling salesman and decision CPM problems. We intend to code these methods and present computational experience with them in the near future.

REFERENCES

[1] Bellmore, M., and J. C. Malone, "Pathology of Traveling Salesman Subtour-Elimination Algorithms," Operations Research, 19, (1971), pp. 278-307.

[2] ________ and G. Nemhauser, "The Traveling Salesman Problem: A Survey," Operations Research, 16 (1968), pp. 538-558.

[3] Charnes, A., and W. W. Cooper, "Some Network Characterizations for Mathematical Programming and Accounting Approaches to Planning and Control," The Accounting Review, 42 (1967), pp. 24-52.

[4] Conway, R. W., W. L. Maxwell, and L. W. Miller, Theory of Scheduling, Addison-Wesley Publishing Company, Inc., Reading, Massachusetts, 1967, pp. 53-67.

[5] Crowston, W., and G. L. Thompson, "Decision CPM: A Method for Simultaneous Planning, Scheduling, and Control of Projects," Operations Research, 15 (1967), pp. 407-426.

[6] Eastman, W. L., "Linear Programming with Pattern Constraints," unpublished Ph.D. dissertation, Harvard University, 1958; also in augmented form: Report No. BL-20, The Computation Laboratory, Harvard University, 1958.

[7] Ford, L. R. and D. R. Fulkerson, Flows in Networks, Princeton University Press, Princeton, N. J. (1963).

[8] Glover, F., D. Karney, D. Klingman, and A. Napier, "A Comparison of Computation Times for Various Starting Procedures, Basis Change Criteria, and Solution Algorithms for Distribution Problems," Research Report CS #44, Graduate School of Business, University of Texas at Austin, (May, 1971).

[9] Hammer, P. L., "Time Minimizing Transportation Problems," Naval Research Logistics Quarterly, 16 (1969), pp. 345-357.

[10] Held, M., and R. M. Karp, "The Traveling-Salesman Problem and Minimum Spanning Trees," Operations Research, 18 (1970), pp. 1138-1162.

[11] ________ and ________, "The Traveling-Salesman Problem and Minimum Spanning Trees: Part II," Mathematical Programming, 1 (1971), pp. 6-25.

[12] Little, J. D. C., K. G. Murty, D. W. Sweeney, and C. Karel, "An Algorithm for the Traveling Salesman Problem," Operations Research, 11, (1963), pp. 972-989.

[13] Shapiro, D., "Algorithms for the Solution of the Optimal Cost Traveling Salesman Problem," Unpublished Sc.D. Thesis, Washington University, St. Louis, 1966.

[14] Srinivasan, V., "A Transshipment Model for Cash Management Decisions," Management Science, forthcoming.

[15] ________ and G. L. Thompson, "An Operator Theory of Parametric Programming for the Transportation Problem, I," Naval Research Logistics Quarterly, forthcoming.

[16] ____________ and ____________, "An Operator Theory of Parametric Programming for the Transportation Problem, II," Naval Research Logistics Quarterly, forthcoming.

[17] ____________ and ____________, "Determining Optimal Growth Paths in Logistics Operations," Naval Logistics Research Quarterly, forthcoming.

[18] ____________ and ____________, "An Algorithm for Assigning Uses to Sources in a Special Class of Transportation Problems," Management Sciences Research Report No. 263, Graduate School of Industrial Administration, Carnegie-Mellon University, November, 1971.

[19] ____________ and ____________, "Benefit-Cost Analysis of Coding Techniques for the Primal Transportation Algorithm," Journal of the Association for Computing Machinery, forthcoming.

[20] Szwarc, W., "The Time Transportation Problem," Zastosowania Matematyki, **8** (1966), pp. 231-242.

SOME EXTENSIONS OF AKERS-FRIEDMAN PRODUCTION SCHEDULING PROBLEM

Ismail B. Turksen and Ravindra Shankar

Department of Industrial Engineering,
University of Toronto, Toronto, Ontario, Canada

The paper introduces a general solution methodology for the deterministic N jobs and M machines scheduling-sequencing problems. In particular, the logical constraints of the problem are transformed into a set of Boolean equations. The reduced set of dominant programs are determined by the solution of these Boolean equations. The optimal program is selected from among these dominant programs.

THE DETERMINISTIC job shop scheduling sequencing problems require a selection of optimal programs which integrate jobs, machines, operations, precedences, and processing times. Akers and Friedman[1] suggested a non-numerical approach for the case of 2 jobs and M machines. The method is limited because the truth table operations are required within the solution technique. Later, Akers[2] introduced a graphical solution method. This graphical solution method was extended by Szwarc[3]. It was also suggested that NxM problem can be decomposed into 2xM problems.

In this report, we state certain improvements on the original approach of Akers and Friedman. These improvements can be realized because we now have a set of computable algorithms for the solution of Boolean equations [4,5,6].

With these algorithms, we have carried out a set of experiments in order to study the decomposition scheme suggested but never investigated in the literature. Our results are encouraging but not conclusive. In the following exposition, we will state the problem, its decompositions, recomposition, and the results of our experiments.

PROBLEM

A job shop process contains N jobs, M machines, and N•M operations. It is assumed here that the precedence relationships among the operations of each job is specified.

Let us define the set of jobs as J, $J=\{J_n | n=1,2,\ldots N\}$, the set of machines as K, $K=\{K_m | m=1,2,\ldots, M\}$, the set of operations as 0, $0=\{0_{nm}\}$ and the set of precedences as R, $R=\{R_n\}$ where the precedence requirements for J_n is defined by R_n, $R_n=(0_{ni} < 0_{nj} < \cdots < 0_{nk})$; or alternatively $R_n=(J_n; K_i < K_j < \cdots < K_k)$. It is also assumed that a real number t_{nm} is known as the processing time of the operation 0_{nm}.

DECOMPOSITION

There is a large number of programs even for small values of N and M, i.e., N=4, M=5, $(N!)^M=(4!)^5= 7,962,624$. However, one can consider the decomposition of these programs in the following manner. Define a pair of jobs to be a subproblem of the original problem.

Let P be the number of subproblems, then $P=\binom{N}{2}$. Assuming that a reduced set of dominant programs can be found for each subproblem, then a recomposition of the subproblems could produce a reduced set of dominant programs for the original problem. If this recomposition determines a rather small set of dominant programs, then an optimal program can be found. There are many questions around this hypothesis. We attempt to answer only a few of these here.

Let Z be the number of programs in the original problem space and S be the number of programs in the decomposed space, then $Z \leq S$ because

$$(N!)^M \leq \left[2^M\right] P.$$

At first, this decomposition of NxM problem in to P subproblems of 2xM type appears undesirable for the enlargement of the computations from Z programs to S programs. However, at present, it is rather difficult to study the Z programs directly, but each of the P programs can be analyzed in terms of Boolean equations as shown below.

Consider a subproblem (J_i, J_j). Define $(K_m; J_i < J_j)$ to mean that the job J_i must precede the job J_j on machine K_m. Associate a Boolean variable y_m as follows: $y_m \in \{0,1\}$, and

$$\left(K_m; J_i < J_j\right) \longrightarrow y_m$$
$$\left(K_m; J_j < J_i\right) \longrightarrow \bar{y}_m$$

Akers and Friedman [1] have shown that optimal programs belong to a set of programs that are technologically feasible and contain no free machines. It can be shown that the technological feasibility conditions and no free machine conditions can be stated as a Boolean equation $w(y)=0$, where $y=(y_1, y_2, \ldots, y_m)$. These Boolean equations can be stated as

$$w(y) = \bigcup_{\eta} \bigcap_{\rho} w_{\eta\rho}(y) = 0$$

in general and they form the constraints of the scheduling problem.

Recently it was shown that there are computable algorithms to find the zero vertices Y^0 of the unit hypercube as $Y^0=\{y|w(y)=0\}$[6]. In particular, the one vertices Y^1 of the unit hypercube can be found as $Y^1=\{y|w(y)=1\}$; and $Y^0=\overline{Y}^1$ can be determined by the complementation operation. We suggest that, in this manner, the dominant programs of each subproblem can be determined.

RECOMPOSITION

Suppose the number of dominant programs in each of the subproblems can be found as stated above. The suggested recomposition scheme requires that a cartesian product of these dominant programs be taken to arrive at the dominant feasible programs of the original problem. When the cartesian product is taken, it can be shown that certain sequences could still be infeasible and these must be eliminated. It was shown that this can be carried out with a construction algorithm assuming a strict precedence relationship among the operations. Finally the resultant dominant feasible programs must be evaluated with a minimum completion time algorithm in order to determine an optimal sequence.

RESULTS

We have carried out twenty-three experiments on the job and machine combination from 2 jobs and 2 machines up to 6 jobs and 6 machines (see Table I). The precedence requirements of operations were generated randomly. The experiments show that for 3x6 problems the computer (CPU) time is less than one second, for 4x6, it is less than 3 seconds, and 6x5, it is around 1 min. It should be pointed out that the computer times are the sum total of reduction, construction, evaluation and interpretation times for scheduling problem.

TABLE I

No. of Jobs	No. of Machines	No. of Programs in the set of Dominant Programs	Execution Time in (CPU) Sec.
2	2	1	0.17
2	3	2	0.27
2	4	2	0.41
2	5	1	0.64
2	6	1	0.31
3	2	2	0.67
3	3	4	2.00
3	4	8	6.01
3	5	1	0.53
3	6	1	0.91
4	2	4	5.21
4	3	8	0.69
4	4	32	2.65
4	5	4	1.19
4	6	8	2.36
5	4	128	11.56
5	5	16	3.20
5	6	128	16.15
6	4	256	30.90
6	5	512	61.75
6	6	*	3.88 (1)

*No. of programs are more than 512

(1) Only one program evaluated.

CONCLUSIONS

The results are encouraging thus far. However, it is observed that the number of dominant programs are rather large at times. This creates the suspicion that for N,M large the number could grow exponentially. Hence making the evaluation rather impossible. This could be remedied if stronger dominancy conditions can be stated as additional Boolean equations. The future work is directed toward a statement of stronger dominancy condition. We suspect that a branch and bound concept together with a Boolean programming approach will lead to some appreciable results.

REFERENCES

1. Akers, Jr., S.E. and J. Friedman, "A Non-numerical Approach to Production Scheduling Problems", Operations Research Journal, Vol. III, 429-442, (1955).

2. Akers, Jr., S.E., "A Graphical Approach to Production Scheduling Problems", Operations Research Journal Vol. 4, No.2, 244-245, (1956).

3. Szwarc, W., "Solution of the Akers-Friedman Scheduling Problem", Operations Research Journal, Vol. 8, No.6, (1960).

4. Turksen, I.B., "Boolean Transformation in a Scheduling Model", M.S. Thesis, University of Pittsburgh, 1962.

5. Holzman, A.G., and I.B. Turksen, "Boolean Isomorphic Structures in an Educational Scheduling Model", Monograph No.1. The Automation of School Information System, Edited by Don D. Bushnell, System Development Corporation, Department of Audio Visual Instruction of the National Education Association of U.S.A., 1964, Chapter 2, pp. 43-48.

6. Turksen, I.B., "The Set of Solutions for the Boolean Equations", Discrete Mathematics (to appear).

RESEARCH ABSTRACT
ON A FEASIBILITY APPROACH TO OPTIMAL SCHEDULES

Victor A. Zaloom

Industrial Engineering Department, Auburn University,
Auburn, Alabama, USA

Optimization algorithms capable of efficiently solving general large scale resource scheduling problems do not exist. There are some specially structured scheduling problems for which optimization algorithms are available; for examples, see Johnson (3), Hu (2), and Chaudhuri (1). A slight extension of the problem discussed by Hu is presented here. First, a portion of Hu's presentation in reference (2) will be discussed. The problem of interest which was solved in (2) had as its objective to minimize schedule duration under the following assumptions:

(1) the project network diagram is a tree, (a tree is a graph containing no closed loops),

(2) each activity requires unit time and a single resource common to each activity,

(3) resources are available in integral numbers.

For this problem Hu presents an algorithm called <u>cutting-the-longest queue</u> which he proves to be an optimization procedure. The elegance of Hu's work is clearly demonstrated in the simplicity of his algorithm. To begin with each node is assigned a label, $\alpha_i = x_i + 1$ where x_i is the length of the path from the root of the tree. The following algorithm is then applied.

> <u>Algorithm</u>: Choose m jobs with labels not less than those not chosen, where m is the number of resources available. In case of a tie, arbitrarily choose any m jobs such that the above algorithm is satisfied. This rule is then repeated until all jobs have been chosen.

An example project is presented in Figure 1. Each node represents a job requiring unit time and unit resource to complete. The labels are placed above each node. If it is assumed that m equals 3, then on the first day any three jobs from the set J_8, J_9, J_{10}, J_{11} may be scheduled. The dotted circles represent an optimal schedule.

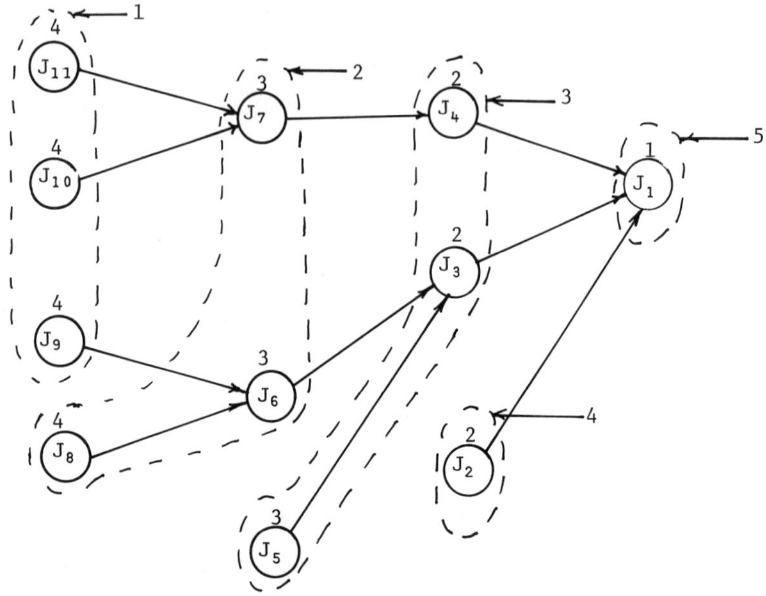

Figure 1. Example Problem - Tree/Diagram

The problem of interest here is again to minimize schedule duration. However, only two of the assumptions of Hu's model are retained. Assumption (1), that the network diagram is a tree is relaxed. General network diagrams are considered. The other assumptions are retained; i.e., each activity is assumed to require a unit of time from a single common resource and resources are available in integral numbers.

Any general activity-on-node network can be converted to a tree by removing all but one arrow leading out of each node. For example, in Figure 2 if (1-3) and (3-5) are removed as shown in Figure 3, the network is a tree. This tree can be

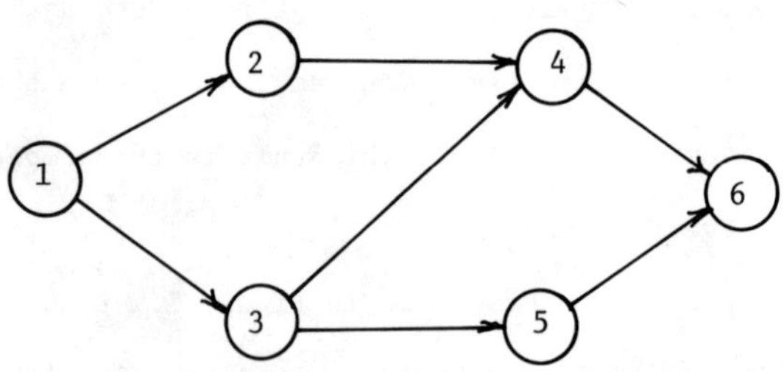

Figure 2. Network Diagram

optimally scheduled using the algorithm presented by Hu. Then the arcs (1-3) and (3-5) may be replaced. If these precedences are satisfied by the optimal tree schedule, then this schedule is optimal for the network.

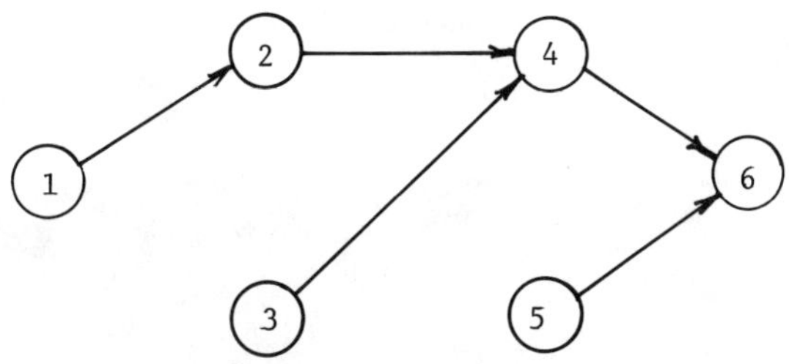

Figure 3. Tree Diagram with Arcs (1-3) and (3-5) Removed

The following theorem is presented in (4).

> Theorem: The minimum time required to complete a reduced network (one obtained from a network by deleting certain precedence arcs) is less than or equal to the minimum time required to complete the original network.

By the theorem, the length of the optimal schedule for the tree is a lower bound on the optimal schedule for the network. Table I gives an optimal schedule for the tree (Figure 3) for m equals 2.

TABLE I

Optimal Tree Schedule

Day	Jobs
1	1,3
2	2,4
3	5
4	6

If arc (3-5) is added, no schedule change is required for feasibility. When (1-3) is added, activity 3 must be delayed. If activity 4 is delayed to day 3 and activity 3 to day 2, then a feasible schedule for the network is obtained. By the theorem, this schedule which is presented in Table II is optimal because it is no longer than the optimal schedule for the tree.

TABLE II

Optimal Network Schedule

Days	Jobs
1	1
2	2,3
3	4,5
4	6

In summary, the feasibility approach is to: (1) convert a network to a tree, (2) find an optimal schedule for the tree, (3) add back the precedence arcs removed, and check to see if the tree schedule is feasible for the network.

If the tree schedule is feasible for the network, then by the theorem it is optimal. If it is not feasible, two problems result. (1) Moving jobs within the scheduled days to satisfy the precedence arcs which were removed is a combinational problem. (2) If there exists no feasible schedule for the original network as short as the optimal schedule for the reduced network, then optimality cannot be established for the best network schedule.

Despite these shortcomings the feasibility approach has the following advantages.

(1) Converting the network to a tree and finding the optimal tree schedule may lead to an optimal schedule for the network.

(2) The optimal schedule for the tree at worst provides a good lower bound on the optimal network schedule.

(3) The criticality of certain precedence restrictions is established.

These advantages combined with the computational savings from determining an optimal tree schedule as opposed to an optimal network schedule appear to make the approach promising.

What is required is an algorithm to optimally replace precedence arcs which had been removed. This is the problem which is currently being worked on. Any suggestions or insights will be well received.

REFERENCES

Chaudhuri, A. R., "On the Problem of Critical Path Method," OPSEARCH, p. 206-214, (1969).

Hu, T. C., "Parallel Sequencing and Assembly Line Balancing, " Operations Research, Vol. 9, p. 841-848, (1961).

Johnson, S. M., "Optimal Two-and-Three Stage Production Schedules with Setup Times Included, " Naval Research Logistics Quarterly, Vol. 1, No. 1, p. 61-68, (1954).

Zaloom, V. A., "An Examination of the Expected Minimum Time to Complete a Resource Constrained Project, " Unpublished Dissertation, University of Houston, p. 54, (1970).

Lecture Notes in Economics and Mathematical Systems

(Vol. 1–15: Lecture Notes in Operations Research and Mathematical Economics, Vol. 16–59: Lecture Notes in Operations Research and Mathematical Systems)

Vol. 1: H. Bühlmann, H. Loeffel, E. Nievergelt, Einführung in die Theorie und Praxis der Entscheidung bei Unsicherheit. 2. Auflage, IV, 125 Seiten 4°. 1969. DM 16,–

Vol. 2: U. N. Bhat, A Study of the Queueing Systems M/G/1 and GI/M/1. VIII, 78 pages. 4°. 1968. DM 16,–

Vol. 3: A. Strauss, An Introduction to Optimal Control Theory. VI, 153 pages. 4°. 1968. DM 16,–

Vol. 4: Branch and Bound: Eine Einführung. 2., geänderte Auflage. Herausgegeben von F. Weinberg. VII, 174 Seiten. 4°. 1972. DM 18,–

Vol. 5: Hyvärinen, Information Theory for Systems Engineers. VIII, 205 pages. 4°. 1968. DM 16,–

Vol. 6: H. P. Künzi, O. Müller, E. Nievergelt, Einführungskursus in die dynamische Programmierung. IV, 103 Seiten. 4°. 1968. DM 16,–

Vol. 7: W. Popp, Einführung in die Theorie der Lagerhaltung. VI, 173 Seiten. 4°. 1968. DM 16,–

Vol. 8: J. Teghem, J. Loris-Teghem, J. P. Lambotte, Modèles d'Attente M/G/1 et GI/M/1 à Arrivées et Services en Groupes. IV, 53 pages. 4°. 1969. DM 16,–

Vol. 9: E. Schultze, Einführung in die mathematischen Grundlagen der Informationstheorie. VI, 116 Seiten. 4°. 1969. DM 16,–

Vol. 10: D. Hochstädter, Stochastische Lagerhaltungsmodelle. VI, 269 Seiten. 4°. 1969. DM 18,–

Vol. 11/12: Mathematical Systems Theory and Economics. Edited by H. W. Kuhn and G. P. Szegö. VIII, IV, 486 pages. 4°. 1969. DM 34,–

Vol. 13: Heuristische Planungsmethoden. Herausgegeben von F. Weinberg und C. A. Zehnder. II, 93 Seiten. 4°. 1969. DM 16,–

Vol. 14: Computing Methods in Optimization Problems. Edited by A. V. Balakrishnan. V, 191 pages. 4°. 1969. DM 16,–

Vol. 15: Economic Models, Estimation and Risk Programming: Essays in Honor of Gerhard Tintner. Edited by K. A. Fox, G. V. L. Narasimham and J. K. Sengupta. VIII, 461 pages. 4°. 1969. DM 24,–

Vol. 16: H. P. Künzi und W. Oettli, Nichtlineare Optimierung: Neuere Verfahren, Bibliographie. IV, 180 Seiten. 4°. 1969. DM 16,–

Vol. 17: H. Bauer und K. Neumann, Berechnung optimaler Steuerungen, Maximumprinzip und dynamische Optimierung. VIII, 188 Seiten. 4°. 1969. DM 16,–

Vol. 18: M. Wolff, Optimale Instandhaltungspolitiken in einfachen Systemen. V, 143 Seiten. 4°. 1970. DM 16,–

Vol. 19: L. Hyvärinen, Mathematical Modeling for Industrial Processes. VI, 122 pages. 4°. 1970. DM 16,–

Vol. 20: G. Uebe, Optimale Fahrpläne. IX, 161 Seiten. 4°. 1970. DM 16,–

Vol. 21: Th. Liebling, Graphentheorie in Planungs- und Tourenproblemen am Beispiel des städtischen Straßendienstes. IX, 118 Seiten. 4°. 1970. DM 16,–

Vol. 22: W. Eichhorn, Theorie der homogenen Produktionsfunktion. VIII, 119 Seiten. 4°. 1970. DM 16,–

Vol. 23: A. Ghosal, Some Aspects of Queueing and Storage Systems. IV, 93 pages. 4°. 1970. DM 16,–

Vol. 24: Feichtinger, Lernprozesse in stochastischen Automaten. V, 66 Seiten. 4°. 1970. DM 16,–

Vol. 25: R. Henn und O. Opitz, Konsum- und Produktionstheorie. I. II, 124 Seiten. 4°. 1970. DM 16,–

Vol. 26: D. Hochstädter und G. Uebe, Ökonometrische Methoden. XII, 250 Seiten. 4°. 1970. DM 18,–

Vol. 27: I. H. Mufti, Computational Methods in Optimal Control Problems. IV, 45 pages. 4°. 1970. DM 16,–

Vol. 28: Theoretical Approaches to Non-Numerical Problem Solving. Edited by R. B. Banerji and M. D. Mesarovic. VI, 466 pages. 4°. 1970. DM 24,–

Vol. 29: S. E. Elmaghraby, Some Network Models in Management Science. III, 177 pages. 4°. 1970. DM 16,–

Vol. 30: H. Noltemeier, Sensitivitätsanalyse bei diskreten linearen Optimierungsproblemen. VI, 102 Seiten. 4°. 1970. DM 16,–

Vol. 31: M. Kühlmeyer, Die nichtzentrale t-Verteilung. II, 106 Seiten. 4°. 1970. DM 16,–

Vol. 32: F. Bartholomes und G. Hotz, Homomorphismen und Reduktionen linearer Sprachen. XII, 143 Seiten. 4°. 1970. DM 16,–

Vol. 33: K. Hinderer, Foundations of Non-stationary Dynamic Programming with Discrete Time Parameter. VI, 160 pages. 4°. 1970. DM 16,–

Vol. 34: H. Störmer, Semi-Markoff-Prozesse mit endlich vielen Zuständen. Theorie und Anwendungen. VII, 128 Seiten. 4°. 1970. DM 16,–

Vol. 35: F. Ferschl, Markovketten. VI, 168 Seiten. 4°. 1970. DM 16,–

Vol. 36: M. P. J. Magill, On a General Economic Theory of Motion. VI, 95 pages. 4°. 1970. DM 16,–

Vol. 37: H. Müller-Merbach, On Round-Off Errors in Linear Programming. VI, 48 pages. 4°. 1970. DM 16,–

Vol. 38: Statistische Methoden I, herausgegeben von E. Walter. VIII, 338 Seiten. 4°. 1970. DM 22,–

Vol. 39: Statistische Methoden II, herausgegeben von E. Walter. IV, 155 Seiten. 4°. 1970. DM 16,–

Vol. 40: H. Drygas, The Coordinate-Free Approach to Gauss-Markov Estimation. VIII, 113 pages. 4°. 1970. DM 16,–

Vol. 41: U. Ueing, Zwei Lösungsmethoden für nichtkonvexe Programmierungsprobleme. VI, 92 Seiten. 4°. 1971. DM 16,–

Vol. 42: A. V. Balakrishnan, Introduction to Optimization Theory in a Hilbert Space. IV, 153 pages. 4°. 1971. DM 16,–

Vol. 43: J. A. Morales, Bayesian Full Information Structural Analysis. VI, 154 pages. 4°. 1971. DM 16,–

Vol. 44: G. Feichtinger, Stochastische Modelle demographischer Prozesse. XIII, 404 Seiten. 4°. 1971. DM 28,–

Vol. 45: K. Wendler, Hauptaustauschschritte (Principal Pivoting). II, 64 Seiten. 4°. 1971. DM 16,–

Vol. 46: C. Boucher, Leçons sur la théorie des automates mathématiques. VIII, 193 pages. 4°. 1971. DM 18,–

Vol. 47: H. A. Nour Eldin, Optimierung linearer Regelsysteme mit quadratischer Zielfunktion. VIII, 163 Seiten. 4°. 1971. DM 16,–

Vol. 48: M. Constam, Fortran für Anfänger. VI, 143 Seiten. 4°. 1971. DM 16,–

Vol. 49: Ch. Schneeweiß, Regelungstechnische stochastische Optimierungsverfahren. XI, 254 Seiten. 4°. 1971. DM 22,–

Vol. 50: Unternehmensforschung Heute – Übersichtsvorträge der Züricher Tagung von SVOR und DGU, September 1970. Herausgegeben von M. Beckmann. VI, 133 Seiten. 4°. 1971. DM 16,–

Vol. 51: Digitale Simulation. Herausgegeben von K. Bauknecht und W. Nef. IV, 207 Seiten. 4°. 1971. DM 18,–

Vol. 52: Invariant Imbedding. Proceedings of the Summer Workshop on Invariant Imbedding Held at the University of Southern California, June – August 1970. Edited by R. E. Bellman and E. D. Denman. IV, 148 pages. 4°. 1971. DM 16,–

Vol. 53: J. Rosenmüller, Kooperative Spiele und Märkte. IV, 152 Seiten. 4°. 1971. DM 16,–

Vol. 54: C. C. von Weizsäcker, Steady State Capital Theory. III, 102 pages. 4°. 1971. DM 16,–

Vol. 55: P. A. V. B. Swamy, Statistical Inference in Random Coefficient Regression Models. VIII, 209 pages. 4°. 1971. DM 20,–

Vol. 56: Mohamed A. El-Hodiri, Constrained Extrema. Introduction to the Differentiable Case with Economic Applications. III, 130 pages. 4°. 1971. DM 16,–

Vol. 57: E. Freund, Zeitvariable Mehrgrößensysteme. VII, 160 Seiten. 4°. 1971. DM 18,–

Vol. 58: P. B. Hagelschuer, Theorie der linearen Dekomposition. VII, 191 Seiten. 4°. 1971. DM 18,–

Vol. 59: J. A. Hanson, Growth in Open Economics. IV, 127 pages. 4°. 1971. DM 16,–

Vol. 60: H. Hauptmann, Schätz- und Kontrolltheorie in stetigen dynamischen Wirtschaftsmodellen. V, 104 Seiten. 4°. 1971. DM 16,–

Vol. 61: K. H. F. Meyer, Wartesysteme mit variabler Bearbeitungsrate. VII, 314 Seiten. 4°. 1971. DM 24,–

Vol. 62: W. Krelle u. G. Gabisch unter Mitarbeit von J. Burgermeister, Wachstumstheorie. VII, 223 Seiten. 4°. 1972. DM 20,–

Vol. 63: J. Kohlas, Monte Carlo Simulation im Operations Research. VI, 162 Seiten. 4°. 1972. DM 16,–

Vol. 64: P. Gessner u. K. Spremann, Optimierung in Funktionenräumen. IV, 120 Seiten. 4°. 1972. DM 16,–

Vol. 65: W. Everling, Exercises in Computer Systems Analysis. VIII, 184 pages. 4°. 1972. DM 18,–

Vol. 66: F. Bauer, P. Garabedian and D. Korn, Supercritical Wing Sections. V, 211 pages. 4°. 1972. DM 20,–

Vol. 67: I. V. Girsanov, Lectures on Mathematical Theory of Extremum Problems. V, 136 pages. 4°. 1972. DM 16,–

Vol. 68: J. Loeckx, Computability and Decidability. An Introduction for Students of Computer Science. VI, 76 pages. 4°. 1972. DM 16,–

Vol. 69: S. Ashour, Sequencing Theory. V, 133 pages. 4°. 1972. DM 16,–

Vol. 70: J. P. Brown, The Economic Effects of Floods. Investigations of a Stochastic Model of Rational Investment Behavior in the Face of Floods. V, 87 pages. 4°. 1972. DM 16,–

Vol. 71: R. Henn und O. Opitz, Konsum- und Produktionstheorie II. V, 134 Seiten. 4°. 1972. DM 16,–

Vol. 72: T. P. Bagchi and J. G. C. Templeton, Numerical Methods in Markov Chains and Bulk Queues. XI, 89 pages. 4°. 1972. DM 16,–

Vol. 73: H. Kiendl, Suboptimale Regler mit abschnittweise linearer Struktur. VI, 146 Seiten. 4°. 1972. DM 16,–

Vol. 74: F. Pokropp, Aggregation von Produktionsfunktionen. VI, 107 Seiten. 4°. 1972. DM 16,–

Vol. 75: GI-Gesellschaft für Informatik e. V. Bericht Nr. 3. 1. Fachtagung über Programmiersprachen · München, 9–11. März 1971. Herausgegeben im Auftrag der Gesellschaft für Informatik von H. Langmaack und M. Paul. VII, 280 Seiten. 4°. 1972. DM 24,–

Vol. 76: G. Fandel, Optimale Entscheidung bei mehrfacher Zielsetzung. 121 Seiten. 4°. 1972. DM 16,–

Vol. 77: A. Auslender, Problemes de Minimax via l'Analyse Convexe et les Inégalités Variationnelles: Théorie et Algorithmes. VII, 132 pages. 4°. 1972. DM 16,–

Vol. 78: GI-Gesellschaft für Informatik e.V. 2. Jahrestagung, Karlsruhe, 2.–4. Oktober 1972. Herausgegeben im Auftrag der Gesellschaft für Informatik von P. Deussen. XI, 576 Seiten. 4°. 1973. DM 36,–

Vol. 79: A. Berman, Cones, Matrices and Mathematical Programming. V, 96 pages. 4°. 1973. DM 16,–

Vol. 80: International Seminar on Trends in Mathematical Modelling, Venice, 13–18 December 1971. Edited by N. Hawkes. VI, 288 pages. 4°. 1973. DM 24,–

Vol. 81: Advanced Course on Software Engineering. Edited by F. L. Bauer. XII, 545 pages. 4°. 1973. DM 32,–

Vol. 82: R. Saeks, Resolution Space, Operators and Systems. X, 267 pages. 4°. 1973. DM 22,–

Vol. 83: NTG/GI-Gesellschaft für Informatik, Nachrichtentechnische Gesellschaft. Fachtagung „Cognitive Verfahren und Systeme", Hamburg, 11.–13. April 1973. Herausgegeben im Auftrag der NTG/GI von Th. Einsele, W. Giloi und H.-H. Nagel. VIII, 373 Seiten. 4°. 1973. DM 28,–

Vol. 84: A. V. Balakrishnan, Stochastic Differential Systems I. Filtering and Control, A Function Space Approach. V, 252 pages. 4°. 1973. DM 22,–

Vol. 85: T. Page, Economics of Involuntary Transfers: A Unified Approach to Pollution and Congestion Externalities. XI, 159 pages. 4°. 1973. DM 18,–

Vol. 86: Symposium on the Theory of Scheduling and Its Applications. Edited by S. E. Elmaghraby. VIII, 437 pages. 4°. 1973. DM 32,–